COSMOGENESIS

The Growth of Order
in the Universe

DAVID LAYZER

New York   Oxford
OXFORD UNIVERSITY PRESS
1990

Oxford University Press

Oxford   New York   Toronto
Delhi   Bombay   Calcutta   Madras   Karachi
Petaling Jaya   Singapore   Hong Kong   Tokyo
Nairobi   Dar es Salaam   Cape Town
Melbourne   Auckland

and associated companies in
Berlin   Ibadan

Copyright © 1990 by Oxford University Press, Inc.

Published by Oxford University Press, Inc.,
200 Madison Avenue, New York, New York 10016

Oxford is a registered trademark of Oxford University Press

Library of Congress Cataloging-in-Publication Data
Layzer, David.
Cosmogenesis : the growth of order in the universe.
Bibliography: p.   Includes index.
1. Cosmology.   2. Evolution.
3. Science—Philosophy.   I. Title.
QB981.L333   1990   523.1   88-12530
ISBN 0-19-505528-4

9 8 7 6 5 4 3 2 1

Printed in the United States of America
on acid-free paper

Preface

No human enterprise seems more fragmented than scientific research. Most scientists devote their working lives to technical problems so specialized that only a tiny fraction of the scientific community understands or cares about them. And more often than not, the solution of an important technical problem creates new, even more specialized technical problems. Science is like the legendary Hydra, which grew two new heads whenever one was lopped off.

Yet at a deeper level, science is becoming increasingly unified and coherent. In the nineteenth century, physics was a collection of small, more or less independent states; now it is a single strong federation based on a few powerful laws. Chemistry and astronomy, autonomous for centuries, have joined the federation. And the physical and biological sciences, which many scientists believed would remain forever separate, became linked midway through the twentieth century by the new molecular science of life.

Oddly enough, the emergence of unifying laws and principles hasn't been accompanied by the emergence of a single, coherent, scientific worldview. Most scientists see the world from the perspective of their own broad disciplines, and the resulting views clash in important ways.

The world of *microscopic physics*—the world of molecules, atoms, and subatomic particles—is a four-dimensional network of events linked by universal mathematical laws. All moments in time are on the same footing. No moment is singled out as the present, and there is no distinction between the directions of the past and the future.

In the world of *macroscopic physics*—the world of everyday experience—time is equipped with an arrow pointing toward the future. The most fundamental law of macroscopic physics, the second law of thermodynamics, asserts that order is continually crumbling into chaos. The Universe is running down.

Biology and *cosmology* also equip time with an arrow pointing toward the future. But the natural processes that define time's arrow for the cosmologist and the biologist are processes that generate order rather than chaos. The worlds of comology and biology are winding up rather than running down. They are worlds in which new forms of order are continually being created.

Yet the worlds of cosmology and biology also differ in a fundamental way. Cosmologists insist that the growth of order is in principle entirely predictable; biologists are largely skeptical of the claim that biological evolution has followed a predictable course.

Finally, some contemporary scientists have challenged the most deeply ingrained of all scientific beliefs, the assumption that science provides, or at least strives for, an objective description of a world that exists independently of the human mind. We are not, they say, mere passive spectators of the passing show, as the etymology of the word *theory* suggests, but participants; we help to make reality.

Each of these conflicting views of the world has at least a plausible scientific basis. (Their advocates would claim much more.) Yet they can't all be true. Are they, perhaps, views of the same mountain from different angles? If so, how do they fit together? Is there a single coherent view of the world that natural scientists study?

I became involved in these questions almost twenty-five years ago, when I set out to try to understand why macroscopic processes like the evaporation and diffusion of perfume from an unstoppered bottle in a still room are irreversible (the perfume will never spontaneously make its way back into the bottle), whereas the motions of individual perfume molecules are perfectly reversible. (There is a standard answer to this question, but like many of my colleagues, I wasn't satisfied with it.) This puzzle led to others, as scientific puzzles have a way of doing, but it was a long time before it dawned on me that I had been working all along on pieces of a much larger puzzle.

This book, the outcome of that process, describes an attempt to construct a coherent scientific view of the world that embraces quantum mechanics, macroscopic physics, cosmology, and biology (including the biology of perception and thought). The final chapter explores in a preliminary way the connection between this worldview and the problem of human freedom.

Because I had to touch on an unusually wide range of scientific topics, and because even professional scientists aren't for the most part as knowledgeable outside their areas of specialization as they would like to be, I have tried to make everything in the book accessible to readers with no specialized scientific knowledge. Accessible but not easy. Although I have omitted many details, especially mathematical details, I haven't omitted or oversimplified any of the important arguments. What has been left out is the kind of detail that nonspecialists won't miss and that specialists should have no trouble filling in for themselves.

A large part of what I have to say in the following pages concerns theories and facts that most scientists regard as well established, but some of my arguments and conclusions are speculative—suggestions that have not yet stood the test of time. A few are even heretical. (Scientific heresies are opinions that contradict what almost all the experts believe but have no very good reason—at least in the heretic's opinion—for believing.) The distinction between what is well established in science and what is speculative is very important. In my opinion, the line is easier to draw than some contemporary philosophers of science are willing to concede. The distinction between speculation and firmly established findings or theories doesn't coincide, however, with the distinction between what scientists currently believe and what they don't believe: current beliefs usually include a large speculative component. I have tried to keep both distinctions clear, and to explain as fully as possible the scientific grounds for current scientific opinions—other people's as well as my own.

This book would have contained more infelicities, obscurities, and downright errors were it not for the wise counsel of several generous friends and colleagues. Chapters 5 and 6 have benefited from Mara Beller's keen historical criticism. Richard Lewontin, Ernst Mayr, and Gunther Stent, who more than a decade ago patiently and skillfully guided my initial efforts to make sense of biological evolution, helped me, just as patiently and skillfully, to improve Chapters 8 to 12, on biological and cultural evolution. (I need hardly add that they bear no blame for the flaws that remain.) Chapters 13 and 14, on neurobiology, owe much to the detailed criticism and expert instruction of John Dowling and J. Allan Hobson. Conversations and correspondence with Lesley Friedman, Alison McIntyre, Thomas Nagel, Mark Rollins, and Gunther Stent helped shape Chapter 15. Prudence Steiner made valuable comments on Chapters 1 and 2. My former students Holly Thomis Doyle and Robert O. Doyle filled the margins of the penultimate draft with queries and suggestions, every one of which proved helpful.

While the book was taking shape, I was able to try out early drafts of some chapters on an exceptionally intelligent and critical readership: students and discussion leaders in two courses in Harvard's Core program. The book owes much to their responses. Invitations to lecture at the van Leer Institute in Jerusalem and at Stockholm University provided further opportunities to get more useful responses. I thank Yehuda Elkana of the van Leer Institute and Per Olov Lindblad and Bertel Laurent of Stockholm University for these kind invitations and for their generous hospitality.

I thank Jacqueline Hartt and Patricia van der Leun for their unflagging and enthusiastic support. Joan Bossert and Irene Pavitt edited the manuscript with great insight, tact, and technical skill. The final version owes much to their excellent suggestions.

George Nichols ably prepared the illustrations from my rough sketches. Judith Layzer prepared the index.

Finally, it is a pleasure to express my gratitude to my wife, Jean, and to Carolyn, Judith, Nicholas, and Jonathan Layzer, who for nearly four years have managed to strike just the right balance between encouragement and impatience.

Cambridge, Mass.   D.L.
August 1988

Contents

I

A
THEORY
OF
ORDER

1

The Unity of Science

The Problem of Order

Around the beginning of the sixth century B.C., in the prosperous trading center of Miletus on the Aegean coast of present-day Turkey, a handful of Greek thinkers made the first recorded efforts to construct a rational or nonmythical account of how order arose in the world. It seemed obvious to them that the world could not have sprung from emptiness. "Nothing," they were fond of saying, "comes from nothing." It seemed equally obvious that the world didn't spring into being ready-made: the *kosmos,* they all agreed, must have evolved from a primordial chaos. (In classical Greek, *kosmos* means both "order" and "world.") But how? Western science and Western philosophy have their roots in the efforts of Thales, Anaximander, Anaximenes, and their successors to answer this question.

Philosophers have long since ceded the question to natural scientists, who, following science's oldest and most fruitful methodological precept, divide and conquer, have separated it into more specific questions:

- How can we account for the permanence, stability, and orderliness of crystals, molecules, atoms, and subatomic particles?
- How did the complex hierarchic structure of the astronomical Universe come into being?
- What is the origin of biological organization in all its manifestations, from DNA to the human mind?
- And why do most kinds of order tend to crumble and decay?

These four questions are central to four great divisions of natural science: quantum physics, cosmology, biology, and macroscopic physics or thermodynamics. These disciplines have yielded a deeper and more detailed understanding of order in its varied manifestations than anyone could have anticipated, even fifty years ago. There are gaps, of course. Theoretical physicists are still striving to unify the laws of physics; the origin of life is still a mystery; and the origin of astronomical systems remains an area of speculation and controversy. These problems lie at the frontiers of modern science, and they are getting a lot of attention.

But another kind of incompleteness in natural science's picture of the world has received less attention: *the four great pieces of the picture don't quite fit together*. Each piece, although still incomplete, is remarkably coherent. Each piece is connected to other pieces. But the connections aren't smooth. There are deep unresolved conflicts between quantum physics and macroscopic physics, between macroscopic physics and cosmology, and between the physical sciences and biology.

Conflicts and Paradoxes

The relation between quantum physics, which describes the invisible world of elementary particles and their interactions, and macroscopic physics, which describes the world of ordinary experience, has perplexed physicists since the birth of quantum physics in 1925. Viewed as a system of mathematical laws, quantum physics includes macroscopic physics as a limiting case. By that I mean that quantum physics and macroscopic physics make the same predictions in the domain where macroscopic physics has been strongly corroborated (the macroscopic domain), but quantum physics also successfully describes the behavior and structure of molecules, atoms, and subatomic particles (the microscopic domain). Yet from another point of view, macroscopic physics seems more fundamental than quantum physics. As we will see later, the laws of quantum physics refer explicitly to the results of measurement. But every measurement necessarily has at least one foot in the world of ordinary experience: it has to be recorded in somebody's lab notebook or on magnetic tape. So quantum physics seems to presuppose its own limiting case—macroscopic physics. This is the mildest of several paradoxes that have sprung up in the region where quantum physics and macrophysics meet and overlap.

The relation between macrophysics and cosmology is also problematic. The central law of macroscopic physics—the second law of thermodynamics—was understood by its inventors, and is still understood by most scientists, to imply that the Universe is running down—that order is degenerating into chaos. How can we reconcile such a tendency with the fact that the world is full of order—that it is a *kosmos* in both senses of the word? Some scientists say, "The contradiction is only apparent. The Second Law assures us that the Universe is running down, so it must have begun with a vast supply of order that is gradually being dissipated." But this way of trying to resolve the difficulty takes us from the frying pan into the fire, because, as we will see, modern cosmology strongly suggests that the early Universe contained far less order than the present-day Universe.

Astronomical evolution and biological evolution are both stories of emerging order. Nevertheless, the views of time and change implicit in modern physics and modern biology are radically different. The physical sciences teach us that all natural phenomena are governed by mathematical laws that connect every physical event with earlier and later events. Imagine that every past and future event was recorded on an immense roll of film. If we knew all the physical laws, we could reconstruct the whole film from a single frame. And in principle there is nothing to prevent us from acquiring complete knowledge of a single frame.

This worldview is epitomized in a much-quoted passage by one of Newton's most illustrious successors, the mathematician and theoretical astronomer Pierre Simon de Laplace (1749–1827):

> We ought then to regard the present state of the Universe as the effect of its previous state and the cause of the one that follows. An intelligence that at a given instant was acquainted with all the forces by which nature is animated and with the state of the bodies of which it is composed would—if it were vast enough to submit these data to analysis—embrace in the same formula the movements of the largest bodies in the Universe and those of the lightest atoms: Nothing would be uncertain for such an intelligence, and the future like the past would be present to its eyes. The human mind offers, in the perfection it has been able to give to astronomy, a feeble idea of this intelligence.[1]

Much the same view of the world was held by Albert Einstein:

> The scientist is possessed by the sense of universal causation. The future, to him, is every whit as necessary and determined as the past.[2]

Most contemporary physical scientists would probably agree with Laplace and Einstein. The world they study is a *block universe,* a four-dimensional net of causally connected events with time as the fourth dimension.[3] In this world, no moment in time is singled out as "now." For Laplace's Intelligence, the future and the past don't exist in an absolute sense, as they do for us.

How does life, regarded as a scientific phenomenon, fit into this worldview? A modern Laplacian might reply:

> Living organisms are collections of molecules that move and interact with one another and with their environment according to the same laws that govern molecules in nonliving matter. A supercomputer, supplied with a complete microscopic description of the biosphere and its environment, would be able to predict the future of life on Earth and to deduce its initial state. Implicit in the present state of the biosphere and its environment are the precise conditions that prevailed in the lifeless broth of organic molecules in which the first self-replicating molecules formed. And implicit in the conditions that prevailed in that broth and its environment is every detail of the living world of today.

If you believe that living matter is subject to the same laws as nonliving matter— and few, if any, contemporary biologists would dispute this assertion—this argument may seem compelling. Yet it clashes with two key aspects of the evolutionary process as described by contemporary evolutionary biologists: randomness and creativity.

Randomness is an essential feature of the reproductive process. In nearly every biological population, new genes and new combinations of genes appear in every generation. Reproduction, whether sexual or asexual, involves the copying of genetic material (DNA). In all modern organisms the copying process is astonishingly accurate. But it isn't perfect. Occasionally there are copying errors, and these have a random character. In sexually reproducing populations there is another source of randomness: the genetic material of each individual is a random combination of contributions from each parent.

The *creative* factor in biological evolution is natural selection, the tendency of genetic changes that favor survival and reproduction to spread in a population, and of changes that hinder survival and reproduction to die out. From the raw material provided by genetic variation, natural selection fashions new biological structures, functions, and behaviors.

A mainstream physicist might reply that the apparent randomness of genetic variation is just a consequence of human ignorance—our inability to understand exceedingly complex but nevertheless completely determinate causal processes—and that evolution is "creative" only in a metaphorical sense. According to this view, evolution merely brings to light varieties of order prefigured in the prebiotic broth.

There is an even more fundamental difference between the physical and the biological views of reality: the physicist's picture of reality seems impossible to reconcile with subjective experience. For there is nothing in the neo-Laplacian picture that corresponds to the central feature of human experience, the passage of time. We humans must watch the film unwind, but Laplace's Intelligence sees it whole. Nor is there anything that corresponds to the aspect of reality (as we experience it) that Greek philosophers called *becoming,* as opposed to the timeless *being* of numbers, triangles, and circles. The universe of modern physics is an enormously expanded and elaborated version of the perfectly ordered but static and lifeless world we encounter in Euclid's *Elements,* of which it is indeed a direct descendant. The biologist's world seems entirely different. Life, as we experience it, is inseparable from unpredictability and novelty.

Freedom and Necessity

What is the relation between being and becoming? Is the future as fixed and immutable as the past? What is chance? These questions bear on one of the perennial problems of Western philosophy, the problem of freedom and necessity.

Each of us belongs to two distinct worlds. As objects in the world that natural science describes we are governed by universal laws. To Laplace's Intelligence we are systems of molecules whose movements are no less predictable and no more the results of free choice than the movements of the planets around the Sun. But as the subjects of our own experience we see the world differently: not as bundles of events frozen into the block universe of Laplace and Einstein like flies in amber, but as the authors of our own actions, the molders of our own lives. However strongly we may believe in the universality of physical laws, we cannot suppress the intuitive conviction that the future is to some degree open and that we help to shape it by our own free choices.

This conviction lies at the basis of every ethical system. Without freedom there can be no responsibility. If we are not really free agents—if our felt freedom is illusory—how can we be guided in our behavior by ethical precepts? And why should society punish some acts and reward others? The Laplacian worldview tends to undermine the basis for ethical behavior.

Judeo-Christian theology faces a similar problem. Although Laplace's Intelligence is not the Judeo-Christian God—Laplace's Intelligence observes and calculates; the Judeo-Christian God wills and acts ("Necessitie and chance approach not mee, and what I will is Fate," says the Almighty in Milton's *Paradise Lost*)—they contemplate similar universes. Nothing is uncertain for an all-knowing God, and the future, like the past, is present to His eyes. But if we cannot choose where we walk, why should those who take the narrow way of righteousness be rewarded in the next life while those who take the primrose path are consigned to the flames of hell?

Theologians have not, of course, neglected this question. Augustine, for example, argued that God's foreknowledge (or more accurately, God's knowledge of what *we* call the future) doesn't *cause* events to happen and is therefore consistent with human free will. Other theologians have embraced the doctrine of predestination and argued that free will is indeed an illusion. Still others have taken the position that divine omniscience and human free will are compatible in a way that surpasses human understanding.

Reconciling the scientific and ethical pictures of the world was a concern of the first scientists. Our scientific picture of the world was foreshadowed by Greek atomism, a theory invented by the natural philosophers Leucippus and Democritus in the fifth century B.C. According to this theory, the world is made up of unchanging, indestructible particles moving about in empty space and interacting with one another in a completely deterministic way. Like modern biologists, Democritus believed that we, too, are assemblies of atoms. Yet Democritus also elaborated a system of ethics based on moral responsibility. He taught that we should do what is right not from fear, whether of punishment or of public disapproval or of the wrath of gods, but in response to our own sense of right and wrong. Unfortunately, the surviving fragments of Democritus's writings don't tell us how or whether he was able to reconcile his deterministic picture of nature with his doctrine of moral responsibility.

A century later, another Greek philosopher with similar ideas about physical reality and moral responsibility faced the same dilemma. Epicurus (341–270 B.C.) sought to reconcile human freedom with the atomic theory by postulating a random element in atomic interactions. Atoms, he said, occasionally "swerve" unpredictably from their paths. In modern times, Arthur Stanley Eddington and other scientists have put forward more sophisticated versions of the same idea. According to quantum physics, it is impossible to predict the exact moment when certain atomic events, such as the decay of a radioactive nucleus, will take place. Eddington believed that this kind of microscopic indeterminism might provide a scientific basis for human freedom:

> It is a consequence of the advent of quantum theory that *physics is no longer pledged to a scheme of deterministic laws. . . .* The future is a combination of the causal influences of the past together with unpredictable elements. . . . [S]cience thereby withdraws its moral opposition to free will.[4]

But neither Epicurus nor Eddington explained what the "freedom" enjoyed by a swerving atom or a radioactive atomic nucleus has to do with the freedom of a human being to choose between two courses of action. Nor has anyone else.

The apparent incompatibility between human autonomy and natural necessity can be used to support either of two opposite conclusions: because we are part of nature, autonomy must be an illusion; or because autonomy is a fact, we can't belong entirely to nature—human nature must have a spiritual or nonmaterial side. The second view was central to the philosophy of Socrates, who taught that mechanistic explanations like those advanced by the atomists don't apply to soul or mind. Socrates's doctrine was taken up and developed by Plato and Aristotle, and it has been dominant in Western philosophy ever since. Immanuel Kant, for example, argued that human autonomy is impossible to understand scientifically, but at the same time impossible to deny. It is the central fact of our existence and an indispensable premise of ethical theory. Whatever the scientific status of human autonomy, said Kant, we cannot help acting "under the idea of freedom."

The view that human beings have a nonnatural or spiritual dimension is, of course, especially congenial to Judeo-Christian religious thought. But it is precisely this view that Charles Darwin's *The Origin of Species,* published in 1859, called into question. Darwin argued that all living organisms have sprung from one or a very few primitive ancestral populations. He also speculated that the first living organisms arose by entirely natural processes from nonliving matter. Modern biology has enormously strengthened these hypotheses, and today they are accepted by virtually every working biologist. But if humankind belongs wholly to the natural world, then human autonomy—no less a fact today than it was in Kant's day or in Socrates's—must be accounted a natural phenomenon. Thus the problem of reconciling human autonomy and natural necessity presents a challenge to natural science as well as to philosophy.

Science and Philosophy

Although the issues I have been discussing are broadly philosophical, I will approach them more in the spirit of science than in the spirit of philosophy. By training and temperament I am a scientist, not a philosopher. More important, I believe that the questions raised by the disunity of the natural sciences and by the conflict between human autonomy and natural necessity are, at their deepest levels, scientific questions. Let me try to make clear what I consider to be the main differences between these two closely related and historically intertwined modes of understanding and explanation.

We may take as exemplars of the two modes two classic Greek texts, Plato's *Republic* and Archimedes's *On the Equilibrium of Floating Bodies*. These books differ in two related ways.

    1. Virtually no modern student of physics reads *On the Equilibrium of Floating Bodies;* no serious student of philosophy can afford not to read the *Republic*.

    2. Archimedes's treatise unequivocally and correctly answers the question it addresses: What conditions determine the equilibrium configuration of a floating body? Plato's dialogue offers unique and valuable insights into a question that has no correct or final answer: What is justice?

The fact that philosophers rarely succeed in answering the questions they ask doesn't reflect unfavorably on philosophers or philosophy. Philosophy seeks to bring clarity, logical consistency, and order into our thinking about questions that by their very nature can never be finally answered, in part because they take on new meanings as the conditions of our lives and the state of our knowledge change. Natural science is different in this respect. A modern Plato wouldn't reinvent the arguments and conclusions of the *Republic,* but a modern Archimedes, faced with the same questions as the historical Archimedes, would in nearly every case invent the same answers, because they are the *right* answers.

The last assertion may shock some readers. Science, we are often told, doesn't supply *right* answers. Theory B may answer a particular kind of question better than theory A, but theories are ephemeral, and tomorrow theory B may be superseded by theory C. And indeed, Archimedes's theory of static equilibrium has been superseded by Newtonian dynamics, which in turn has been superseded by Einstein's general theory of relativity.

Yet if you want to predict how far a ship can roll without capsizing, you may rely on Archimedes's theory of floating bodies. If you want to predict whether a framework of steel girders will be stable, you may rely on Archimedes's theory of statics. Newton's theory makes the same predictions. So does Einstein's, and—we may be confident—so will the theory that supersedes Einstein's.

How can this be? How can theories as different as Archimedean statics, Newtonian dynamics, and Einstein's general theory of relativity give the same, correct answers to questions that lie in their common domain? And what can we infer about scientific theories and their evolution from the fact that they do agree?

The philosopher Willard Van Orman Quine argues that theories are *underdetermined* by the requirement that their predictions agree with experience—that is, that many different theories are compatible with this requirement.

> We have no reason to suppose that man's surface irritations [sensory information about the external world] even unto eternity admit of any one systematization that is scientifically better or simpler than all possible others. It seems likelier, if only on account of symmetries or dualities, that countless alternative theories would be tied for first place. Scientific method is the way to to truth, but it affords even in principle no unique definition of truth.[5]

Thus, according to Quine, scientific truth has many forms. The historian of science Thomas Kuhn goes further. He concedes that new theories work better than the theories they succeed, but denies that science progresses toward objective truths about the world. In some respects, says Kuhn, Einstein is closer to Aristotle than to Newton.[6]

Many working scientists disagree with both Quine and Kuhn. They maintain that well-established scientific theories are *overdetermined,* rather than underdetermined, by experience; that scientific theories are approaching a nucleus of objective truth about the external world; and that, in the words of Einstein, "nature is the realization of the simplest conceivable mathematical ideas."[7] These opinions are closely connected. Let's take a closer look at what they mean and at the evidence on which they rest.

Underdetermined or Overdetermined?

Because we can usually make up many stories to fit a given set of facts, it seems plausible that scientific theories should be underdetermined by the observational and experimental data that can be brought to bear on them at any given moment. But let's look at some historical evidence.

Every working scientist knows how hard it is to frame a plausible hypothesis that explains an unexpected experimental or observational finding without coming into conflict with other experimental or observational findings. For example, there is not yet (in April 1989) *any* widely accepted explanation of the recently discovered phenomenon of high-temperature superconductivity (the observation that certain compounds lose their electrical resistance at temperatures well above absolute zero). In practice, explaining a new phenomenon seems to be less like making up a plausible story to fit a small set of facts than like putting together a large jigsaw puzzle when you don't know what the picture will turn out to be like, or whether you have all the pieces.

The history of physics does furnish examples of theories that have the same scope and make the same predictions. But these theories have always turned out to have the same underlying structure, in the way that Euclid's "synthetic" geometry and Descartes's "analytic" geometry have the same underlying structure. (In Descartes's geometry, points in a plane are represented by pairs of numbers; lines and circles are represented by algebraic equations. Each of Euclid's theorems has an algebraic counterpart, and each of Descartes's theorems has a geometric counterpart. Although the two theories look entirely different, they are really the same theory in different mathematical clothing.) For example, in 1926 there were two competing theories of quantum physics: Erwin Schroedinger's "wave mechanics" and the "matrix mechanics" of Werner Heisenberg, Max Born, and Pascual Jordan. Schroedinger showed that these theories have exactly the same underlying structure, just as Euclidean and Cartesian geometry have the same underlying structure. The same thing happened in the 1940s when three apparently quite different theories of quantum electrodynamics were shown to have the same underlying structure.

New phenomena like high-temperature superconductivity usually stimulate many rival hypotheses. But if one of these hypotheses thrives, by acquiring strong experimental or observational support, the others usually wither away. For example, at the beginning of the nineteenth century, physicists were divided between the hypothesis that heat is an indestructible substance and the hypothesis that it is interconvertible with mechanical energy. Each hyothesis enjoyed some experimental support, and each had passionate advocates. During the 1840s, however, James Joule and Robert von Mayer demonstrated experimentally that heat and mechanical energy are interconvertible at a fixed rate of exchange in a variety of physical contexts. The rival hypothesis promptly collapsed.

Again, during the opening years of the twentieth century, physicists were divided between two competing theories of electromagnetic phenomena in moving bodies. The two theories accounted equally well for the available data. In choosing sides, most physicists were guided by philosophical or aesthetic prejudices. Some preferred H. A. Lorentz's theory because it posited a medium in which light waves could propagate as vibrations. Others preferred Albert Einstein's theory because it *didn't* posit such a medium and was mathematically simpler than Lorentz's theory. Einstein's theory won out not because it was simpler but because it proved to be more accurate and broader in scope than Lorentz's theory. (Of course, the success of Einstein's theory strengthened the prejudice in favor of mathematical simplicity.)

In an ideal world, scientists would promptly accept strongly corroborated hypotheses and would reserve judgment on hypotheses that have not yet been strongly corroborated. But scientists, like other people, are guided by prejudices as well as by precepts. Strongly corroborated hypotheses aren't always quickly accepted by the scientific community. Einstein's photon hypothesis, which we discuss later, is a famous example. In spite of Einstein's immense personal authority, the hypothesis wasn't widely accepted until long after it had been strongly corroborated.

Some hypotheses, though, win widespread acceptance *before* they are strongly corroborated. The hypothesis of an all-pervading ether that acts as a carrier for light, heat, and gravitation was never strongly corroborated, but for over two centuries it was accepted by every leading physicist, from Newton to Lorentz. Light and heat seemed to *need* a carrier. Eventually the ether hypothesis fell before Einstein's special theory of relativity. Today the cosmic-fireball hypothesis enjoys equally widespread support within the scientific community, although the evidence in its favor is still rather meager.

As long as they are not held too dogmatically (and sometimes even when they are), prejudices play a constructive role in the growth of science. They motivate scientists to test and develop new hypotheses. Steven Weinberg makes the point very well:

> I do not believe that scientific progress is always best advanced by keeping an altogether open mind. It is often necessary to forget one's doubts and to follow the consequences of one's assumptions wherever they may lead—the great thing is not to be free of theoretical prejudices, but to have the right theoretical prejudices. And always, the test of any theoretical preconception is where it leads.[8]

By the "right" prejudices, Weinberg means those that are eventually confirmed by experiments or observations.

In the light of the evidence we have been discussing, what can we say about Quine's assertion that scientific theories are underdetermined by experimental and observational data? It is true that the theories of Archimedes, Newton, and Einstein are equally consistent with experimental data on the equilibrium configurations of floating bodies. *But in their common domain, they are not really different.* Archimedes's theory is a limiting case of Newton's, and Newton's theory is a limiting case of Einstein's. This example is typical. The history of physics shows that when a strongly corroborated mathematical theory is superseded by a more comprehensive and accurate theory, it persists as a limiting case of that theory. Or so most natural scientists believe. But what, precisely, do scientists mean by the phrase *limiting case?*

Limiting Cases and the Evolution of Physics

In a good street map of Kansas City, the distance between any two points bears a fixed ratio to the actual distance; the map has the same scale everywhere. It isn't possible to make a map of the world with this property. The most common world maps greatly exaggerate distances between points lying in the polar caps, but every world map necessarily distorts some regions. To make a faithful map of a large portion of the Earth's surface we have to use a globe. Yet the part of a large globe that represents a city is virtually indistinguishable from a map of the same small region drawn on a flat sheet of paper. The city map's geometry (plane geometry) is a *limiting case* of the globe's geometry (spherical geometry).

Let's take a closer look at this example. On the surface of a sphere, the shortest route between two points is an arc of the great circle that passes through these points. (A great circle is a circle whose center coincides with the center of the sphere.) The sides of a spherical triangle are arcs of great circles. The properties of such triangles are different from the properties of triangles in a plane. For example, the sum of the angles of any plane triangle is exactly 180°; but the sum of the angles of a spherical triangle is greater than 180° and increases with the area of the triangle. However, the geometry of a patch the size of Kansas City on a sphere the size of the Earth is nearly indistinguishable from plane geometry, and the differences between the two geometries diminish with the size of the patch, approaching zero as the area of the patch (or more precisely, the fraction of the sphere's surface that it occupies) approaches zero. This is what is meant by the statement that plane geometry is a limiting case of spherical geometry, approximately valid for sufficiently small patches.

As its name suggests, geometry began as the science of measurements on the surface of the Earth. Euclid's axioms for plane geometry were the first scientific laws. Eventually they were superseded by the axioms of spherical geometry, which apply to arbitrarily large regions on the surface of the Earth. But plane geometry remains of great practical importance as a limiting case, approximately valid for small tracts. In the same way and in the same sense, Archimedes's laws of statics and Galileo's laws governing the motions of falling and thrown objects near the surface of the Earth became limiting cases of Newton's theory, which in turn became a limiting case of Einstein's theory, which itself may one day become a limiting case of a still more comprehensive theory.

In this progression, each theory comprehends more phenomena than its predecessor and is more accurate than its predecessor. Archimedes's and Galileo's theories apply near the surface of the Earth; Newton's theory applies not only to the entire Earth but to all astronomical systems, from planets and their satellites to great clusters of galaxies. It doesn't, however, apply to the Universe as a whole. Einstein's theory does. Einstein's theory, moreover, gives a slightly more accurate account than Newton's theory of the motions of the planets.

Simplicity

Physics and astronomy grew out of efforts to understand certain kinds of shared experience. These shared experiences define the phenomenal world, the world of appearances. Until the beginning of the seventeenth century, the phenomenal world was coextensive with the world of human sensation. When Galileo turned his telescope toward the heavens and saw sights previously unseen——the mountains of the Moon, the phases of Venus, the Milky Way as a swarm of individual stars—he destroyed the comfortable illusion that what we see is what there is. By creating new kinds of shared experience, Galileo's observations not only extended the boundaries of the phenomenal world, but also demonstrated that from the standpoint of physical science, there is nothing fundamental about the limitations imposed by human senses—a conclusion that profoundly shocked Galileo's contemporaries. Since then, the phenomenal world not only has expanded enormously but also has sprouted a multitude of new dimensions—colors we can't see, odors we can't smell, sounds we can't hear. Observation and experiment have revealed new worlds outside the Milky Way and inside the atom. But at the same time, the system of mathematical laws and axioms underlying our theoretical description of the phenomenal world has become—in spite of the conflicts we discussed earlier—increasingly unified, increasingly accurate in its predictions, and even, in a sense, increasingly simple.

Simplicity, as physicists use the word, is not a simple concept. By any ordinary standards, Einstein's theory of gravitation is far more complex than Newton's. Newton's theory uses one function of position to describe a gravitational field; Einstein's theory uses ten functions of position and time. And the equations satisfied by Einstein's ten functions are far more complicated and vastly more difficult to solve than the single equation of Newton's theory.

Yet in another way Einstein's theory is simpler than Newton's. In Newton's theory there are two kinds of mass: inertial mass, a measure of an object's resistance to acceleration by an applied force; and gravitational mass, a measure of the strength with which an object attracts and is attracted by other objects. Experiments of a kind first carried out by Galileo have shown that if these two kinds of mass are measured in appropriate units, they are numerically equal. Although this remarkable finding is consistent with Newton's theory, it isn't an integral part of the theory. It has to be tacked on as an amendment. If new experiments were to show that the ratio between gravitational and inertial mass isn't quite constant, the amendment could easily be amended. The rest of Newton's theory would be unaffected. In Einstein's theory inertial and gravitational mass are identical in principle. Or rather, the distinction never arises. It is in this way that Einstein's theory is simpler—as well as more vulnerable to experimental refutation—than Newton's.

During the period when Einstein was working out and publishing his theories of space, time, and gravitation, most physicists and philosophers of science considered physical theories to be nothing more than economical systematizations of experience. Einstein's theory of gravitation didn't fit this description. Not only was it formulated in abstract and difficult mathematical language, but the quantities that figured in the theory were not directly measurable. A systematization of experience obviously has to treat measurable quantities, not mathematical abstractions like the curvature of spacetime. Ernst Mach and other positivists accordingly rejected Einstein's theory.

Experimental verification of a few key predictions—a small correction to Mercury's orbit, the bending of starlight by the Sun, a slight reddening of the light emitted by compact stars—vindicated not only Einstein's theory of gravitation but also his view of the nature of physical theories. It seemed that Nature shared Einstein's prejudice in favor of the kind of simplicity he had built into his theory.

Einstein's theories set the tone for twentieth-century physics. The history of quantum physics is the story of a continuing and highly successful search for simpler, deeper, more comprehensive, and more abstract unifying principles. These principles are not systematizations of experience. They are, as Einstein was fond of saying, "free creations of the human mind." Yet they capture, in the language of abstract mathematics, regularities that lie hidden deep beneath appearance. Why do these regularities have a mathematical form? And why are they accessible to human reason? These are the great mysteries at the heart of humankind's most sustained and successful rational enterprise. As Einstein once remarked, in an essay called "Physics and Reality," "the eternal mystery of the world is its comprehensibility."

An Overview

Natural science has two main projects. One is to understand the laws that underlie natural phenomena. The other is to understand the processes that have shaped the world as we know it. Here, too, unity beckons, although unity of a different kind from the unity of the fundamental laws.

The two projects address two kinds of natural order. I will refer to the first kind as timeless, or Pythagorean, order. It resides in the unchanging mathematical laws that underlie all natural phenomena. (Pythagoras and his followers believed that numbers, geometric shapes, and mathematical harmony underlie the phenomenal world.) The second kind resides in structures that come into being, evolve, and eventually decay. I will refer to it as timebound, or Heraclitean, order. (The worldview of the pre-Socratic philosopher Heraclitus, epitomized in the saying "You cannot step twice into the same river," stressed change and flux.)

The distinction between timeless and timebound order reflects a more fundamental distinction, between *laws* and *initial conditions*. The laws of physics define the realm of the possible; initial conditions define the realm of the actual. Laplace's Intelligence knows all the initial conditions as well as all the laws. So first we reexamine and reinterpret this distinction, which goes back to Galileo and Newton. I will argue that every set of initial conditions contains a regular component and a random component. The regular component represents a form of timebound order. In principle, it is predictable. I will argue that all forms of timebound order consist in the absence of randomness and that they can be subsumed under a single mathematical formula invented by Ludwig Boltzmann over a century ago.

Then I will argue that, contrary to current opinion, the random component of initial conditions doesn't arise entirely from human ignorance—that certain kinds of initial conditions are unknowable in principle, even by a Laplacian Intelligence. This argument will help us to understand why the future is different from the past, despite the fact that the laws governing atoms and molecules don't discriminate between the two directions of time.

From randomness, we pass to another of the book's major themes: discreteness. A discrete aggregate is one whose members can be listed or numbered. (The list may be infinitely long.) The chemical elements, the stars in our Galaxy, and the integers are all discrete aggregates. By contrast, continuous aggregates, such as the points in a line segment, cannot be listed or numbered.[9] The world of classical physics is continuous; the world of quantum physics is discrete. For example, classical physics would allow a hydrogen atom, which consists of a massive positively charged particle (the proton) and a much lighter negatively charged particle (the electron), to have any radius whatever and any energy whatever. But quantum physics predicts—and experiments confirm—that the hydrogen atom has a discrete aggregate of possible physical states, including a state of lowest energy and least radius.

We will see that discreteness lies at the root of timeless, or Pythagorean, order. We will also see that Boltzmann's definition of randomness—a definition that will enable us to give a unified account of timebound, or Heraclitean, order—applies only to systems whose possible states form a discrete aggregate. The two kinds of order therefore have a common basis.

Searching for a deeper understanding of the link between randomness and discreteness, we will be led to a new (and highly speculative) interpretation of quantum indeterminacy and to an equally speculative resolution of certain paradoxes that quantum physicists have been discussing for well over half a century. We will also be led to a deeper understanding of time's arrow as it manifests itself in macroscopic physics and ordinary experience.

The second part of the book explores the varieties of timebound order. We will see how our earlier discussions of randomness and discreteness make it possible to resolve two of the conflicts mentioned at the beginning of this chapter: between macroscopic physics, which tells us that the Universe is running down, and cosmology, which tells us that the Universe is winding up; and between the mechanistic and deterministic worldview of physics and the evolutionary world-view of biology.

Human thought and action are the most prolific sources of order in the world as we know it. Speech, writing, music, painting and sculpture, dance, and the ordinary activities of everyday life are at once the most familiar and the most mysterious of all order-generating processes. An examination of these processes paves the way for a discussion of our final question: Is there room for human freedom in a scientific picture of the world? Philosophers from Immanuel Kant to Thomas Nagel have argued that the subjective and the scientific views of the world—the view from within and the view from without—are irreconcilable. I will argue that the two ways of looking at the world are compatible and that the future is as open as our intuition tells us it is.

2

Order and Randomness

Natural science is a quest for order. But if you were to ask a particle physicist, a thermodynamicist, a cosmologist, and a biologist to define *order,* you would probably get four quite different answers. The particle physicist might identify order with the regularities expressed by fundamental physical laws and symmetry principles. The thermodynamicist might say that order consists in deviations from the uniform featureless state of thermodynamic equilibrium toward which all physical systems, and indeed the Universe as a whole, are tending. The cosmologist might point to the orderly structure of the Solar System and the Milky Way, perhaps adding that the thermodynamicist's homily about the decay of order apparently doesn't apply to self-gravitating systems. (An isolated self-gravitating gas sphere, for instance, grows continually hotter and denser as it radiates energy into empty space.) Finally, the biologist might argue that biological organization is qualitatively different from the kinds of order that concern her three colleagues in the physical sciences. All these views of order are valid. Our problem isn't to choose among them but to integrate them into a coherent picture, to bring order to the subject of order.

Laws and Initial Conditions

Scientists believe that all natural phenomena are governed *but not fully determined* by a handful of universal laws.

Consider, for example, Galileo's theory of motion, which describes the motion of projectiles near the surface of the earth. It rests on two mathematical laws:

1. A projectile's horizontal velocity is uniform in speed and direction.

2. Its vertical velocity decreases at a steady rate (about 10 meters per second each second).

These laws apply to objects dropped from towers, to cannonballs shot from cannons, and to stones thrown by small boys; they hold in Pisa, in Paris, and in Paducah; in our century as well as in Galileo's. But they don't tell us where or in what direction or with what speed a particular projectile will be launched. And unless we have these data, we can't predict the trajectory. The position, speed, and direction of a projectile at a given moment are called *initial conditions*. Once these data have been specified, Galileo's laws enable us to predict the projectile's position, speed, and direction of travel at both earlier and later times, as long as it is traveling freely.

This example illustrates four widely held beliefs about laws and initial conditions:

1. Laws and initial conditions are perfectly distinct. Thus Galileo's laws of motion are distinct from the initial conditions that define specific trajectories.

2. Laws express *universal* and *necessary* aspects of phenomena; initial conditions express their *particular* and *contingent* aspects. Thus Galileo's laws describe features shared by all possible trajectories, while each set of initial conditions defines a particular trajectory.

3. Laws express the *regularities* underlying phenomena; initial conditions have a *random* character.[1]

4. The initial conditions needed to determine all future states of the world are in principle knowable in complete detail.

I will argue that these beliefs are valid only in restricted contexts and that they don't apply to the Universe as a whole.

To see why, we have to take a closer look at Galileo's theory. It predicts that a projectile rises and falls in a parabolic arc. But this prediction has limited validity. If its initial speed is great enough, the projectile will leave the Earth and never return (Figure 2.1). Thus *Galileo's laws are valid for only a certain range of initial conditions* (those in which the projectile's initial speed lies below a certain value). In addition, the projectile must be large enough and heavy enough for the effects of air resistance to be negligible—another initial condition. Thus a precise statement of Galileo's laws must refer to initial conditions. In this example, then, laws and initial conditions are not distinct.

But suppose we replace Galileo's laws by Newton's. The class of permitted initial conditions is then much wider. Newton's laws allow for air resistance, and they apply to projectiles whose initial speed is great enough to overcome the Earth's gravitational pull. But Newton's laws, too, include a tacit reference to initial conditions: they apply only to projectiles whose speed is much less than the speed of light.

Einstein's theory of gravitation overcomes this limitation. It applies to projectiles moving with arbitrary speeds. Have we, then, finally succeeded in separating the laws from the initial conditions?

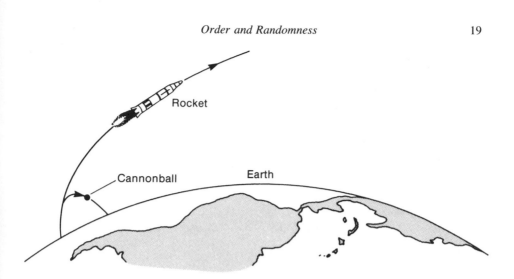

FIGURE 2.1 Galileo's theory predicts that a projectile rises and falls along a parabolic arc. But this prediction has limited validity. A projectile whose initial speed is great enough will leave the Earth and never return.

Not quite. Newton's laws of motion hold only in certain reference frames. The surface of the Earth, for example, is *not* one of these privileged reference frames. Because the Earth rotates about its axis, a hockey puck sliding on a perfectly flat, horizontal, and frictionless ice rink traces a slightly curved path instead of the straight line predicted by Galileo's law of inertia or Newton's first law of motion (Figure 2.2). If hockey games were played on rapidly rotating ice rinks, the curvature would be obvious. But what determines the frames of reference in which Newton's laws hold, frames in which the trajectories of hockey pucks are straight lines?

Newton answered this question in a way that modern scientists consider unscientific. He said that *space itself* provides an absolute standard of rest. An object is at rest if it is at rest relative to space, it is in motion if it is in motion relative to space, and it is accelerated if it is accelerated relative to space. But how can we tell whether an object is at rest relative to space?

We can't. Space itself, considered apart from its contents, is unobservable. Nowadays we would say that Newtonian space is a mathematical construct with no physical counterpart. Newton himself took a different view; he identified space with the "sensorium of God." In effect, he asserted that the question "What defines an unaccelerated frame of reference?" has no scientific answer.

Ernst Mach (1838–1916), equally distinguished as a physicist, a psychologist, and a philosopher of science, attempted to supply a scientific answer. In *The Science of Mechanics and Its Development* (1883), he argued that Newton's laws of motion—and indeed all physical laws—are nothing more than economical descriptions of what has or can be observed. Because space, as Newton conceived it, is unobservable in principle, it ought not to figure in the laws of motion. Mach suggested that the role assigned by Newton to absolute space should be given to the fixed stars. All references to accelerated and unaccelerated motion would then

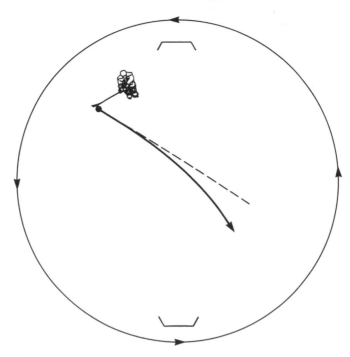

FIGURE 2.2 The arrow shows the curved trajectory of a hockey puck sliding on a rotating ice rink. The dashed straight line describes the trajectory relative to a nonrotating floor beneath the rink.

be interpreted as references to motion accelerated or unaccelerated *relative to the fixed stars.*

Since Mach's time, the "fixed stars" have been found to belong to a huge self-gravitating stellar system with complex internal motions. Moreover, this system, the Galaxy, is accelerated by the gravitational attractions of neighboring galaxies. It isn't hard, however, to modify Mach's suggestions to take these findings into account. We have only to replace the "fixed stars" with a large sample of distant galaxies, and stipulate that "unaccelerated" means "unaccelerated with respect to a sufficiently large sample of distant galaxies."

But there is a more serious difficulty with Mach's proposal and with its underlying philosophy. Consider first Mach's view of physical laws as economical representations of data. Mach could argue that physical concepts like mass, velocity, force, tension, temperature, heat, and electric charge are directly related to specific measurements, as are more abstract constructs like electric and magnetic fields. Only the atomic theory of matter couldn't be reconciled with Mach's view, because atoms hadn't yet been observed directly, and Mach warded off this threat to his philosophy by declaring the atomic hypothesis to be unnecessary and unscientific. But twentieth-century physics is another story. Relativity and quantum physics are emphatically not "economical representations of data." They are far too abstract, far too removed from direct experience, to qualify for such a description. At the same time, they seem to many modern physicists to express the deep structure of an objective physical reality.

Mach's criticism of Newton's ideas about absolute space made a deep impression on Einstein, who didn't, however, consider that Mach had solved the problem by simply decreeing that the "fixed stars" provide a standard of rest. Einstein believed that an adequate theory of gravitation must furnish a *causal link* between *local* unaccelerated reference frames—frames in which frictionless hockey pucks move with constant speed in perfectly straight lines—and a *cosmic* reference frame, determined by the cosmic distribution of matter, energy, and motion. Such a theory would enable one to *predict* (rather than merely stipulate) the hockey puck's unaccelerated motion relative to a standard of rest defined by the structure of the Universe as a whole. This ambitious requirement has come to be known as "Mach's principle," an appellation that manages to be unfair to both Mach, who didn't believe that physical laws express causal links, and Einstein, who actually formulated the principle.

Einstein not only formulated the requirement, but also succeeded in constructing a theory of gravitation that meets it: general relativity. According to general relativity, the geometric structure of spacetime determines its physical contents, and vice versa. To apply Einstein's theory to the Universe as a whole, we need an additional assumption. Following Einstein, most cosmologists postulate that *at a given moment in time the density of mass, averaged over a sufficiently large volume, has the same value everywhere in the Universe.* Supplemented by this assumption, Einstein's theory does indeed satisfy Mach's principle: it predicts that the hockey puck will move in a straight line with constant speed relative to a frame of reference in which the large-scale cosmic distribution of matter, energy, and motion is the same everywhere and in all directions.

What is the current status of Einstein's cosmological postulate? Some cosmologists argue that cosmic uniformity is a phenomenon that ought to be explained rather than posited. What would such an explanation be like? I think it would have to show that *any* universe satisfying Einstein's law of gravitation and the other laws of physics must evolve into a universe that is uniform on a sufficiently large scale, at least in our neighborhood. Any less sweeping demonstration would have to invoke an initial condition that would necessarily be more complex than the postulate of cosmic uniformity.

But it certainly isn't true that *any* initial conditions would lead to a universe as uniform as the one our telescopes show us. It may be possible to account for the large-scale uniformity of the Universe by invoking a weaker—but less simple—initial condition than Einstein's postulate, but it isn't possible to formulate a theoretical description of the Universe as a whole that dispenses entirely with initial conditions.

The postulate of cosmic uniformity isn't like other initial conditions. In all other contexts, initial conditions serve to distinguish the particular system one wishes to study from other systems of the same kind—the Earth from other planets, the Sun from other stars, the Galaxy from other galaxies. Does the postulate of cosmic uniformity distinguish our Universe from other possible universes? Surely talk about other possible universes, unobservable in principle, belongs to science fiction rather than science. In any case, there is no reason to suppose that other universes (whatever that may mean) obey the same physical laws as our own. Mathematicians may study "model universes," but in the lexicon of science, the word *Universe* has no plural form. The initial conditions that define the Universe are no less unique than the laws. Thus one of the key distinctions between laws and initial conditions—the uniqueness of laws as against the multiplicity of initial conditions—breaks down at the cosmological level of description.

Another of the conventional distinctions between laws and initial conditions hinges on the supposed randomness of initial conditions. Laws are supposed to express regularities, while initial conditions are supposed to have a random character. This distinction unquestionably holds for the initial conditions that define ordinary physical and astronomical systems. Consider a star. Among the conditions that serve to define it is its birthweight. Astronomical observations, interpreted in the light of theories of stellar structure and evolution, tell us that the initial masses of stars span a wide range. Some stars are one-tenth as massive as the Sun; others, 100 or more times as massive. And there is no evident reason why a particular star—the Sun, for example—should be born with a particular mass.

A theory of star formation might, however, predict the fraction of stars with initial masses in a specified range—within 10 percent of the Sun's present mass, say, or between 10 and 100 times the Sun's present mass. In other words, it might predict the *statistical profile* or *frequency distribution* of initial stellar masses (Figure 2.3). Thus the initial conditions that relate to initial stellar masses have a regular component as well as a random component. The random component arises from the fact that nothing in any present or foreseeable theory of how stars form would enable us to assign a definite initial mass to a particular star such as the Sun. The regular component resides in the statistical profile of stellar masses (Figure 2.3). It tells us the fraction of initial masses in any specified interval, and an adequate theory of how stars form would predict it.

To construct such a theory, we would need to understand the *process* of star formation, and we would need to know the relevant *initial conditions*—the conditions that paved the way for star formation. In practice these two tasks are intertwined. To arrive at an understanding of the physical process, one must know something about the initial conditions; to discover the initial conditions, one must understand the processes that link them to their effects. In principle, however, physical processes are distinct from initial conditions. Once the latter are specified, the former are determined by basic physical laws. Thus the problem of predicting the statistical profile of initial stellar masses reduces in principle to the problem of getting information about the conditions under which stars form.

These conditions, in turn, have a random component and a regular component. To predict the regular component, we would have to understand the antecedent physical processes and their associated initial conditions. And so on. Every set of initial conditions is linked by physical processes to another set of initial conditions. This causal chain is neither circular nor endless. Because the regular component of each set of initial conditions is explained by antecedent initial conditions

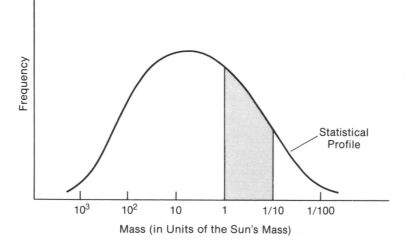

FIGURE 2.3 Hypothetical statistical profile of initial stellar masses. If the total area under the profile is taken to be unity, the shaded area represents the fraction of stars whose initial masses lie between one-tenth the Sun's mass and the Sun's mass.

that refer to a more inclusive physical system, we must eventually reach a set of conditions that describes the initial state of the Universe as a whole. This is the end of the line. Like Euclid's axioms, these cosmic initial conditions can't be deduced from anything else. They contain information about the world that we will never be able to explain.

Most scientists hope that this residue of the unexplained and unexplainable is small or even nonexistent. Although many people—mystics, for example—don't share this hope, it doesn't rest on prejudice alone. It rests in part on past experience. Science has repeatedly come up against apparently unexplainable phenomena. Nearly always it has succeeded, sooner or later, in explaining them. There is also a methodological reason for assuming that the cosmic initial conditions are as simple as possible: the simplest hypotheses don't always turn out to be right, but they have a strong claim to be tested before less simple hypotheses. This claim is expressed by the precept known as Ockham's razor: "What can be done with fewer assumptions is done in vain with more."

To sum up the discussion so far: we have scrutinized three nuggets of conventional wisdom concerning physical laws and initial conditions. We have seen that the distinction between these two determinants of natural phenomena isn't as clean as it appears at first sight and that it may disappear entirely at the cosmological level of description. The contrast between the universality of laws and the specificity and contingency of initial conditions also fades away at the cosmological level. Finally, we have seen that initial conditions have a regular as well as a random component. Understanding the regularities in initial conditions is indeed the central concern of cosmologists and biologists. These regularities are generated and destroyed by processes that obey universal and timeless physical laws, but they can't be deduced from these laws alone, as can the properties of a hydrogen atom or a diamond.

There remains the question of whether the initial conditions of the Universe determine its future states in the manner envisioned by Laplace and Einstein. We have seen that these initial conditions have a large random component. Our next task is to make more precise the notion of randomness.

Randomness

Many scientific terms are implicitly defined by the theories in which they figure. The terms *point* and *line* are implicitly defined by axioms, which tell us not what points and lines "really are" but what properties of the objects we call "points" and "lines" we are allowed to use in discourse bounded by those axioms. The meaning of *randomness,* like the meaning of *line,* is defined by its theoretical context.

The conventional scientific view distinguishes two kinds of randomness: one appropriate to quantum physics and the other to classical physics. According to quantum physics, certain elementary events are irreducibly random—the decay of a radioactive atomic nucleus, for instance. The precise moment when an unstable atomic nucleus will decay is inherently unpredictable. Quantum physics tells us only the probability that a given nucleus will decay during a given time interval, and in this respect no future theory will ever do better. At least that is what most contemporary physicists believe (correctly, in my opinion).

Randomness also figures in macroscopic physics. A physicist studying how the pressure, temperature, and density of air in a room are interrelated assumes that individual air molecules are moving with random speeds in random directions. But this assumption is made only for the sake of convenience. The physicist takes it for granted that the randomness of molecular motion, unlike the randomness of radioactive decay, is *reducible.* A complete description of the air in the room would assign every molecule a definite speed and direction of motion, within the limits imposed by quantum physics. For the physicist's purpose, all this microscopic detail is irrelevant. But in principle it is knowable. At least that is what most contemporary physicists believe.

Einstein, who believed strongly in the unity of physics, refused to accept the prevailing view that macroscopic and microscopic randomness are fundamentally dissimilar. He believed that the kind of randomness exemplified by radioactive decay is in fact reducible, just like the randomness that physicists attribute to molecular motions.

Fortunately, the question of reducibility can be separated from the question of how randomness should be defined and measured. That question was asked and answered in 1873 by Ludwig Boltzmann. Boltzmann was looking for a statistical property of a collection of molecules that would mimic *entropy,* a macroscopic property of gases. Entropy was invented (and named) by the physicist Rudolf Clausius in 1854. Unlike such quantities as pressure and temperature, it has no counterpart in experience. We can calculate the entropy of the air in a room as a function of temperature, pressure, and chemical composition, but nature hasn't equipped us with the means for sensing changes in entropy, as it has equipped us with heat sensors, pressure sensors, and a sense of smell. Nevertheless, entropy is enormously important in all sorts of practical contexts, from chemical engineering to biochemistry. Multiplied by the absolute temperature, it measures the quantity of energy that is *not* available for doing useful work under conditions of constant temperature and pressure (as in a living cell).

Physical processes that waste energy create entropy. Friction generates entropy. So does the direct flow of heat from a warmer to a cooler body, because two bodies at different temperatures can be used to run a heat engine, which converts part of the heat flow into work. So also do chemical reactions in living cells. Clausius postulated that *all* natural processes generate entropy unless the entropy is already as large as it can be. This postulate is one way of stating the second law of thermodynamics. The First Law says that no natural process creates or destroys energy; the Second Law implies that all natural processes *degrade* energy.

Although Clausius's definition enabled physicists to calculate the entropy of a gas, the physical meaning of what they were calculating remained obscure. They would have liked to understand entropy in the way they understood temperature and pressure. Temperature had been shown to be a measure of the average energy of molecular motion. The pressure that a gas exerts on the walls of a vessel had been shown to be a measure of the recoil the walls experience when gas molecules bounce off them. What is the molecular interpretation of entropy? How is entropy related to the statistical distribution of molecular positions and velocities? These are the questions Boltzmann set out to answer.

He discovered that an isolated volume of gas always evolves in such a way that the motions and spatial distribution of the gas molecules become increasingly random. Eventually the gas settles down into a state in which the molecules are as randomly distributed as they can be, both in space and in velocity. This state is known as *thermodynamic equilibrium.* More precisely, Boltzmann proved that molecular interactions always cause a specific measure of randomness to increase, unless it is already as large as it can be. Randomness, as Boltzmann defined it, turned out to have exactly the same mathematical properties as Clausius's entropy.

To illustrate Boltzmann's conception of randomness, let's consider a simple analogy. A collection of molecules viewed at a single moment may be compared to a hand of cards. A card corresponds to a molecule in a particular place, moving with a particular speed. Suit (clubs, diamonds, hearts, spades) corresponds to position; face value (one to thirteen), to speed:

$$card \leftrightarrow molecule$$

$$suit \leftrightarrow position$$

$$face\ value \leftrightarrow speed$$

Consider hands of five cards drawn from a standard deck of fifty-two. (The number 52 corresponds to the number of molecules in the collection.)

Every hand corresponds to what I will call a *microstate* of the collection of molecules. To specify a microstate, we have to specify the position and speed of every molecule in the collection. The fact that the number of distinct five-card hands is very large corresponds to the fact that the number of microstates of the collection of molecules is very large.

A *macrostate* is a class or category of microstates that have some specified property in common—for example, the property that all the molecules are (momentarily) in the left half of the box that houses the collection, or the property that the average speed of the molecules in the top half of the box is twice that of the molecules in the bottom half. The analogue of a macrostate is a class or category of five-card hands that have some specified property in common. The game of poker provides a convenient set of categories: four of a kind (four of the five cards have the same face value), three of a kind, flush (all five cards belong to the same suit), straight (the face values of the five cards form an unbroken sequence), and so on. Thus

$$\text{poker hand} \leftrightarrow \text{microstate}$$

$$\text{flush, straight, and so on} \leftrightarrow \text{macrostate}$$

We are now ready to define randomness. Boltzmann's key idea was to measure the randomness of a macrostate by the number of its member microstates. (Boltzmann called the microstates that belong to a given macrostate its *complexions*. Thus the randomness of a macrostate is measured by the number of its complexions.) Analogously, the randomness associated with a category of poker hands, such as a straight or a flush, is measured by the number of distinct hands in that category. The best poker hands belong to categories of low randomness, the worst to categories of high randomness. The most random category (because more hands belong to it than to any other category) consists of hands in which the cards are not all of the same suit, in which their face values don't form an unbroken sequence, and in which no two cards have the same face value. The least random category is the straight flush: five cards of the same suit whose face values form an unbroken sequence.

Boltzmann defined the randomness of a macrostate not as the number of its complexions, but as the logarithm of that number. You may recall that logarithms are a device for converting products into sums. The logarithm of the product of two numbers is the sum of their logarithms. In the days before cheap electronic calculators, people used tables of logarithms or slide rules—devices for adding logarithms mechanically—to do long multiplication and long division. This property of logarithms accounts for its appearance in Boltzmann's formula, as the following example will make clear.

Consider two distinct physical systems, A and B. We may think of them together as a composite system AB. To specify a macrostate or a microstate of the composite system we have to specify a macrostate or a microstate of each system separately. Suppose that A is in a macrostate X with N_X microstates, and B is in a macrostate Y with N_Y microstates. Then the composite system AB is in a macrostate (which we may call XY) with $N_X \times N_Y$ microstates, because to specify a microstate of the composite system we have to specify the microstates of each of its component systems and there are just $N_X \times N_Y$ distinct ways of doing this. According to Boltzmann's definition, the randomness of the macrostate XY is the logarithm of this number, $N_X \times N_Y$. This logarithm is equal to the sum of the logarithms of N_X and N_Y. Thus the randomness of the macrostate XY is the sum of the randomnesses of the macrostates X and Y. Boltzmann wanted randomness to have this property because he hoped to identify randomness with entropy, and the entropy of a composite system is the sum of the entropies of its component systems. (Other physical quantities that have this property are volume and energy.)

I haven't yet mentioned the main ingredient in the intuitive notion of randomness: chance. The connection between randomness and chance is easy to make. Suppose we don't know which microstate a system is in. If we *postulate* that all microstates are equally likely to be realized—if we assign them equal probabilities—then macrostates that contain a relatively large number of microstates and hence have relatively high randomness are more likely to be realized than macrostates with relatively few microstates. For example, a poker hand that belongs to a category with high randomness—a category that contains a relatively large number of distinct hands—is more likely to be dealt from a well-shuffled deck than a hand that belongs to a category with low randomness; pairs are more common in honest poker games than flushes, which in turn are more common than straight flushes. This follows from the condition that defines a well-shuffled deck: that every distinct hand is equally likely to be dealt from it. The probability of being dealt a hand that belongs to a given category is then proportional to the number of distinct hands in that category.

In scientific contexts, it isn't so easy to assign probabilities to microstates. It may seem plausible that the positions and velocities of molecules in a gas have been "well shuffled" by interactions of the molecules with one another and with the outside world. And indeed the assumption that they are well shuffled nearly always works. But justifying it presents a serious theoretical problem. (I will explain the problem and sketch a possible solution in Chapter 3.)

Order

I propose to define *order* as absence of randomness. At first sight, this definition may seem topsy-turvy. Surely, you may say, it ought to be the other way around: order is a positive quality, and randomness is the absence of this quality. Fair enough. But the varieties of order differ greatly from one another. Structural and chemical order have little in common; biological and physical order, even less. And the processes that generate different varieties of order are as dissimilar as their outcomes. Disorder, by contrast, has a universal quality, and the processes that generate it are governed by a single grand generalization, the second law of thermodynamics. To paraphrase Tolstoy: disorderly systems are all alike; every orderly system is orderly in its own way.

Randomness is a numerical measure of disorder. In the same way, *information*, defined by the formula

$$\text{information} = \text{maximum randomness} - \text{actual randomness}$$

is a numerical measure of order. Information is thus potential randomness. It is the amount by which the randomness of a macrostate falls short of its greatest possible value. Processes that increase randomness destroy information; processes that decrease randomness generate information.

In defining order as absence of randomness, we have focused attention on the common, quantifiable core of all varieties of order. But we haven't suppressed the qualities that make each variety specific. Recall our definition of randomness: the logarithm of the number of microstates, or complexions, belonging to a given macrostate. *The qualities that make each variety of order specific are just the defining properties of the macrostates.* A category of poker hands isn't orderly in the same way as the macrostate of a gas—the defining properties of the two kinds of macrostates are completely different—but the same formula assigns a definite quantity of information to both kinds of order.

Biological Order

The preceding definitions of randomness, information, and order allow us to resolve a long-standing controversy between physicists and biologists about the nature of biological order. The terms of the controversy were clearly set forth by Joseph Needham almost half a century ago. Needham contrasts two views of order:

> The development of modern science has led to a curious divergence of world-views. For the astronomers and the physicists the world is, in popular words, continually "running down" to a state of dead inertness when heat has been uniformly distributed through it. For the biologists and sociologists, a part of the world, at any rate (and for us a very important part) is undergoing a progressive development in which an upward trend is seen, lower states of organisation being succeeded by higher states. For the ordinary man the contradiction, if such it is, is serious, because many physicists, in expounding the former of these principles, the second law of thermodynamics, employ the word "organisation" and say it is always decreasing. Is there a real contradiction here? If so, how can it be resolved?

Needham then mentions two opinions about how the contradiction might be resolved:

1. The concepts of organisation as held by physicists and biologists are the same; but all biological, and hence social, organisation is kept going at the expense of an over-compensating degradation of energy in metabolic upkeep.
2. The two concepts are quite different and incommensurable. We should distinguish between *Order* and *Organisation*.[2]

The first opinion is still held by most physical scientists as well as by some biologists. In an essay published at about the same time as Needham's, Erwin Schroedinger, one of the founders of quantum mechanics, remarked that

> according to what the physicist calls "order," the heat stored up in the sun represents a fabulous provision for order . . . a small fraction [of which] suffices to maintain life on earth by supplying the necessary amount of "order."[3]

Three years later, in an influential little book called *What Is Life?*, Schroedinger went even further. He argued that living organisms maintain their biological organization in the face of the universal tendency for order to decay by "sucking orderliness from [their] environment."

Needham defended the second opinion. His argument, later endorsed by the biologist Peter Medawar, runs as follows: "Biological organization is only the extrapolation of patterns to be found in the non-living world." But even in physical contexts, an increase in organization may be accompanied by an increase in entropy. For example, when ice crystals form in water carefully cooled below its freezing point, entropy is generated because according to the Second Law, every spontaneous process generates entropy. Yet crystallization obviously causes the spatial distribution of water molecules to become more highly organized. So organization has nothing to do with order as physicists define it (that is, in terms of entropy).

Both parts of this argument are flawed. It is true that the crystallization of a supercooled liquid increases its entropy, but entropy isn't a measure of spatial disorder alone. Disorderly molecular *motions* also contribute to entropy. When supercooled water freezes, the increased disorder of molecular motions more than offsets the increased order of crystalline organization.

More important, although biological order is *founded* on the order exemplified by crystals and molecules, it isn't, as Needham maintained, an "extrapolation" of that kind of order. Unlike the structure of a crystal, the structure of a leaf is not implicit in any physical law; it is the outcome of a process, and the outcome is inseparable from the process.

Information and entropy were first linked by James Clerk Maxwell in a famous thought experiment invented to illustrate the limitations of the Second Law. Maxwell imagined a vessel filled with air and divided into two parts, *A* and *B*, by a partition with a small hole. "A being, who can see the individual molecules, opens and closes this hole, so as to allow only the swifter molecules to pass from *A* to *B*, and only the slower ones to pass from *B* to *A*," Maxwell wrote.[4] Thus *A* grows progressively cooler while *B* grows progressively hotter. This conclusion seems to contradict the Second Law, which predicts that temperature differences cannot arise in an isolated system at uniform temperature. Maxwell concluded that the Second Law, along with other laws based on macroscopic observations and experiments, may not be "applicable to the more delicate observations and experiments which we may suppose made by one who can perceive and handle the individual molecules which we deal with only in large masses."

Leo Szilard, in 1929, drew a different conclusion from Maxwell's thought experiment. Maxwell's demon (as the tiny doorkeeper came to be known) causes the entropy of a gas to decrease by obtaining and exploiting information about its microscopic state. Szilard proved that any device—animate or inanimate—that does this must generate more entropy in the process of gaining the information than it causes the gas to lose. Thus the combined entropy of the gas and the demon (or its inanimate counterpart) must increase with time.

Szilard had demonstrated that the process of acquiring information generates entropy. In 1946, Claude Shannon elucidated the structure of information itself. Shannon detached Boltzmann's definition of entropy from its original physical setting (the theory of gases) and applied it to *messages*—strings of letters or other symbols. Armed with this definition, he went on to construct an elegant and consequential theory of communication. But it was the initial step that captured the imagination of scientists in many fields. A "message," as defined by Shannon, is a highly abstract structure, susceptible to an endless variety of concrete interpretations. All kinds of orderly structures can be "digitized"—encoded in strings of numbers—including human speech, music, even pictures (astronomical data are nowadays *gathered,* as well as recorded and analyzed, in this form). And by the mid-1950s it was clear that nature had chosen to encode biological organization in strings of symbols; the strings are molecules of DNA, and the symbols are molecular groups known as nucleotides. Thus information theory seemed to hold the key to an understanding of biological organization.

Yet this bright promise was not fulfilled. As Medawar has written,

> In my opinion the audacious attempt to reveal the formal equivalence of the ideas of biological organization and of thermodynamic order, non-randonmess, and information must be judged to have failed. We still seek a theory of Order in its most interesting and important form, that which is represented by the complex functional and structural integration of living organisms.[5]

Medawar's objection to the use of information as a measure of biological order has been voiced by many biologists. It is brought out by the following imaginary dialogue:

PHYSICIST: A protein is an intricately twisted and folded chain, sometimes with two or more intertwined strands. Each strand is made up of chemical groups called amino acids, linked together in a linear sequence. Only twenty distinct amino acids appear in proteins encoded by DNA. A protein's intricate structure is entirely determined by the linear sequence of amino acids in each of its strands. Suppose we represent the amino acids by the first twenty letters of the alphabet. Then we may regard the linear sequence of amino acids in a strand as a "message." Such "messages" are typically several hundred characters long. The information content of a message is the logarithm of the number of possible messages of the same length. Since there are twenty possibilities for each letter of the message, there are $20 \times 20 \times \ldots \times 20$, or 20^n, distinct messages of length n. The logarithm of this number is $n\log 20$. That is the information content of a message of length n.

BIOLOGIST: Thus all "messages" of the same length contain the same quantity of information. You are entitled to define *information* in that way, but your definition doesn't capture the essential ingredient of biological order. Biological order resides in the *meaning* of a message, not in its length. The meaning of a protein is its specific biological function. Changing one or two letters in a message may greatly impair, or even destroy, the protein's biological function and hence the property that biologists call "order." Yet it would not alter its information content, as physicists define *information*.

The biologist's point is well taken. It is true that the usual definition of information doesn't capture the essential ingredient of biological order. The reason isn't hard to discover. Recall that the randomness of a macrostate is the logarithm of the number of its constituent microstates, and that the information of a macrostate is the difference between its largest possible randomness and its actual randomness. *The conventional recipe for calculating the information of a protein implicitly identifies macrostates with microstates.* Because it assumes that every macrostate consists of just one microstate, it assigns every macrostate zero randomness and the same (very large) quantity of information.

This way of defining macrostates is appropriate in communication engineering, the context in which Shannon developed information theory. Because the cost of transmitting a message doesn't depend on its meaning but only on its length, the telephone company doesn't concern itself with the meanings of the messages it transmits. Biologists, however, care about the meaning of information encoded in DNA. Accordingly, they are justified in rejecting definitions of genetic information that fail to take into account its biological meaning. But Boltzmann's definition of randomness, like every mathematical definition, is a pure form, devoid of content. Let's see if we can't find an appropriate biological content for it.

Why is an insulin molecule an orderly structure? There are many reasons. The building blocks of insulin, as of every protein, are amino acids. Every amino acid is an orderly structure of atoms, which are themselves orderly structures. The amino acids are linked in a chain—another orderly structure. Moreover, only twenty different amino acids—the same twenty in every protein—appear in the chain. These are regularities that pertain to the whole class of proteins, and the fact that the chemical strategy of life features this class of molecules is a profound aspect of biological order. But the specific aspect of biological order peculiar to insulin is its *sequence of amino acids*. Let us for the moment focus attention on this single aspect and ignore all the rest.

How shall we assign values of randomness and information to this sequence? Let's consider an analogous problem. Imagine that a manuscript containing a previously unknown sonnet by Shakespeare has been discovered. Unfortunately, one line has been eaten by mice. A contest is announced to replace the missing line, entries to be judged by a panel of distinguished Shakespeare scholars and poets.

What criteria should the judges use to evaluate the entries? It is probably impossible to lay down explicit criteria beyond the obvious ones concerning meter, rhyme, and diction. Yet experience suggests that there would be a high degree of consensus about which entries were the best and which were not worthy of serious consideration.

The biological analogue of selection by a panel of judges is natural selection. The "fittest" variant of a protein is the variant most likely to maximize its possessor's contribution to the next generation's gene pool. A gene that codes for a defective protein decreases its possessor's chances of surviving to reproductive age or of having healthy offspring. Hence carriers of that gene will tend to become less common in the population. Fitness is usually a difficult quantity to assess—almost as difficult as literary quality. Biologists can rarely be certain that they understand all the functions of a given protein, or how that protein behaves in all relevant internal and external environments. Population geneticists assume, however, that in a given population and a given range of environments every protein has a definite fitness, whether or not it can be measured.

Accepting this assumption, I propose to assign every protein to a class consisting of all the variants of that protein that have equal or greater fitness. Highly fit proteins will then belong to very small classes; unfit proteins, to large classes. Nonfunctional proteins belong to the class of all possible variants: every variant is at least as effective as one that is completely ineffective. We now identify these fitness classes with macrostates. Because the information of a macrostate is equal to the logarithm of the total number of variants minus the logarithm of the number of microstates belonging to that macrostate, our definition assigns high values of information to fit proteins and zero information to nonfunctional proteins.

This definition lends precision to the intuitive notion of biological order (at least in the case of individual proteins and the genes that code for them). Biological order, as we have defined it, contributes to an organism's expectation of reproductive success in a given range of environments. Conversely, natural selection always tends to generate biological order, because it increases the proportion of relatively fit variants of a given protein and decreases the proportion of relatively unfit variants. Thus the aspect of biological order we have been considering is shaped by natural selection.

The Hierarchy of Order-Generating Processes

Biological order isn't confined to the adaptedness of individual proteins to specific biological functions. The chemical reactions underlying the development of an organism are highly coordinated and exquisitely regulated. Some fraction—probably a large fraction—of an organism's genetic material is used to encode the molecular mechanisms that accomplish the miracles of coordination and regulation that even the least complex forms of life exhibit. Just as we did for individual proteins, we can consider the part of the gene script that encodes a specific regulatory task, and define the information associated with it. This information, too, must have been generated by natural selection.

But natural selection is not the only process that generates biological order, and not all forms of biological order reside in genes. No one is born with the ability to speak Chinese or play the piano. These and other kinds of behavior are learned during our lifetimes. As we will see later, learned skills and behaviors probably have their physiological basis in patterns of connections between nerve cells, patterns that come into being gradually during the learning process and whose structure, considered abstractedly, mirrors the structure of the corresponding skill or behavior. The more complex a skill or behavior, the more complex must be the underlying pattern of neural connections and the greater its information content.

Human culture in all its varied manifestations—language, the arts, social and political institutions, religions and ideologies, to name a few of the most obvious—represents still another variety of biological order.

How are all the varieties of order that we have discussed related?

The fundamental distinction between laws and initial conditions entails an equally fundamental distinction between the timeless (Pythagorean) order exemplified by crystals, molecules, atoms, nuclei, and subnuclear particles and the timebound (Heraclitean) order embodied in astronomical and biological structures. Pythagorean order is a direct consequence of physical laws and the entities that these laws posit. Heraclitean order is generated and destroyed by physical processes.

These processes and their outcomes rely on and exploit various aspects of Pythagorean order. For example, the cyclic character of seasonal change, which has profoundly influenced the course of biological evolution, is a consequence of the fact that the Earth moves in a closed orbit around the Sun, which in turn is a consequence of the inverse-square law of gravitational attraction. Again, all forms of biological order depend on the properties of DNA, a molecule that consists of two threadlike chains twisted together in a double helix. Each chain is a linear sequence of molecular units that function as characters in a script that encodes a developmental program. The DNA in a fertilized egg directs a series of chemical reactions that result in the development and growth of an organism. The precise execution of this program depends on the permanence and stability of molecular structure. So do the processes by which DNA is passed from generation to generation. Life could not have arisen or evolved without a language written in permanent characters.

These characters are permanent in the sense that their structure is prescribed by the timeless mathematical laws of quantum physics, but not in the material sense envisioned by Democritus and Newton. In living cells, chemical reactions create new alliances between atoms and reshuffle electrons between molecules. These structural transformations are as essential to the business of life as the structural invariance mandated by the laws of quantum physics.

In Greek, *atom* means ''uncuttable.'' We now know that atoms are not truly uncuttable. But are they composed of elementary (the modern term for ''uncuttable'') particles? The answer depends, at least in part, on how much energy can be mobilized for the task of demolition.

Molecules begin to break up into their constituent atoms and atoms begin to lose electrons at absolute temperatures of a few thousand degrees. Such temperatures are reached in electric furnaces and in the surface layers of the Sun and other stars. Close to the center of a star, where temperatures are measured in millions or tens of millions of degrees, atoms are fully dissociated into free electrons and nuclei. At these temperatures nuclei are just beginning to lose the character of elementary particles. Just as chemical reactions rearrange the atomic constituents of molecules, usually releasing energy in the process, nuclear reactions rearrange the nucleons (neutrons and protons) in nuclei, releasing energy in the process. Such reactions supply the energy that stars radiate in the form of light.

Nuclei dissociate into their constituent nucleons at temperatures around 10,000 million degrees (10^{10} K). Even the cores of stars don't get that hot. But physicists have built particle accelerators that produce electrons and nuclei with the energies they would have at these temperatures. This has allowed physicists to study the structure of atomic nuclei in somewhat the same way as chemists study the structure of atoms and molecules. In this range of experimentally produced energies, the world seems to be is made up of just five kinds of particles: protons, neutrons, electrons, neutrinos, and photons.

But just as the phenomena of chemistry made it plain that atoms and molecules are composite particles, certain nuclear phenomena suggested to physicists that protons and neutrons are composite particles. A new generation of accelerators, capable of producing particles with energies corresponding to temperatures around 10^{13} K—a thousand times greater than the energies needed to split the atomic nucleus—allowed physicists to verify this prediction. Experiments at these energies brought forth a copious, varied, and seemingly chaotic assortment of extremely short-lived particles. Gradually physicists developed theories that brought order to this newly discovered zoo. Although no particle physicist is satisfied with the present state of the theory, it represents a tremendous advance from the state of knowledge and understanding of twenty years ago.

According to current theory, both the neutron and the proton are composed of three elementary particles called quarks. The quarks that compose matter in its ordinary forms—matter as it exists outside high-energy physics labs—are of two kinds, labeled u (for "up") and d (for "down"). The proton contains two u quarks and one d quark; the neutron, one u quark and two d quarks. Since the proton carries one unit of positive electric charge and the neutron is electrically uncharged, the u quark must carry two-thirds of a unit of positive charge; the d quark, one-third of a unit of negative charge. Short-lived particles called mesons are composed of a quark and an antiquark. (A fundamental symmetry of physical laws ensures that for every charged particle there exists an oppositely charged antiparticle with otherwise identical properties. The electron's positively charged antiparticle is the positron, the proton's negatively charged antiparticle is the antiproton, and so on.[6]) The electric charge carried by a meson is either zero ($= \frac{2}{3} - \frac{2}{3} = -\frac{1}{3} + \frac{1}{3}$) or +1 ($= \frac{2}{3} + \frac{1}{3}$) or −1 ($= -\frac{2}{3} - \frac{1}{3}$). Free quarks, which would carry a fractional electric charge, haven't yet been observed, and current theories say they won't be observed because they can't exist in isolation.

Do electrons and quarks mark the end of the road? Are they truly elementary particles? Nobody knows. There are theoretical arguments on both sides of the question. Experiments with the next generation of particle accelerators (superconducting supercolliders) may bring us closer to an answer.

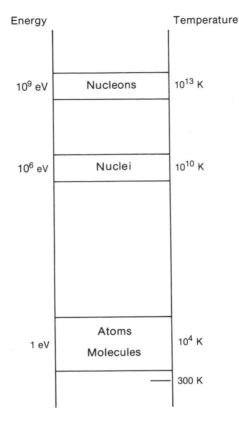

FIGURE 2.4 The quantum ladder. The composite structure of molecules and atoms manifests itself at temperatures around 10,000 degrees absolute (10^4 K). The composite structure of atomic nuclei manifests itself at temperatures 1 million times greater, and that of nucleons (protons and neutrons), 1,000 times greater still. The corresponding energies, measured in a unit called the electron volt (abbreviated eV), are shown at the left. Notice that the temperature and energy scales are logarithmic; equal vertical intervals correspond to equal powers of ten.

Figure 2.4 sums up an important aspect of the preceding discussion. The energy levels of molecules and atoms, atomic nuclei, and nucleons occupy three widely separated bands. Physicist Victor F. Weisskopf refers to them as rungs of the quantum ladder. The lowest rung is occupied by the energy levels of molecules and atoms, the next rung by the energy levels of atomic nuclei, and the highest rung (to date) by the energy levels of protons and neutrons. Within each rung, there is additional hierarchic structure. For example, the energy needed to make or break a link in one of the two intertwined chains that compose a molecule of DNA is much greater than the energy needed to separate the two chains at a given point. This fact has great biological importance. Reproduction depends on reliable replication of genetic material. Because the energy needed to break links between adjacent units in a chain is so large, genetic information, which is encoded in the linear sequence of the units that make up a single chain, is highly stable; because the energy needed to make or break bonds between opposed and complementary units of the intertwined chains is so small, this information is readily copied.

Astronomical and biological systems also exhibit hierarchic structure, as we will discuss in detail in later chapters. These structural hierarchies are outcomes of a more fundamental *hierarchy of initial conditions and order-generating processes.* Every order-generating process requires initial conditions created by processes that occupy earlier hierarchic levels and creates initial conditions required by order-generating processes that occupy later hierarchical levels. This scheme is represented in Figure 2.5, in which only the main levels of the hierarchy are shown:

- Cosmogenic processes in the early Universe gave rise to self-gravitating systems, including the primordial Solar System. They also produced *chemical order,* a concept explained in Chapter 8.
- Within the primordial Solar system, processes of a qualitatively different kind from those that shaped primordial self-gravitating systems produced the planets and their satellites.
- Then physical and chemical processes in the Earth's atmosphere, oceans, and landmasses created conditions that allowed the first self-replicating molecules to emerge, thereby initiating the process of biological evolution.
- Finally, in *Homo sapiens* biological evolution produced the initial conditions needed for cultural evolution, the most rapid and prolific of all order-generating processes.

Each of these hierarchic levels is itself a hierarchically organized aggregate of finer levels. Consider the level called "biological evolution." Evolution isn't a unidirectional process. On the contrary, it is omnidirectional. Yet if we focus attention on those exceedingly rare (but also exceedingly interesting) evolutionary changes that created new levels of biological organization, we see a more detailed hierarchy of processes and initial conditions with the same abstract structure as the coarse hierarchy illustrated in Figure 2.5:

- The emergence of self-replicating molecules paved the way for the evolution of self-reproducing molecular communities.
- These evolved into the ancestors of modern bacterial cells. Bacterial cells evolved into higher, nucleated cells, perhaps by forming symbiotic communities.
- Communities of higher cells became simple multicellular organisms like sponges.
- Increased cellular differentiation led to the emergence of higher plants and of animals like flatworms, with specialized tissues and organs.
- The concerted evolution of sense organs, a central nervous system, and limbs enabled animals to acquire, process, and act on specific kinds of information about their environment.
- As brains evolved further, they gave their possessors the ability to

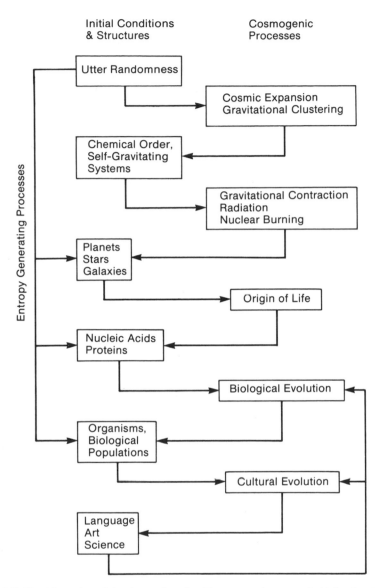

Initial Conditions
& Structures

Cosmogenic
Processes

Utter Randomness

Cosmic Expansion
Gravitational Clustering

Chemical Order,
Self-Gravitating
Systems

Gravitational Contraction
Radiation
Nuclear Burning

Planets
Stars
Galaxies

Origin of Life

Nucleic Acids
Proteins

Biological Evolution

Organisms,
Biological
Populations

Cultural Evolution

Language
Art
Science

Entropy Generating Processes

FIGURE 2.5 The hierarchy of timebound order. Time runs downward. At the beginning of the cosmic expansion, there was little or no order in the Universe. As the Universe expanded, chemical and structural order come into being. Geologic processes on Earth (not explicitly indicated in the diagram) created the conditions under which the first self-replicating molecules formed. Every process creates structures and initial conditions that make possible new processes. Notice the feedback loops that allow human culture to influence both its own evolution and biological evolution itself.

construct and manipulate images, memories, and ideas, to hypothesize possible courses of action, and to choose among them.

Each level and sublevel of the hierarchy of initial conditions embodies a qualitatively distinct variety of order. Yet all these varieties of order consist in the absence of the same kind of disorder. They are like different figures sculpted from the same block of marble. Although the manifestations of order are endlessly varied, the processes that generated and are generating them belong to a small number of distinct categories, and the randomness that order annihilates is of just one kind. Our next task is to probe the nature of randomness, the top rung of the hierarchy illustrated in Figure 2.5.

3

The Strong
Cosmological Principle
and Time's Arrow

According to Laplace and Einstein, chance is a product of human ignorance. This view of chance implies that timebound forms of order, which consist in the absence of randomness, are as subjective as chance itself. For, as we discussed in Chapter 2, microscopic randomness and macroscopic order are two sides of the same coin. It also implies that the openness of the future is illusory. To Laplace's Intelligence, whose knowledge of the external world is perfect and complete, the future differs from the past only in the way that right differs from left.

Does quantum physics offer an escape from Laplacian determinism? Quantum physics tells us that the exact moment when a radioactive atomic nucleus will decay is unpredictable in principle. Yet the laws of quantum physics are just as deterministic as those of classical physics. Given a complete quantal description of the world at a single moment of time, a supercomputer programmed with the laws of quantum physics could calculate any of its past or future states. Why, then, is the exact moment of radioactive decay unpredictable?

Strange as it may seem, quantum physicists disagree about the answer to this fundamental question. Some of them say that when we measure or observe a quantal system, its physical state changes discontinuously and unpredictably. Others assert that the physical state of a quantal system is unknowable. Quantum physics, they say, is just a computational device for calculating the probabilities of observable events. This group of physicists, like the first group, simply postulates that measurements of quantal systems normally have unpredictable outcomes. A third, much smaller, group believes that every measurement causes the entire Universe to split into infinitely many parallel universes, in each of which the measurement has a definite outcome. Which of these universes we happen to live in is a matter of chance. These aren't the only ways in which physicists have tried to explain why measurements of quantal systems have unpredictable outcomes, but they are the most popular.

All of them, however, have serious flaws, as we will see later. This chapter lays the groundwork for what will (I hope) prove to be a better answer. I will argue that, contrary to conventional wisdom, there is a vast amount of genuine, irreducible randomness in the world.

A Toy Universe

Let's begin with a simple model, a "toy universe," that exhibits irreducible randomness. Imagine an infinite straight line divided into cells of equal width, as in Figure 3.1. Some cells contain a counter, represented in the figure by a small filled circle; some are empty. We may represent the distribution of counters by a sequence of zeros and ones:

$$. . . 00101100010 . . .$$

Suppose the counters have been distributed according to the following rule. For each cell, the decision whether to put a counter in the cell or leave it empty has been made by the toss of an unbiased coin. Thus if zero stands for tails, and one for heads, the sequence of zeros and ones also represents the outcome of an infinite sequence of coin tosses. What kind of information does such a sequence contain?

By measuring the fraction of occupied cells in long sub-sequences, we can discover that the average number of counters per cell is close to one-half. The more sub-sequences we measure, the more precisely we can estimate this average value. We can also answer questions like: Does the chance of finding a counter in a given cell depend on whether the adjacent cells (or any other cells) are occupied? Again, our answer will not be exact, but by carrying out enough measurements we can make it as precise as we please. Thus we can recover, with arbitrary precision, the information that characterizes the procedure used to construct the

FIGURE 3.1 A toy universe, shown as an infinite straight line partitioned into equal compartments, each of which contains one counter or none. The state of this "universe" is described by an endless series of zeros and ones.

sequence. I will call this information *statistical information*, because it refers to average values and probabilities: the average number of counters per cell is one-half, and the probability of finding a counter in a given cell doesn't depend on how neighboring cells are occupied.

The actual sequence of zeros and ones represents *nonstatistical* information. At first sight, the sequence seems to contain an infinite quantity of such nonstatistical information. This would certainly be the case if the sequence had a beginning or an end—that is, if it looked like

$$10101100010 \ . \ . \ .$$

or

$$. \ . \ . \ 10101100010$$

instead of like

$$. \ . \ . \ 10101100010 \ . \ . \ .$$

Given two sequences that have a beginning or two sequences that have an end, we can try to decide whether they are identical by matching them. Of course, we can't match two infinite sequences along their entire lengths, but we can at least formulate a criterion for identity:

> *Two sequences that have a beginning (or and end) are identical if, and only if, the sub-sequences consisting of their first (or last)* N *digits are identical for every value of* N.

For if two infinite sequences of the same kind are *not* identical, they must differ in at least one place—say the millionth place. Then the sub-sequences consisting of the first (or last) million digits will differ in that place.

Now consider two sequences, both constructed in the manner described above through tosses of an unbiased coin, that have neither a beginning nor an end. How can we tell whether the two sequences differ? Because neither sequence has an end or a beginning, there are infinitely many ways of trying to match them. The most straightforward generalization of the preceding criterion to this case seems to be the following:

> *Two sequences that have no beginning or end are identical if, and only if, every sub-sequence of the first sequence, no matter how long, has an identical counterpart in the second sequence.*

This criterion is obviously *necessary;* that is, if we could find a single sub-sequence of the first sequence that had no counterpart in the second sequence, the sequences could certainly be said to be different. The criterion tells us, for example, that the sequence

$$. \ . \ . \ 01010101 \ . \ . \ .$$

in which zeros and ones alternate throughout, differs from a sequence in which zeros and ones occur at random. It is harder to decide whether the criterion is *sufficient*—that is, whether two sequences that satisfy the criterion can still be said to be different. Is there a test that would distinguish two such sequences? I can't prove that such test doesn't exist, but I haven't been able to think of one.

Thus endless and beginningless sequences of independent tosses of an unbiased coin are identical. The sequence of zeros and ones is completely determined by its statistical properties; a complete description of the sequence contains only statistical information.

The world contemplated by Laplace's Intelligence is like an infinite sequence of zeros and ones that has a beginning or an end. A complete description of it contains an infinite quantity of information, and a statistical description of it is very incomplete. I will argue that the real world is like an infinite sequence of zeros and ones that has no beginning and no end. A complete description of such a world contains a relatively small quantity of information, all of it statistical.

Can a complete description of the real Universe contain only statistical information—information that refers only to average values and probabilities? It turns out that the answer is yes if the Universe has the following properties:

1. It is *infinite.*
2. It is *statistically uniform.* That is, its average properties are the same at every point in space.
3. Local nonuniformities have a maximum scale.
4. The microstates of finite systems form *discrete,* rather than continuous, arrays.

A universe with these properties has a "statistically repetitive" structure. All possibilities that are consistent with the statistical description (in the toy universe, all finite sub-sequences of zeros and ones) are realized. In fact, they are realized over and over again because the possible realizations are limited (by properties 2 to 4) and the opportunities for their realization are unlimited (by property 1). Thus a single realization of the statistical description contains all possible structures.

The role of discreteness (property 4) is crucial. In the toy universe, each cell represents a system whose physical state is represented by the occupancy of the cell. Thus every system has just two possible states, represented by the occupancies zero and one. According to quantum physics, finite systems do in fact have discrete arrays of possible states, so our toy universe is a quantum universe. In the classical analogue of the toy universe, every counter is at a definite point. To specify a microstate of this classical toy universe we would have to specify not an endless sequence of zeros and ones but an endless sequence of real numbers, represented by nonterminating decimals. In such a sequence, there is zero probability that any single number will be repeated, and zero probability that it will occur in another sequence with the same statistical properties. The microstates of this classical toy universe are thus easily distinguishable, at least in principle.

(These assertions depend on the fact, proved by Georg Cantor at the end of the nineteenth century, that the real numbers, or nonterminating decimals, are infinitely more numerous than the integers. Both sets of numbers are infinite, but the infinity of real numbers is of a higher order than the infinity of integers. In a list of real numbers generated by a procedure that contains some random element, there is zero probability that any of the numbers will be duplicated. See Chapter 1, note 9.)

Thus irreducible or objective randomness—randomness that does not arise from human ignorance—can exist only in a universe whose most fundamental laws are quantal rather than classical. We will explore the link between discreteness and quantum physics in Chapter 4.

Cosmological Observations

The notion that stars are sprinkled more or less evenly throughout an infinite space was held by the Greek atomists. They also conjectured that the stars are distant suns, each with its family of planets. Aristotle, however, held that the Earth is at the center of the Universe. The conflict between these opposing worldviews persisted for well over 2,000 years.

Copernicus adopted the Pythagorean view that the Earth is just one of five planets circling the Sun, but considered the Sun itself to be at the center of Creation, as did Kepler a century later. Giordano Bruno, at the end of the sixteenth century, maintained that the Universe is infinite and unbounded ("the center is everywhere, the circumference nowhere"), that the stars are suns, and that there is nothing special about either the Solar System or its environs. For spreading these and other heretical opinions, he was burned at the stake.

Huygens and Newton were convinced that the stars are suns uniformly sprinkled throughout an infinite space. Nevertheless, the heliocentric hypothesis remained popular among observational astronomers until well into the twentieth century. Harlow Shapley, in 1918, found convincing astronomical evidence that the Sun is far from the center of a vast stellar system, but concluded that this system, the Galaxy, lies at the center of the astronomical Universe.

Newton and Huygens were probably influenced more by a preconceived idea of what the Universe ought to be like than by observational evidence. On any clear, dark night, you can see that the stars are not sprinkled evenly over the sky, but are concentrated toward a bright irregular band of light, the Milky Way. In 1750, Thomas Wright, an amateur astronomer and theologian, suggested that the appearance of the Milky Way could be explained by supposing that the stars are "all moving the same way and not much deviating from the same plane, as the planets in their heliocentric motion do round the solar body." Immanuel Kant developed this inspired suggestion into a convincing scientific argument that became the cornerstone of a comprehensive theory of the astronomical Universe and its origin, which he published in 1755. More than a century earlier, Galileo had looked at the Milky Way through his newly constructed telescope and seen that its light came, at least in part, from a multitude of stars too faint to be seen individually with the naked eye. Kant inferred from this observation that the stars are neither uniformly sprinkled throughout an infinite space nor distributed in a spherical shell centered on the Sun. Instead, he argued, they must belong to a vast disk-shaped system whose members revolve around a distant center under the influence of gravity in the same way that the planets revolve around the Sun.

Kant's speculations about the shape and structure of the astronomical Universe did not end here. Astronomers had noticed here and there among the stars small, diffuse patches of light, some of them vaguely S-shaped. Kant conjectured that these luminous patches are not tenuous clouds of incandescent gas, as his contemporaries believed them to be, but distant stellar systems comparable in all respects with our own Milky Way and distributed more or less evenly throughout infinite space. In this way Kant was able to reconcile the hypothesis of cosmic uniformity with astronomical observations.

Curiously enough, Kant's picture of the astronomical Universe was not taken up by astronomers. Observational evidence seemed to indicate that the Sun was at the center of our stellar system. As for the "spiral nebulae" (the name eventually adopted for Kant's nebulous patches), astronomical opinion remained divided between Kant's view that they are remote stellar systems comparable with our own and the neo-Aristotelean view that they are relatively small and nearby satellites of the Milky Way. The issue was decisively settled in the 1920s by Edwin Hubble, using observations made with what was then the world's largest telescope, a 100-inch reflector on Mount Wilson in southern California. Hubble's observations established that the spiral nebulae are, as Kant had conjectured, stellar systems comparable in size and mass with our own, and that their spatial distribution, viewed on a sufficiently large scale, is the same in all directions and shows

> no evidence of a thinning-out, no trace of a physical boundary. There is not the slightest suggestion of a supersystem of nebulae isolated in a larger world. . . . We may assume that the realm of the nebulae is the universe and that the observable region is a fair sample.[1]

Hubble's studies of the galaxies (the modern term for nebulae) confirmed the prejudices of Democritus, Huygens, Newton, and Kant. They also revealed a new and startling feature of the astronomical Universe: the cosmic expansion. Hubble found that the galaxies are moving away from one another at a rate that, if maintained, causes the distance between any two of them to double in a fixed period of time, provided they don't belong to the same self-gravitating group or cluster. The relative motions of well-separated galaxies are like those of printed spots on an expanding balloon.

A theoretical interpretation of Hubble's discovery already existed in the scientific literature. In 1916, Einstein applied his newly constructed theory of gravitation to the simplest model of the Universe he could imagine: a static, uniform, unbounded medium filling all space. To construct a self-consistent theory, Eistein had to modify his law of gravitation, a step he later described as the greatest scientific mistake of his career. But six years later, in 1922, Alexander Friedmann showed that Einstein's original law does apply to a uniform, unbounded medium, provided the medium is not static but expands uniformly from an initial state of infinite density in the finite past. Hubble's discovery of the cosmic expansion corroborated Friedmann's prediction and supported Einstein's and Hubble's postulate of cosmic uniformity.

Modern studies of the spatial distribution and motions of distant galaxies have enabled astronomers to sample a vastly greater portion of "the realm of the nebulae" than Hubble was able to do, but they have left his main conclusion unaltered: *at any given moment, the average properties of the astronomical Universe are the same everywhere in space, and they are the same in all directions from any given point.*

Strong and unexpected confirmation for this conclusion (which cosmologists call the Cosmological Principle) came in 1965 with the discovery that we are immersed in a sea of blackbody radiation (Chapter 5) whose temperature is about 3 degrees above absolute zero. The photons that make up this sea originated at distances far beyond those of the most distant galaxies, at a time before galaxies could have existed. Their distribution in direction provides by far the most sensitive test we have of the hypothesis that the average properties of the astronomical Universe are the same in all directions from the Earth.

Thus astronomical evidence strongly supports the Cosmological Principle. But what is its scientific status? Is it a phenomenon in need of explanation? Or a postulate?

Theoretical Arguments Bearing on the Cosmological Principle

We touched on this question in Chapter 2, where I argued that the conditions that define the initial state of the Universe can only be posited; they can't be derived from antecedent initial conditions. Many cosmologists believe, however, that it is neither necessary nor desirable to posit statistical uniformity as a property of the initial state of the Universe. They argue that statistical uniformity could have been produced by physical processes in the early Universe, and that what can be explained need not be posited. Because this argument is widely accepted, it deserves careful consideration.

Statistical uniformity is the usual outcome of protracted interactions, which tend to smooth out nonuniformities of density, temperature, and motion. The air density and temperature in a still room are almost perfectly uniform. Can't we attribute the remarkable isotropy of the cosmic microwave background, as well as the less accurately established but still remarkable large-scale isotropy and uniformity in the spatial distribution of galaxies, to interactions analogous to those responsible for the macroscopically uniform distribution of air molecules in a room?

No. The microwave photons we observe today are very old. They have been traveling undisturbed for nearly the entire history of the Universe. Variations in the brightness distribution of microwave radiation over the sky are therefore directly related to *spatial* variations in the temperature and energy density of radiation in the early Universe. Thus the observed lack of variation in the brightness distribution of the radiation over the sky implies a certain calculable degree of uniformity in the spatial distribution of temperature and energy in the early Universe. And this degree of uniformity could not have been produced by interactions, in the way that the macroscopically uniform distribution of air molecules in a room is produced by interactions.

The reason has to do with *spheres of influence*. Imagine a sealed room containing nothing but air and an inflated balloon. Suddenly the balloon bursts. How long will it take the air to settle down to its equilibrium state of uniform density? The expansion of the suddenly released air generates a pressure wave that travels in all directions, carrying news of the altered conditions in the room and initiating local responses to this news. No region can begin to respond until the news has reached it. The time that must elapse between the bursting of the balloon and the attainment of a new, uniform equilibrium state must therefore exceed the time it takes a pressure wave, moving with the speed of sound, to cross the room. The same reasoning tells us that the radius of a region within which equilibrium can be achieved within a given time cannot exceed the distance that sound would travel during that time.

In the early Universe, news about local irregularities traveled with the speed of light. The radius of a region within which uniformity could have been produced when the Universe was t seconds old (a "sphere of influence") is therefore ct, where c is the speed of light. If we calculate the *mass* of such a region, we find that it is proportional to the age t (because, according to Friedmann's theory, the mass density is inversely proportional to the square of t). Interactions strong enough to smooth out large density contrasts occurred only at very early times, when the mass density itself was very high. At these early times, the mass within a sphere of influence was very small—a tiny fraction of the mass of the currently observable Universe. This argument suggests that we can't account for the isotropy of the cosmic microwave background and the isotropy and uniformity of the spatial distribution of galaxies by invoking interactions in the early Universe.

There is a loophole, however. According to an ingenious theory proposed by physicist Alan Guth in 1981, there may have been a period near the beginning of the cosmic expansion, lasting less than 10^{-35} second, during which the Universe expanded by a factor of 10^{50} or more. During this "inflationary" episode, the radius of a sphere of influence would have expanded by a similar factor. If this theory is correct, the whole of the present observable Universe would have been contained within a preinflationary sphere of influence. Interactions in the very early Universe could then have produced the uniformity and isotropy of the present observable Universe, even if the initial state was far from uniform.

At first sight, a theory that *explains* the uniformity and isotropy of the present Universe seems preferable to one that *postulates* these properties. This would certainly be the case if the initial conditions needed to define the Universe were strictly analogous to those needed to define stars or galaxies. We have seen, however, that the initial conditions needed to define the Universe differ in two important ways from those needed to define other physical systems. The cosmological initial conditions define a unique system, and they are irreducible: they can't be deduced from antecedent initial conditions. These properties make the cosmological initial conditions more like laws than like ordinary auxiliary conditions. It is advantageous to have a theory of stellar structure and evolution that embraces a wide range of initial conditions, because the objects the theory describes embody a wide range of initial conditions. It is not advantageous to have a theory of the Universe that embraces a wide range of initial conditions because there is just one Universe. A cosmological theory that deduces rather than posits statistical uniformity must posit more complicated initial conditions. Since this information would form the first link in a causal chain, it could never be explained. In fact, only an insignificant fraction of it could ever be known. Scientific knowledge about the Universe could never be more than a tiny island in a vast sea of invincible ignorance.

Perhaps that is the way the world is. But why assume at the outset that it is so? A follower of Karl Popper might even argue that such an assumption would be methodologically unsound. According to Popper, the distinctive feature of scientific hypotheses is testability. The hypothesis that the Universe is largely unknowable is untestable, because if taken seriously it guarantees its own truth.

Instead of assuming that a complete description of the initial state contains a vast quantity of information, nearly all of it unknowable, I will posit that *a complete description of the initial state contains no information at all that would serve to define a preferred position in space or a preferred direction at any point.* I call this postulate the *Strong Cosmological Principle,* because it explicitly asserts that *no* statistical property of the Universe singles out a preferred position or direction. Earlier in this chapter I argued that a statistical description of an infinite universe satisfying the Strong Cosmological Principle is also a *complete* description: it can't be supplemented by detailed microscopic information because any two realizations of the same statistical description are identical.

If this is true at one moment of time, it is true always. The initial state of the Universe is connected to later states by physical laws. None of our present physical laws discriminates between different positions or between different directions at a given point. For example, planets revolving around the star Sirius (if there are any) obey the same law of gravitation as planets revolving around the Sun, and that law makes no distinction between different directions at a given point. Thus the rate of a pendulum clock (which depends on the weight of the bob) would be exactly the same everywhere on the surface of a uniform, spherical, nonrotating planet. (On the surface of the Earth, the rate of a pendulum clock varies with latitude because of the Earth's rotation and its slightly nonspherical shape.) Since the initial state of the Universe treats all points in space and all directions at every point in a perfectly democratic manner, and since the laws that govern all subsequent changes in the state of the Universe can't introduce preferred positions or directions if there are none to start with, it follows that a complete description of *any* state of the Universe contains no information that would serve to define a preferred position in space or a preferred direction at any point.

This doesn't mean that we can't *gather* detailed, nonstatistical information about any finite portion of the Universe—the Earth and its environs, for example. Nevertheless, a complete description of the Universe satisfying the Strong Cosmological Principle couldn't contain a statement such as "In this part of space, there is a star of a certain mass surrounded by nine planets with the following characteristics. . . ." A description that doesn't discriminate between positions has no way to define the word *this,* just as a description that doesn't discriminate between unaccelerated reference frames (special relativity) has no way to define the word *rest.* A complete statistical account of cosmic evolution would predict that a certain fraction of the mass of the Universe has condensed into stars, that a certain fraction of these stars have properties similar (in some well-defined sense) to those of the Sun, that a certain fraction of these stars have planetary systems, and so on. In other words, it would assign a planetary system like ours some frequency of occurrence, and that is all that can be demanded of a description satisfying the Strong Cosmological Principle. Obvious as the distinction between one place and another may seem, the Strong Cosmological Principle implies that it is merely a prejudice, just as the Principle of Relativity implies that the obvious distinction between rest and uniform motion is merely a prejudice.

Like any scientific postulate, the Strong Cosmological Principle must be judged not by how well it agrees with our prejudices, but by how well it agrees with the evidence and how it fits into the web of physical laws. I will argue that the Strong Cosmological Principle resolves a fundamental difficulty with Boltzmann's statistical interpretation of the second law of thermodynamics. First, however, I have to address a philosophical issue.

Information and Objectivity

"A complete description of the Universe contains only statistical information." Is this a hypothesis about the Universe itself, or about a *description* of the Universe? There are really two issues here. One has to do with the relation between reality and descriptions of reality; the other, with the status of *information*. It will be helpful to separate them.

In his opening lecture to freshman physics students at the California Institute of Technology in 1961, Richard Feynman asked, "How do we *know* there are atoms?"

> By one of the tricks mentioned earlier: we make the hypothesis that there are atoms, and one after the other results come out the way we predict, as they ought to if things *are* made of atoms.[2]

I think that most scientists who have given the matter some thought would agree with this view of the relation between theoretical descriptions and physical reality. To say that atoms (or quarks or genes) exist is to say that these objects are posited or predicted by a highly successful (that is, strongly corroborated) theory.

As theories evolve, the entities they posit or predict change, sometimes radically. The electron posited by J. J. Thomson differs radically from the electron defined by Schroedinger's wave equation, which in turn differs just as radically from the electron defined by Dirac's relativistic theory of the electron. These objects are called by the same name, despite their differences, because they have common characteristics (they are pointlike particles of a certain mass that carry a unit of negative electric charge) and play analogous roles in the descriptions that contain them. Similarly, the gene of Mendelian genetics, the gene named in the slogan "one gene, one enzyme," and the fragmented gene of modern molecular genetics (see Chapter 11) are radically different versions of the unit of heredity.

When scientists assert that electrons or genes aren't just theoretical constructs but "really exist," they mean—or should mean—that recognizable versions of these constructs will continue to figure in successful theories. And because scientific theories evolve by incorporating rather than rejecting earlier theories, the answers that scientists give to the ancient philosophical question "What is there?"[3] change less rapidly and less radically in form than in content.

The assertion that only statistical information is present in a complete description of the Universe is therefore the same sort of statement as the assertion that galaxies are the building blocks of the Universe. It *seems* to be a different sort of statement partly because the word *information* refers explicitly (rather than implicitly) to a theoretical description and partly because the word has subjective connotations: information is something *we* know (or don't know) about something out there. In fact, information, as here defined and used, is as objective as quantities like mass and energy. It differs from mass and energy in being a property of the Universe as a whole rather than of individual subsystems. The same is true of probability. Recall that in the toy universe, the probability of finding a counter in a given cell can be measured with arbitrary precision. But to increase the precision of a measurement, we must examine more cells. The probability of finding a counter in a single cell is a property of an infinite assembly of cells: it is the fraction of cells in the assembly that contain a counter. Because the cells are inherently indistinguishable (by virtue of the Strong Cosmological Principle), their properties are implicitly properties of the entire infinite assembly.

Time's Arrow and the Strong Cosmological Principle

Many everyday phenomena exhibit time's arrow. A cup of hot coffee left to itself gradually cools to room temperature; thereafter, it doesn't spontaneously extract energy from the surrounding air and grow warmer. If I have stirred cream into the coffee, it doesn't spontaneously reseparate. The motions of the coffee produced by stirring die out and don't spontaneously regenerate themselves. These and all other spontaneous macroscopic processes have a built-in temporal signpost: this way to the future.

And yet individual molecular collisions, as well as other microscopic events that make up any macroscopic process, *are* reversible. They don't define a preferred direction of time. The laws that govern atomic and subatomic processes are also reversible. Where, then, does macroscopic irreversibility come from? How can a law that defines a preferred direction of time (the second law of thermodynamics) be derived from more basic laws that accord equal status to the directions of the past and the future? Let's see if we can answer these questions.

Consider a simple experiment in which time's arrow is evident at the macroscopic level but disappears when we examine the behavior of individual molecules. An opened bottle of perfume is left standing in a sealed and perfectly still room. The perfume evaporates. Its molecules diffuse, and eventually they become evenly mixed with air molecules. This process is obviously irreversible; the perfume will never make its way back into the bottle. To see why, let's look at the process in microscopic detail.

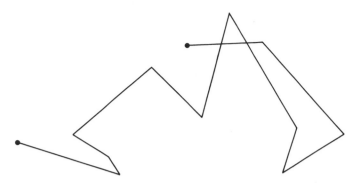

FIGURE 3.2 Path of a molecule in a gas. At each encounter between molecules, the directions of both molecules change sharply.

Each perfume molecule moves along a zigzag path, altering direction sharply whenever it encounters another molecule (Figure 3.2). A reversed film of any encounter represents another possible encounter. Since a molecule's history is a sequence of reversible free flights punctuated by reversible encounters, it, too, is reversible. So if collisions with the walls of the room were also reversible (an unrealistic but not preposterous assumption), the history of the entire collection of air and perfume molecules would be reversible. A reversed film of the experiment, showing the perfume molecules making their way back into the bottle, would make perfect microscopic sense. It would violate no law of microscopic physics. Laplace's Intelligence, which could apprehend, store, and manipulate information about the position and velocity of each molecule, would have no fault to find with it. Of course, *we* would recognize at once that the film had been reversed. Experience, supported by *macroscopic* physics, tells us that such things don't happen. Why not?

The standard answer is: they *could* happen; it's not impossible for the perfume to find its way back into the bottle, just highly unlikely. But aren't physical laws supposed to tell us what *does* happen, not what is *likely* to happen? Aren't statements about probability—the kind of statements meteorologists make about tomorrow's weather—based on incomplete information? Wouldn't Laplace's Intelligence be able to predict tomorrow's weather with complete certainty?

To begin with, we have to be clear about the meaning of *probability* or *chance*. Some people, including a few physical scientists, define the probability of an event, given certain data, as the degree of confidence that a perfectly rational person would have in the occurrence of that event, given those data. Physical scientists, by and large, interpret chance differently. When they say that a radioactive atom has a probability of one-half of decaying within the next hour, they mean two things: first, that half the members of a very large—in principle, infinitely large—sample of identical radioactive atoms will decay within an hour; second, that there is no way of knowing which atoms will decay. These two statements define the *statistical* view of chance. My state of mind regarding the likelihood of an individual event—or of anything else—doesn't come into it. The probability of a given outcome is the frequency of that outcome in a very large (theoretically infinite) collection of identical trials. That's how we defined the probability of finding a counter in a given cell of the toy universe we discussed earlier in this chapter.

Interpreted in this spirit, the statement "The perfume is highly unlikely to find its way back into the bottle" means something like this: if we consider a vast assembly of identical rooms in which identical copies of a bottle of perfume have been allowed to evaporate and diffuse, and if we inspect these rooms after a given interval of time, we will find that the perfume has found its way back into the bottle in only an absurdly small fraction of them.

Why is that fraction so small? For two reasons. First, only an absurdly small fraction of the molecular configurations that are consistent with the way I have prepared the experiment (sealed the room, polished its walls, uncorked the bottle of perfume) lead to that outcome. This seems plausible and isn't hard to prove rigorously. Second, I *assume* that all the molecular configurations that are consistent with the way I have prepared the experiment are equally likely.

This assumption is an essential part of the standard explanation of irreversibility. It appears in all the textbooks. Yet its meaning is far from clear. After all, Laplace's Intelligence, who at a single glance takes in the precise configuration of all the molecules in a room, would not agree that all the configurations consistent with given macroscopic conditions are equally likely. Because the Intelligence *knows* the initial molecular configuration, it would assign a probability of one (certainty) to that configuration and a probability of zero to all the rest. In assigning equal probabilities to each of the initial molecular configurations that are consistent with the way we have prepared the experiment, aren't we describing the state of our knowledge rather than the state of the gas? And if that is so, isn't irreversibility a subjective phenomenon caused by our severely limited capacity for handling information?

Some physicists—I think Einstein was one of them—have bitten the bullet and said, "Yes, irreversibility *is* a subjective phenomenon, a product of the human condition." To a Laplacian Intelligence (or to God), there would be no becoming, only being in a four-dimensional spacetime manifold.

If we reject this point of view, asserting instead that the irreversible behavior of the perfume molecules in our experiment is an objective phenomenon, then we must also maintain that a complete description of the room doesn't fully specify the positions and velocities of the molecules. (If it did, the probabilities that figure in a statistical description of irreversibility would refer to the state of our knowledge rather than to the state of the gas.) But how can a complete description fail to specify the microscopic state of a gas? How could the microscopic state be unknowable even by Laplace's Intelligence?

One plausible answer is this: our sealed room can't be perfectly isolated. For example, we can't shield the molecules in it from gravitational disturbances. Nor can we control interactions between the molecules and the walls, which themselves consist of molecules whose positions and velocities are indirectly influenced by events in the outside world. Thus the assumption of perfect isolation is an unattainable idealization.

But isn't it, nevertheless, a *useful* idealization? Don't *all* theoretical descriptions depend on idealizations of one kind or another?

The answer isn't quite straightforward. Consider friction. Before Newton, friction was regarded as an irrelevant complication and was therefore ignored. Archimedes studied ideal (that is, frictionless) liquids and frictionless machines. Galileo theorized about blocks of wood sliding down perfectly smooth planes and about pendulums that encountered no air resistance (a form of friction). Newton departed from this tradition and studied friction—in an idealized form. In the same spirit of ignoring irrelevant complications to get a clearer view of underlying regularities, students of thermodynamics consider systems cut off from all interaction with the outside world. An ideal isolated system exchanges no mass, no energy, *and no information* with its surroundings.

The neglect of mass exchange and energy exchange can be justified by arguments much like those Galileo would have used to justify his neglect of air resistance in studying the motion of a pendulum. A system that is *almost* perfectly insulated against mass and energy flow behaves, in the short run, very much like an ideal system that is perfectly insulated against these flows. A sufficiently clever experimenter can ensure that the effects of mass and energy leakage are small and uninteresting. *He cannot, however, insulate a gas effectively against the leakage of microscopic information; he cannot shield it from external disturbances that render its microscopic state highly uncertain.* Let a single molecule deviate infinitesimally from its prescribed course, and molecular collisions amplify the deviation and transmit it to molecules in every corner of the room.

A similar phenomenon occurs in the parlor game Telephone, in which a message is whispered from one person to another. If the number of players is large, the final form of the message may be strikingly different from its initial form; errors tend to get amplified in the transmission.

There is an easy and useful way to visualize how tiny disturbances of a gas propagate and grow. Consider first a single molecule constrained to move along a straight line. Its physical state is defined by its position and its velocity (direction and speed). We can represent any combination of position and velocity by a point in a two-dimensional diagram, as in Figures 3.3 and 3.4. The horizontal axis in the diagram represents the line along which the molecule is moving, and the horizontal coordinate of a point in the diagram specifies the particle's position on the line. The vertical coordinate represents the molecule's velocity. Points with positive vertical coordinates (points above the horizontal axis) represent particles moving from left to right; points with negative vertical coordinates (points below the horizontal axis) represent particles moving from right to left.

Let's now generalize this representation by considering a particle constrained to move not along a line but in a plane. To specify its state, we now have to specify two position coordinates and two velocity components. We therefore need a four-dimensional diagram. Analogously, to represent the state of a particle that can roam freely in three dimensions, we need a six-dimensional diagram. Of course we can't draw four-dimensional—let alone six-dimensional—diagrams, but fortunately there is no need to. Everything that's useful to visualize can be represented by a two-dimensional diagram like Figure 3.3 or Figure 3.4, in which the hori-

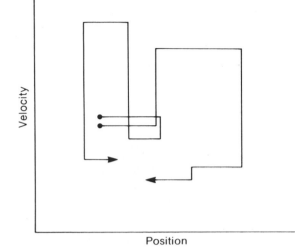

FIGURE 3.3 Trajectories in the 6N-dimensional position–velocity space representing the evolution of two initially similar microstates of a gas consisting of N particles. The 3N position coordinates are represented by the horizontal coordinate; the 3N velocity components, by the vertical coordinate. Between collisions, the 3N velocity components are constant. Whenever two molecules collide, their six velocity components change abruptly. The two trajectories diverge very rapidly.

FIGURE 3.4 If the initial microstate of a gas is uncertain, it may be represented by a small probability cloud in the 6N-dimensional state space. As the gas evolves, the cloud rapidly expands to fill the whole state space.

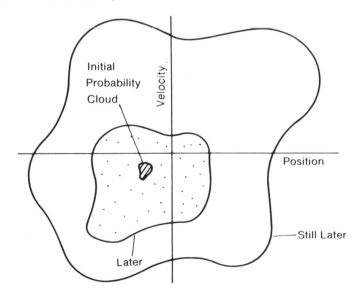

zontal coordinate of a point stands for all three position coordinates and the vertical coordinate stands for all three velocity components.

To represent the state of a pair of molecules, we need a space of twelve dimensions. A point in this space is specified by twelve coordinates. The first three represent the position coordinates of the first molecule; the next three, the position coordinates of the second molecule; the next three, the velocity components of the first molecule; and the final three, the velocity components of the second molecule. But once again we may use a two-dimensional diagram to visualize what is going on. In this diagram the horizontal coordinate stands for all six position coordinates; the vertical coordinate, for all six velocity components. Proceeding in this way, we can represent the state of a collection of an arbitrary number, N, of molecules by a single point in a space of $6N$ dimensions, but we can still use a two-dimensional diagram to visualize what is going on. Let's now do that.

As the molecules in a gas move about and collide, the point in $6N$-dimensional space that represents their collective state traces an exceedingly complicated, zigzag path. Every time two molecules collide, six of the coordinates (the three velocity coordinates of each of the colliding molecules) change drastically, and such collisions occur at an enormous rate. Consider two nearly identical initial states of the gas, represented by two nearby points in the $6N$-dimensional space. Because molecular collisions amplify the initial difference between the two states, the two paths diverge rapidly, however close their starting points (Figure 3.3). But because external disturbances are unavoidably present, the initial state of the gas must be slightly uncertain. So instead of representing it by a point, we should represent it by a small *cloud*, as in Figure 3.4. The probability that the gas's representative point lies in a given region of the cloud is proportional to the ($6N$-dimensional) volume of the region. Because any two nearby points in the cloud diverge in the course of time, every portion of the cloud must expand with time. The whole cloud therefore expands, and eventually fills the accessible part of the $6N$-dimensional space uniformly.

I think this argument is correct and relevant, but incomplete. It assumes that the external disturbances have a random character. No doubt they do—to us. But to Laplace's Intelligence, for whom the Universe is an open book, the microscopic state of our not-quite-isolated roomful of molecules evolves in a perfectly determinate way. The randomness we attribute to external disturbances is illusory, and so is the irreversibility of change. I think we can't avoid this conclusion if we admit the possibility of a Laplacian Intelligence—that is, if we admit the possibility of a complete microscopic description of the world.

The Strong Cosmological Principle supplies the missing ingredient in the preceding account of irreversibility: it makes randomness an objective property of the Universe. The Strong Cosmological Principle implies that nonstatistical information about the state of the Universe as a whole is objectively nonexistent. The probabilities that figure in a complete statistical description of the Universe reflect an objective absence of microscopic order rather than incompleteness in the state of our knowledge.

If we accept the Strong Cosmological Principle, an explanation of why we shouldn't expect the dispersed perfume to reassemble in its bottle runs as follows. We may imagine that the sealed room in our imaginary experiment is one member of an infinite assembly of macroscopically identical rooms sprinkled more or less uniformly throughout the Universe. (A description in which there was only one such room would define a preferred position in space). In all but an exceedingly small fraction of these rooms, the perfume will behave "normally"; it will evaporate and remain dispersed. Because we can't tell which room is ours, we conclude that the perfume will "almost certainly" behave normally. This is exactly the conclusion drawn in every modern textbook. The Strong Cosmological Principle supplies an objective basis for the conclusion and a concrete interpretation of the phrase "almost certainly."

This account of irreversible behavior presupposes that the initial state of the system we are considering doesn't contain *hidden order*. Hidden order and its consequences are illustrated by a popular classroom demonstration.

The demonstrator enters carrying a large glass cylinder like that shown in Figure 3.5. Inside the cylinder is a slightly smaller cylinder. The space between the two cylinders is filled with a colored fluid, and the inner cylinder is fitted with a crank and mounted in such a way that it can rotate inside the outer cylinder in either direction. The demonstrator now begins to turn the inner cylinder slowly counterclockwise. As she does so, the fluid between the cylinders gradually becomes *less* uniform in color. Eventually, the dye concentrates itself into a narrow, straight strip. Order has evolved from chaos, in apparent defiance of the second law of thermodynamics!

The behavior of the dye becomes less mysterious when we learn more about the initial conditions. The fluid between the cylinders (glycerol is a common choice) is highly viscous. If it were not stirred, the dye would diffuse very slowly—hardly at all during the time occupied by the demonstration. Before the demonstrator entered the classroom, she injected a narrow strip of dye into the fluid between the cylinders. She then turned the inner cylinder slowly and carefully in the *clockwise* direction, spreading the strip of dye into a much thinner, very tight spiral sheet. Then she walked into the classroom and proceeded to crank the inner cylinder in the opposite direction, thereby unwinding the spiral and eventually restoring a slightly degraded version of the true initial conditions. The public part of the proceedings didn't create macroscopic order; it merely revealed the presence of macroscopic order that had been created earlier.

This experiment teaches us that macroscopic systems may contain hidden order. Can we be sure that the initial state of our sealed room with its open bottle of perfume doesn't contain hidden order analogous to that of the dye in the fluid between the glass cylinders—hidden order that would cause the dispersed perfume molecules to reassemble in their bottle? I think we can. Even if a demon trickster had, unbeknown to us, injected hidden order into the initial microscopic state of the gas, external disturbances (which, thanks to the Strong Cosmological Principle, must have a genuinely random component) would frustrate his plan. There are other macroscopic systems, however, whose evolution is sensitive to the presence of hidden order in the initial state. In the preceding example, the hidden order, although invisible to students in the classroom, was on a macroscopic scale.

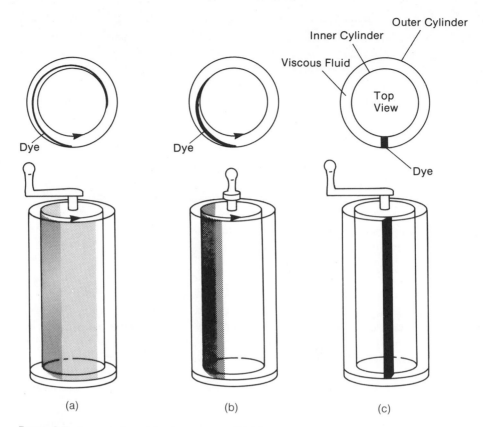

FIGURE **3.5** An experiment to demonstrate "hidden order." The apparatus consists of two concentric cylinders with a viscous fluid, such as glycerol or corn syrup, between them. By turning a crank, the experimenter can rotate the inner cylinder relative to the outer cylinder. Before entering a lecture hall, the demonstrator injects a thin vertical ribbon of dye into the fluid between the cylinders. She then rotates the inner cylinder, winding the ribbon into a thin spiral sheet, as shown in (a). To the audience, the dye appears uniformly distributed between the cylinders. The top view shows that the dye actually occupies a small fraction of the region between the cylinders; its orderly distribution is present but hidden. By rotating the inner cylinder in the opposite direction, the demonstrator restores a slightly degraded version of the thin ribbon of dye, as shown in (b) and (c). The spectators, who did not witness the initial winding up of the ribbon, think that they are seeing order evolve from chaos. They are actually seeing hidden order being made manifest.

But physicists have been able to devise analogous experiments in which the order hidden in an initial state is truly at the molecular level.[4] A macroscopic system with such an initial state may evolve "anti-entropically"; that is, its macroscopic order may increase with time. This is illustrated by the following thought experiment, which features a mutant of Maxwell's demon.

Recall that in Maxwell's thought experiment a minuscule doorkeeper regulates the passage of individual molecules through a tiny door in a partition between two compartments of an isolated box of gas, letting only fast molecules pass from left to right, only slow molecules from right to left. As a result, the right-hand compartment gradually heats up; the left-hand compartment gradually cools down. An observer unaware of the demon's presence would record a steady decline in the entropy of the gas, in apparent violation of the Second Law. But to acquire the information needed to discriminate between fast and slow molecules, the demon would have to generate more than enough entropy to offset this decline. Thus the combined entropy of the gas and the demon would always increase, as required by the Second Law.

Suppose, however, that we replace the demon by a computer that has been programmed by a Laplacian Intelligence to open and shut the door at predetermined times, calculated from an exact knowledge of the gas's initial state and of the relevant physical laws governing the motion of the gas molecules. Unlike Maxwell's demon, the computer doesn't have to acquire information about individual molecules; it has all the information it needs at the outset. Of course, the computer and the mechanism that opens and shuts the door generate some entropy, but at a rate that is not directly linked to the rate at which the gas loses entropy. In principle, there is no reason why the total entropy of the system couldn't decline, at least for a time, thereby violating the Second Law.

In this thought experiment, an enormous quantity of microscopic information—the information represented by a description of the gas's initial microstate—is present to begin with. Some of this information becomes manifest in the form of macroscopic order (a difference in temperature between the two compartments) in the course of the experiment. This thought experiment illustrates once more that the decline of macroscopic order (that is, the growth of entropy) in an isolated system is contingent on the initial absence of certain kinds of *microscopic* order. The Strong Cosmological Principle ensures that microscopic order *may* be objectively absent, but doesn't rule out the possibility that the initial state of a particular system could contain hidden microscopic order.

Even if hidden microscopic order is initially present, however, it may be destroyed too quickly by external disturbances to give rise to a noticeable increase in macroscopic order. This is always the case for macroscopic samples of gases. But there are other systems with a large number of particles for which anti-entropic initial conditions *can* be realized. Such systems may not obey the laws of macroscopic physics, which presuppose the irrelevance of microscopic information and which accordingly predict entropic behavior—the diffusion of dye in a liquid, the flow of heat from warmer to cooler regions. To predict how such a system will behave, we must know something about its history.

Thus the Second Law has an entirely different character from the laws that govern elementary particles and their interactions. These laws are independent of initial conditions. They are true expressions of timeless order. The Second Law rests on a *historical generalization:* it holds for macroscopic systems that were formed with little or no microscopic order. The Second Law owes its enormous—but incomplete—generality to the fact that nearly all naturally occurring systems were indeed formed with little or no microscopic order.

4

The Importance of Being Discrete

The world of Galileo, Newton, and Maxwell—the world of classical physics—is fundamentally continuous. In Chapters 2 and 3, we saw that Boltzmann's definition of randomness and the Strong Cosmological Principle, the twin pillars supporting the theory of timebound order sketched there, depend on the assumption that the world is fundamentally discrete. In this chapter and the next, we will try to gain insight into what is arguably the most important scientific discovery of the twentieth century, the discovery that physical reality is at bottom discrete.

"The Point at Which Science Must Stop"

Newton's *Mathematical Principles of Natural Philosophy* (the *Principia*) appeared in 1687. During the following century, scientists devoted most of their efforts to working out its implications. They constructed mathematical theories that described the motions of particles, of solid bodies, of fluids, and of the planets and their satellites, all based on Newton's three laws of motion and his theory of universal gravitation. With these branches of science well secured, nineteenth-century scientists set out to conquer new territories. Experimental and theoretical investigations of light, electricity, and magnetism culminated in James Clerk Maxwell's electrodynamics, a theory that expanded the framework of Newtonian physics and brought all three groups of phenomena under a common roof. Studies of heat culminated in the science of thermodynamics, with its two great laws: the law of conservation of energy (formulated by Robert Mayer and James Joule) and the law of entropy growth (formulated by Rudolf Clausius). Unlike electromagnetism, however, thermodynamics had only a tenuous link with Newtonian physics, through the concept of energy. Maxwell, Clausius, and Ludwig Boltzmann immediately set out to construct a dynamical basis for the science of heat. Their point of departure was the atomic hypothesis. Their aim was to show that the macroscopic properties of gases, liquids, and solids could be deduced from the properties and behavior of their constituent molecules, moving and interacting according to Newton's laws of motion. In its initial stages, this program was spectacularly successful. By 1880, many scientists had come to believe that all the laws of physics had been discovered and that the remaining unsolved problems—the problem of radiation, for example—wouldn't hold out much longer.

Not all physicists were so sanguine, however. In a popular lecture published in 1880, Maxwell began by explaining why he and other physicists were convinced that gases were made up of molecules moving and interacting according to the laws of Newtonian physics. But, he went on, molecules can't be just miniature billiard balls:

> The molecule, though indestructible, is not a hard rigid body, but is capable of internal movements, and when these are excited, it emits rays, the wave-length of which is a measure of the time of vibration of the molecule.

Less concisely: if we look at a glowing cloud of gas through a spectroscope, whose prism spreads the light into a rainbow, or spectrum, we see a set of bright lines, each a definite pure color. A sodium lamp, for example, emits most of its visible light in two closely spaced yellow lines. The colors (or wavelengths) of these lines are characteristic of the molecules that make up the cloud. Every atom and molecule has its own pattern of bright lines, as distinctive as a thumbprint. What is more—and this is the crux of Maxwell's argument—careful measurements have shown that the patterns of lines emitted by samples of atomic hydrogen obtained from different terrestrial sources are identical with one another and with the patterns imprinted on the light we receive from the Sun, the stars, and interstellar gas clouds.

> We are thus assured that molecules [we would now call them atoms] of the same nature as those of our hydrogen exist in those distant regions, or at least did exist when the light by which we see them was emitted. . . . A molecule of hydrogen, . . . whether in Sirius or Arcturus, executes its vibrations in precisely the same time.

It follows, said Maxwell, that the vibrations in a hydrogen "molecule" don't obey Newton's laws. This becomes obvious when we compare them with the motions of planets in the solar system, which we know obey Newton's laws:

> The form and dimensions of the orbits of the planets, for instance, are not determined by any law of nature, but depend upon [initial conditions]. . . . [N]atural causes . . . are at work, which tend to modify, if they do not at length destroy, all the arrangements of the earth and the whole solar system. But though in the course of ages catastrophes have occurred and may yet occur in the heavens,

though ancient systems may be dissolved and new systems evolved out of their ruins, the molecules out of which these systems are built—the foundation stones of the material universe—remain unbroken and unworn.

Maxwell concluded that physics had reached an impasse:

> We have been led, along a strictly scientific path, very near to the point at which Science must stop.[1]

By "Science" Maxwell meant not just the physics of his own day but any theory of the kind that modern physicists call "classical." A classical theory pictures the physical world as made up of *events,* each referring to a definite point in space and a definite instant of time. The mathematical laws of a classical theory link events that refer to a given moment of time with events that refer to earlier and later moments.

What are these events? Newton, like Democritus, pictured the world as made up of particles moving in empty space. In Newtonian physics, every event is associated with a particle and is defined by the particle's position, velocity, and mass. Around the middle of the nineteenth century, Michael Faraday added an important new element to this picture, the electromagnetic field. Faraday thought of electric and magnetic fields as permeating the space between electrically charged particles, the sources of the field. The electric field at any given point in space is equal to the force that would be experienced by a stationary particle at that point carrying one unit of positive electric charge, but the field is there whether or not a charged particle is there to experience it. Analogously, a magnetic field may be defined by its action on a piece of wire carrying an electric current or on a moving electric charge.

Although the introduction of the electromagnetic field greatly enriched classical physics, it didn't alter its basic character. Neither did Einstein's two great theories of space, time, and energy: special relativity (1905) and general relativity (1915); they continued to describe the world in a "classical" way—that is, as consisting of events in a spacetime continuum, albeit a spacetime continuum with radically different properties from those attributed to it by earlier scientists.

The Quantum World

Maxwell was perfectly correct in his belief that science, in the only sense that would have been intelligible to a scientist of the late nineteenth century, would never be able to explain why the "foundation stones of the material universe" are forever "unbroken and unworn." Quantum physics, which emerged during the first quarter of the twentieth century, *did* explain this, and much more besides. But quantum physics is not a classical theory. Its picture of physical reality differs so radically from that of classical physics that the term *picture* has to be stretched to apply to it. How do you picture a pointlike particle that can be in many places at the same time? According to quantum physics, the electron is such a particle.

Or rather, that is the kind of statement about electrons you get when you try to translate the language of quantum physics into English. The result doesn't make sense—and certainly can't be illustrated by a picture.

The basic concepts of classical physics are not far removed from experience and ordinary language. (At least they don't seem so today.) Words like *velocity, acceleration, mass,* and *force* may not mean quite the same things to a poet as to a physicist. Still, the differences in meaning aren't likely to give rise to gross misunderstandings. The basic concepts of quantum physics refer to mathematical notions that are far less directly connected with experience. Yet the quantum world does have a coherent and understandable structure, some of whose most important features can, I believe, be communicated in largely nonmathematical language.

The quantum world differs from the classical world in four important ways.

1. The classical world is continuous; the quantum world is discrete. The terms *continuous* and *discrete* apply to collections. The members of a discrete collection are separate from one other; each one has some space, real or metaphorical, around it. In a continuous collection, or continuum, there are no gaps. The points on a line form a continuum; the divisions on a ruler form a discrete set.

Maxwell's comparison between atoms and planetary systems illustrates this contrast between the classical and quantum worlds. An artificial satellite has a continuous range of possible periods and orbital diameters, but a hydrogen atom must have a discrete set of possible physical states, because it emits and absorbs light only at particular wavelengths. The chemical elements themselves, arranged in order of mass or of the electric charge carried by the atomic nucleus, form a discrete sequence. There is no element whose mass lies between the mass of hydrogen (1 unit) and the mass of deuterium (2 units), or whose nuclear electric charge lies between those of hydrogen (1 unit) and helium (2 units). In classical physics, mass and electric charge have continuous ranges of possible values.

2. The classical world is a world of flux and irreversible change. But crystals, molecules, atoms, and subatomic particles—the objects that populate the quantum world—are permanent and immutable.

3. In the classical world, change is both continuous and, in principle, completely predictable. The observable effects of change in the quantum world are discontinuous and, in principle, partly unpredictable. When a hydrogen atom is observed to emit a photon, its physical state changes abruptly. According to quantum physics, it is impossible to predict exactly when such an event will occur. Theory tells us only what fraction of the atoms in a cloud of excited hydrogen atoms will emit a photon during a given interval of time.

4. The classical world is populated by individuals; the quantum world is populated by clones. Two classical objects—a pair of ball bearings, for example—can't be precisely alike in every respect. But according to quantum physics, all electrons are exact replicas of one another. They are indistinguishable not only in practice but also in principle, as are all hydrogen atoms, all water molecules, and all salt crystals (apart from size). This is not simply a dogma, but a testable and strongly corroborated hypothesis. For example, if electrons weren't absolutely indistinguishable, two hydrogen atoms would form a much more weakly bound

molecule than they actually do. The absolute indistinguishability of the electrons in the two atoms gives rise to an ''extra'' attractive force between them. The indistinguishability of electrons is also responsible for the structure of the periodic table—that is, for the fact that elements in the same column of the table (inert gases, halogens, alkali metals, alkali earths, and so on) have similar chemical properties.

These four nonclassical aspects of the quantum world are closely related. Perhaps the most fundamental of them is discreteness, because it is implicated in all the others. For example, consider the second property, permanence. The fact that atoms and molecules have fixed properties depends on the fact that their possible physical states are discrete rather than continuous. When the physical state of an atom or a molecule changes, it changes from one well-defined state to another.

Indistinguishability, the fourth property, also depends on discreteness. Electrons wouldn't be indistinguishable if their electric charge could assume a continuous range of possible values.

Finally, discreteness entails unpredictability, the third property. This connection is a little subtler than the others. Consider a collection of free neutrons. Outside the atomic nucleus, the neutron is an unstable particle, decaying spontaneously into a proton, an electron, and an antineutrino (a massless particle that interacts very weakly with other kinds of matter). After about ten and a half minutes, half the neutrons in a large sample will have decayed. Quantum physics doesn't, however, predict when an individual neutron will decay. Now suppose that neutrons had some as-yet-unknown property that, if it could be observed, would enable us to predict their individual lifetimes. Since the observed lifetimes of individual neutrons form a continuum, the possible values of this hypothetical property would also have to form a continuum. But if the internal states of atomic and subatomic objects form discrete aggregates, this is impossible. Thus discreteness is incompatible with the predictability of certain kinds of atomic events.

Discreteness and Continuity in Greek Mathematics

The first generations of Greek mathematicians recognized that discreteness and continuity are complementary and contradictory aspects of the world of mathematics (which they believed underlies the world of experience). Arithmetic is the realm of the discrete; geometry, of the continuous. We can represent the integers by line segments whose lengths are multiples of an arbitrarily chosen unit segment, but we can't always represent line segments by integers. For example, it isn't possible to represent a side and a diagonal of a square simultaneously by integers. Pythagoras, or one of his disciples, discovered this by the following argument:

Suppose it *was* possible to represent a side and a diagonal of a square by two integers, s and d, so that the side contained s appropriately chosen units of length and the diagonal, d units. We may assume that s and d have no common divisor; if they did have a common divisor, c, we could choose a new unit of length c

times as long as the original one, and then the square's side and diagonal would be represented by integers without a common divisor. Pythagoras's theorem tells us that $d^2 = 2s^2$. But that is impossible, because it implies that d is an even number, which implies that d^2 is divisible by 4, which, in turn, implies that s^2 is even, and hence that s itself is even, contrary to our assumption that s and d have no common divisor. So our original premise, that there exists a common unit of measurement for a side and a diagonal of a square, must be false.

We may think of geometry, with its continuously variable figures and magnitudes, as a metaphor for the world of everyday experience, and of arithmetic, the science of integers, as a metaphor for the discrete, unchanging quantum world that underlies experience. But the relations between geometry and classical physics and between arithmetic and quantum physics are more than metaphorical. Classical physics is a direct descendant of Greek geometry. Archimedes widened the net of Euclid's geometry by adding axioms about weight, force, and the balance of forces. Archimedean statics forms the basis of the modern mechanical engineering curriculum. Archimedes himself invented many practical applications of his theories, among them a variety of devices for redirecting and multiplying forces, such as pulleys, winches, and the hydraulic screw. Galileo began where Archimedes had left off. His earliest efforts were the solutions of problems that Archimedes might have assigned to a graduate student, had he had one. Galileo went on, however, to extend Greek mathematical physics in a direction that no Greek, not even Archimedes, had anticipated: the description of motion—the motion of bodies sliding without friction down inclined planes and curved surfaces, falling from towers, being shot out of cannons. Finally, Galileo's new science of motion, along with Johannes Kepler's theory of planetary motion, itself a direct continuation of the work of a long line of Greek mathematicians, paved the way for the central achievement of classical physics: Newton's theory of motion and universal gravitation.

Quantum physics has no such direct link with the Greek past. Yet its worldview has much in common with that of Pythagoras and his followers, who believed that mathematical objects and relations are the building blocks of physical reality. Aristotle, who was not well disposed toward this worldview—he considered mathematics to be an idealized representation of the surface appearance of things—tells us in the *Metaphysics* that the Pythagoreans "supposed the whole heaven to be a *harmonia* and a number." The Pythagorean notion of harmony will repay a closer look because, as we will see, it also figures in the quantum physicist's picture of physical reality—not, of course, in a way that the Pythagoreans could have anticipated, but in a way they surely would have appreciated.

Pythagoras's Theory of Musical Harmony

Pythagoras discovered that consonant musical intervals—in his day they would have been octaves, fifths, and fourths—are produced by vibrating strings that have the same thickness, density, and tension and whose lengths are in the ratio of small integers. Halving the length of a vibrating string raises its pitch an octave. Reducing its length by a third raises its pitch a fifth. Thus octaves correspond to the ratio 2/1; fifths, to the ratio 3/2. The interval of a fourth is a fifth below the octave, so it corresponds to the ratio $2/1 \times 2/3 = 4/3$. Moving upward and downward in fifths from an arbitrary starting point (middle C, say) and calling notes that differ by an octave by the same name, we obtain a sequence that includes

. . . , E-flat, B-flat, F, C, G, D, A, E, B, F-sharp, C-sharp, G-sharp, . . .

Each set of seven consecutive notes in this sequence yields a diatonic scale, the scale defined by seven consecutive white keys on a piano. For example, the sequence beginning with F contains the seven tones of the C major scale; the sequence beginning with E-flat contains the seven tones of the B-flat major scale; and so on. Each set of five consecutive notes yields a pentatonic scale, the scale defined by five consecutive black keys on a piano. The pentatonic scale is common in folk music.

Pythagoras's scale isn't quite identical with the modern one, though. A sequence of perfect fifths goes on endlessly in both directions. E-sharp is a little sharper than F; C-flat is a little flatter than B; and so on. But Pythagoras's theory suggests a way to close up the sequence, to make it into a *circle* of fifths containing twelve and only twelve distinct tones. The theory makes two distinct assertions: (1) that equal musical intervals correspond to equal ratios of the lengths of vibrating strings of given density, thickness, and tension; and (2) that *consonant* intervals correspond to ratios between small whole numbers. To close the circle of fifths, we have to divide the octave into twelve equal intervals. Pythagoras's first assertion implies that the ratio corresponding to the interval between successive tones in a scale containing twelve equal intervals must be the twelfth root of 2. Since the fifth tone of the diatonic scale (G in the C major scale) is the seventh tone of the twelve-tone scale, the interval of a fifth must correspond to the seventwelfths power of 2, or 1.498307. This differs from a perfect Pythagorean fifth, corresponding to the ratio 1.5, by slightly more than one part in a thousand.

In one of his Norton Lectures, on music of the golden age from Bach to Beethoven, Leonard Bernstein refers to the interplay between the "two forces of chromaticism and diatonicism, forces that were equally powerful and presumably contradictory in nature. This point of delicate balance is like the still center in the flux of musical history." The preceding calculation shows that this delicate balance is a consequence of the fact that the number 2 raised to the seven-twelfths power is very close to 3/2.

Nowadays we regard music and physics as being at opposite poles of the cultural spectrum, but to Pythagoras and his contemporaries the properties of musical sounds were no less a part of the natural world than the properties of solids. The rule that every musical interval corresponds to a definite ratio between the lengths of vibrating strings of the same density, thickness, and tension must have seemed to them just as objective as the rule that the volumes of similar solids are proportional to the cubes of corresponding edges: both would have been regarded as natural laws. And the connection between *consonant* musical intervals and whole numbers might easily have been seen as the tip of an iceberg, affording the first glimpse of a deep correspondence between natural order and mathematical order.

Although the Pythagorean belief in an underlying mathematical order is shared by most contemporary theoretical physicists, some historians and philosophers of science dismiss it as a form of mysticism akin to numerology and astrology. The difference between the Pythagoreans and modern physicists, according to these writers, is that modern physicists, although they may be mystics at heart, submit the laws they devise to rigorous testing, whereas the ancient Greeks, as everybody knows, were long on speculation but short on testing. The evidence usually adduced to support this view is that Plato consistently disparaged experience as a source of knowledge.

Plato wasn't a practicing scientist, however. Even though documentary evidence is lacking, I think we can safely assume that experiments and observations played important roles in the development of all branches of Greek mathematical science: geometry, harmony, optics, astronomy, mechanics, and hydrostatics. Why? Because Greek science is highly veridical. The theorems of Euclidean geometry express verifiable properties of rigid bodies. The theorems of Greek optics correctly describe how light-rays are reflected from plane and spherical mirrors. Pythagoras's theory of musical harmony works. So does Aristarchus's heliocentric model of the Solar System, which Kepler took as his starting point. And so does Archimedes's theory of statics, which Galileo, Huygens, and Newton took as theirs. Unless we assume that Greek mathematical scientists were clairvoyant, we must assume that in formulating their theories they relied heavily on experiments and observations, although, of course, their experiments may have been less systematic than those of seventeenth-century scientists.

Overtones

To understand the connection between musical harmony and quantum physics, we have to dig a little deeper into the question of why certain intervals, such as octaves and fifths, are pleasing to the ear, while others—octaves and fifths played slightly out of tune, for instance—are not. The answer (insofar as it falls within the domain of physics) has to do with *overtones*. A well-made tuning fork or an electronic oscillator emits a pure tone, wholly lacking in overtones. Musical instruments, however, produce tones that are mixtures of pure tones: the fundamental (which determines the pitch we hear) and its overtones. The relative strengths of these pure tones determine the quality or timbre of the sound. The full sound of a fine violin is rich in overtones, while the pure sound of a recorder is weak in overtones.

The strongest overtones in most musical tones are the first two: the octave and the fifth above the octave. The pitch of the next strongest overtone is two octaves above the fundamental, and of the following one, two octaves and a major third above the fundamental.

Experiments show that two *pure* tones played together produce a dissonant sound only if their pitches are just close enough to be distinguishable. Two *musical* tones played together produce a dissonant sound whenever any of their constituent pure tones are just close enough in pitch to be distinguishable. Octaves and fifths played in tune on musical instruments are consonant because the pitches

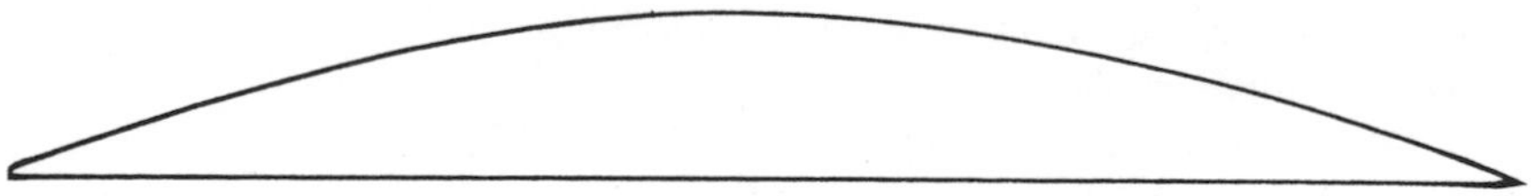

FIGURE 4.1 An ideal string vibrating in its fundamental mode. The waveform is half a sine wave.

of their constituent pure tones either coincide or are well separated. Unisons, octaves, and fifths played slightly out of tune are especially dissonant because their strongest pure constituents are close but distinguishable in pitch.

The fundamental and its overtones correspond to different ways (modes) in which an ideal string with fixed endpoints can vibrate. When an ideal string vibrates in the fundamental mode its waveform is half a sine wave (Figure 4.1). The waveform of the first overtone, whose pitch is an octave higher than that of the fundamental, is a full sine wave (Figure 4.2). The midpoint of the string is a node of the vibration; that is, it remains at rest. You can make a bowed string vibrate in this mode by touching it lightly at this point. The waveform of the next overtone has two nodes (Figure 4.3).

In each of these modes, every point on the string except the endpoints and the nodes (if there are any) moves up and down periodically. If we plot the vertical displacement of a single point on the string against time, we get another sine curve (Figure 4.4). The curves that represent the vertical motion at different points differ only in amplitude. This kind of vibration is called *harmonic,* because a sound wave in which the air pressure varies in this way carries a pure musical tone.

The duration of a single cycle of a vibration is called the period. The periods of the various modes are simply related. The period of the first overtone is one-half the period of the fundamental; the period of the second overtone is one-third the period of the fundamental; and so on. Thus the period of a mode is proportional to its wavelength.

Up until now, we've discussed *possible* modes of vibration of a perfect violin string. In practice, however, a plucked or bowed violin string never vibrates in one of these modes. Plucking or bowing excites a *composite mode*—a *superposition* of simple harmonic modes, each vibrating with its own period as though the other modes weren't there. The meaning of this statement is illustrated in Figure 4.5. To construct the waveform of a composite mode, we add the displacements

FIGURE 4.2 An ideal string vibrating in its first overtone. The waveform is a full sine wave.

FIGURE 4.3 An ideal string vibrating in its second overtone. The waveform is a sine wave and a half.

associated with its constituent simple harmonic modes, treating upward displacements from the string's unperturbed position as positive quantities and downward displacements as negative quantities, so that equal and opposite displacements cancel.

The contribution of each simple harmonic mode to the superposition is proportional to the mode's *amplitude*. The amplitude of a mode is a positive or negative number equal in magnitude to the height of the highest point of the waveform. If the amplitude is positive, that mode's contributions to the net displacement at each point are added to the contributions of the remaining modes; if it is negative, they are subtracted. Thus in Figure 4.5, the amplitudes of two modes (the fundamental and the fourth overtone) are positive, and the amplitude of the third mode (the second overtone) is negative. To specify a composite mode, we have to specify the amplitude of each of its constituent harmonic modes.

The vibrating string exemplifies two important properties shared by all vibrating systems.

1. *Simple harmonic modes.* Every vibrating system has simple harmonic modes, in which the displacement at every point is represented, as in Figure 4.4, by a sine curve with a fixed period. In general, the ratios between the periods of the simple harmonic modes are not whole numbers, as they are for the vibrating string. And the waveforms of the simple harmonic modes have less simple shapes

FIGURE 4.4 The vertical displacement of a single point on a vibrating string plotted against time. The plotted points lie on a sine curve. The width of the curve represents the period of the vibration.

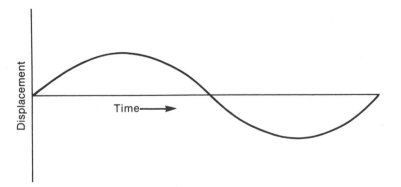

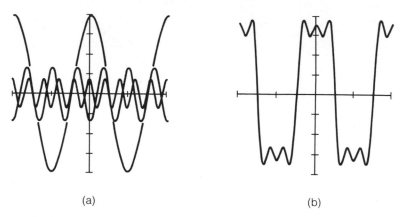

(a)   (b)

FIGURE 4.5 Constructing a composite mode by adding simple modes. (a) Three sinusoidal waves whose wavelengths are in the ratios 1:3:5. (b) The composite wave formed from these simple waves. The three waves interfere constructively at the center of the illustrated region and interfere destructively on either side of this region. (From *The Science of Musical Sound,* by John R. Pierce. Copyright © 1983 Scientific American Books Inc. Reprinted with permission)

than those of a vibrating string. Figure 4.6 shows the waveforms of some harmonic modes of a vibrating drumhead of uniform thickness. The modes illustrated here are all symmetric about the drumhead's axis of symmetry, a line perpendicular to the drumhead through its center. Although the periods of the overtones are not submultiples (1/2, 1/3, 1/4, and so on) of the period of the fundamental, as they are for the vibrating string, they do get smaller as the number of nodes in the waveform gets larger.

2. *Superposition.* A vibrating system can vibrate in several modes at once, each mode vibrating as though the others weren't there. The displacement at any point is the algebraic sum of the displacements resulting from the individual modes—upward displacements being counted as positive, negative displacements as negative.

Physical States of the Hydrogen Atom

Now consider a hydrogen atom, which consists of a single electron bound by electrostatic attraction to a proton (a particle about 2,000 times as massive as an electron, carrying 1 unit of positive electric charge). Because the proton is so massive, we may treat it as a fixed center of attraction, just as we treat the Sun as a fixed center of attraction for the planets. If the electron behaved like a planet revolving around the Sun, its possible physical states would be represented by orbits. According to quantum physics, however, electrons bound in atoms don't behave like planets.

The possible physical states of a hydrogen atom are analogous to the modes of vibration of a vibrating system. States in which the energy of the hydrogen

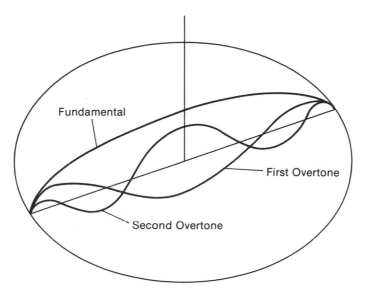

FIGURE 4.6 Vertical cross section of a drumhead of uniform thickness, vibrating in its three lowest axially symmetric modes (schematic). The waveforms are *not* sine curves, and the periods of the first and second overtones are not submultiples of the period of the fundamental. The frequency (reciprocal period) of the first overtone is 2.3 times the frequency of the fundamental (instead of 2 times the frequency of the fundamental, as it is for a vibrating string). The frequency of the second overtone is 3.6 times (instead of 3 times) the frequency of the fundamental.

atom has a definite value correspond to simple harmonic vibrations (which have a definite period). The most general state of a hydrogen atom corresponds to a superposition of simple harmonic vibrations.

The vibrations that represent physical states of a hydrogen atom are three dimensional. Figure 4.7 shows the waveforms of some harmonic modes in which the displacement has the same value at every point on any given sphere centered on the nucleus. These vibrations are analogous to the pulsations of a rubber ball that has been compressed and released. (Certain stars whose brightness varies periodically pulsate in the same way.)

As in vibrating strings and drumheads, the harmonic vibration with the longest period—the fundamental—has a nodeless waveform, and the period of the vibration decreases as the number of nodes increases. The periods are given by a remarkably simple rule. Let P_1 stand for the period of the fundamental, P_2 for the period of the first spherically symmetric overtone, and so on. Then the ratio between any two periods is equal to the ratio of the squares of the corresponding whole numbers. For example, the period of the first overtone, P_2, is 2^2, or 4, times the period of the fundamental, P_1; the period of the second overtone, P_3, is 3^2, or 9, times the period of the fundamental; and so on. This is the connection I hinted at earlier between the discreteness of the physical states of a hydrogen atom and the Pythagorean theory of musical harmony. I think the Pythagoreans would

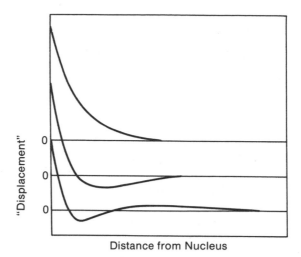

FIGURE 4.7 Waveforms of three spherically symmetric "vibrations" that represent physical states of the hydrogen atom. "Displacement" is plotted against distance from the nucleus.

have found it appropriate that the kinds of vibrations that produce pure musical tones also underlie the structure of hydrogen, the most abundant constituent of the Universe, and that the pitches of these vibrations—audible only to the mind's ear, like the music of the heavenly spheres—should be connected to the integers by a simple rule.

If we accept the analogy between states of a hydrogen atom and vibrations of a vibrating system, we can understand why the structure of a hydrogen atom, unlike the structure of a planetary system, is immune to change. Waveforms, like the spheres and cubes of Euclidean geometry, are mathematical objects, endowed with permanent and immutable properties. We can disrupt a hydrogen atom, but we can't alter the waveforms and pitches of the simple vibrations that represent its possible states; and when the proton and the electron of the disrupted atom recombine to form a new atom, it will be indistinguishable from every other hydrogen atom in the Universe. But what is the basis for the analogy? In what way do the physical states of a hydrogen atom resemble vibrations? How are the waveforms and periods of these vibrations connected with quantities that physicists can actually measure?

The best way to understand the quantum physicist's picture of a hydrogen atom is to retrace the route that led to it. We will see how and why physicists were forced to jettison their cherished belief in the possibility of picturing atomic and subatomic structures and processes. Classical physics had populated the world with two distinct classes of energy-carrying objects: waves and particles. A wave is in many places at the same time; a particle is always in a particular place. Light consists of waves; matter, of particles. Quantum physics fused these two concepts, endowing light with particlelike properties and material particles with wavelike properties. The quantum picture of physical reality is not a picture at all in the conventional sense. We can grasp it, but we can't visualize it.

5

Seven Steps to Quantum Physics

The first quarter of the twentieth century was a period of transition for theoretical physics. Classical physics had broken down, but there was as yet no coherent theory to replace it, only disconnected fragments. Gradually, in the minds of a few people, these fragments began to fit together into a new and strange shape. We can best understand the novel features of the picture that finally emerged by retracing the path that led to it—not the actual historical path, with all its twists, turns, branchings, and detours, but the shorter and simpler path that emerges when we look at the past through the eyes of the present. Historians rightly disparage this way of viewing the past (''Whig history''), but our concern here is not so much with history as with genealogy—the genealogy of ideas.

Step 1: Boltzmann's Theory of Entropy

The story as it is usually told begins in 1900 with Max Planck's theory of black-body radiation. It should begin almost thirty years earlier with Ludwig Boltzmann's theory of entropy. I described that theory in Chapter 2, but in so doing I glossed over an important difficulty. Boltzmann's solution to this difficulty set Planck on the path that eventually led to quantum physics.

Boltzmann identified the entropy of a physical system in a given macrostate with the randomness of that macrostate. He defined the randomness of a macrostate as the logarithm of the number of its microstates. Let's try to apply this definition to a gas. In Chapter 3 we represented the state of a single molecule by a point in a six-dimensional space. Three of the point's six coordinates specify

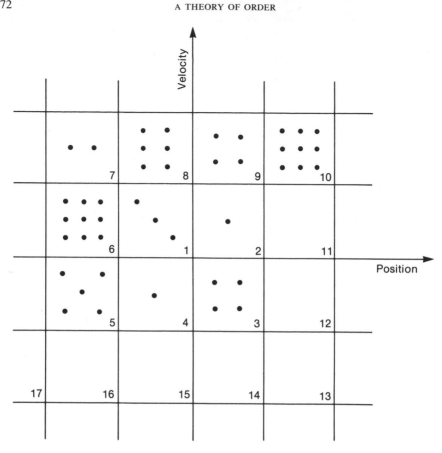

FIGURE 5.1 Numbered macrocells in the two-dimensional state space of a particle moving along a line. Each macrocell represents a range of possible positions and velocities. To specify the macrostate of a gas consisting of such particles, we have to specify the number of particles in each macrocell. There are three particles in cell 1, one in cell 2, four in cell 3. The macrostate of the gas is specified by a series of numbers that begins 3, 1, 4, 1, 5, 9, 2, 6, 4, 9.

the molecule's position; the other three specify its velocity. To describe the state of a pair of molecules, we used a twelve-dimensional space. But we could equally well have represented the state of a pair of particles by a pair of points in a common six-dimensional position–velocity space. For our present purpose the second representation is more convenient. We represent the *microstate* of a gas consisting of N molecules by a set of N points in the six-dimensional space.

To define the *macrostates* of a single molecule, we divide the six-dimensional space into equal compartments, which I will call *macrocells* (Figure 5.1). Every macrocell represents a possible macrostate of a single molecule. The points in a macrocell represent the microstates that make up the macrostate. When we say that a molecule is in a certain macrostate, we mean that its position and velocity lie within the limits that define the corresponding macrocell.

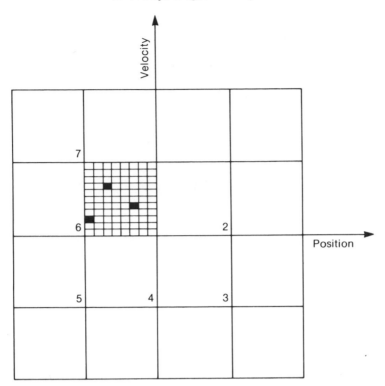

FIGURE 5.2 Each macrocell is divided ito a large number of equal microcells small enough to ensure that none contains more than one particle.

The macrostate of a *gas* is defined by the way in which the points that represent molecular microstates are distributed among the macrocells. To specify a macrostate of the gas, we have to specify how many representative points (points representing the microstates of individual molecules) lie in each macrocell.

So far, so good. Now let's try to calculate the randomness of a macrostate. The first step is to count the microstates belonging to that macrostate. Consider, for simplicity, a gas consisting of a single molecule. We specify its macrostate by pointing to a particular macrocell. The points belonging to that macrocell represent its microstates. But the number of points in any macrocell is infinite, so the calculation breaks down. (Actually, the difficulty is even more severe. The set of points in a macrocell is not only infinite, but also *uncountable*. See Chapter 1, note 9.)

Boltzmann got around this difficulty by assuming that a molecule's possible microstates are represented not by points in the six-dimensional position–velocity space but by identical, very small cells *(microcells)*, as in Figure 5.2. He stipulated that these cells should be so small that they almost never contain more than one representative point. It is then relatively easy to figure out how many microstates belong to any given macrostate, and thus to calculate the randomness of that macrostate.

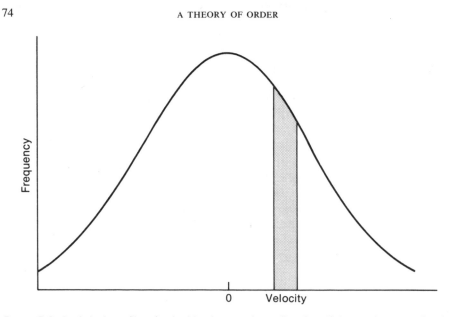

FIGURE 5.3 Statistical profile of velocities in any given direction. If the total area under the curve is unity, the area of a vertical slice like that indicated in the figure is equal to the fraction of the molecules that have velocity components within the limits that define the slice.

But doesn't the result of the calculation depend on the size of the microcells? Yes, but in what proves to be a "harmless" way. Changing the size of the microcells causes the number of microstates in any given macrostate to be multiplied by a certain number—the same number for every macrostate. Because the randomness of a macrostate is the logarithm of the number of its microstates and the logarithm converts products into sums, changing the size of the microcells increases the randomness of every macrostate by the same amount. Thus the *difference* between the randomnesses of any two macrostates doesn't change. In nearly all scientific applications, we are interested in how the randomness (or entropy) of a system changes, not in its absolute value. For example, chemical reactions in a living cell generate entropy and thereby decrease the cell's store of potentially useful energy. The entropy generated is the difference between the entropies of the final and the initial states, so we can calculate it without specifying the size of the microcells.

Let's take a closer look at Boltzmann's way of calculating the randomness of a gas. Its molecules are moving in all directions with a wide range of speeds. Boltzmann's aim was to predict the *statistical profile* of molecular velocities in a given direction (Figure 5.3). Assume that the gas is in a box whose walls are kept at a fixed temperature. This temperature determines the average kinetic energy of the gas molecules and hence their average squared velocity. (The kinetic energy of a particle is proportional to the square of its velocity.) Stipulating that the temperature of the gas has a given value therefore imposes a constraint on the statistical profile of molecular velocities: the profile must have a shape that gives

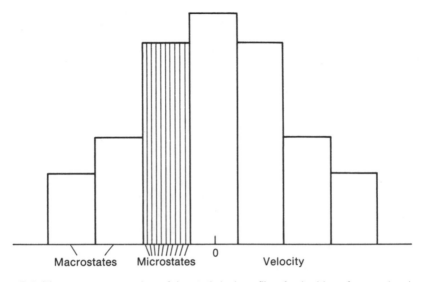

Macrostates   Microstates   0   Velocity

FIGURE 5.4 Discrete representation of the statistical profile of velocities of gas molecules in a given direction. The large velocity intervals, greatly exaggerated in this diagram, represent velocity macrostates. The small intervals, whose size is even more greatly exaggerated, represent velocity microstates.

the average squared velocity the value appropriate to that temperature. Physically, this constraint means that there is a fixed quantity of energy to be distributed among the molecules. The combined kinetic energy of the molecules is proportional to both the absolute temperature and the number of molecules. A vast number of velocity profiles satisfy this constraint. Which is the right profile?

We may approximate the statistical profile of molecular velocities in a given direction by a bar graph (Figure 5.4). The velocity intervals in this graph represent velocity macrocells, and the graph itself defines a possible *macrostate* of the gas. (Distinct macrostates have distinct graphs.) Next we divide each velocity interval into smaller intervals, as illustrated in Figure 5.4. These represent velocity microcells. Every way of distributing the molecules among microcells represents a possible *microstate* of the gas. Every microstate belongs to a definite macrostate, because every way of distributing the molecules among microcells defines a way of distributing the molecules among macrocells.

A gas molecule undergoes a huge number of collisions every second. Because molecular collisions cause random changes in the molecular velocities, it seems plausible that the macrostate of a gas that has been left to itself for a long time should be as random as it can be, subject to the constraint that the combined kinetic energy of the molecules has a fixed value. Now according to Boltzmann's definition, the randomness of a macrostate is equal to the logarithm of the number of its microstates. Hence the most random macrostate is the one with the greatest number of microstates. The velocity profile that satisfies this condition isn't difficult to calculate. It is the bell-shaped curve illustrated in Figure 5.3.

Boltzmann's theory was spectacularly successful. By reproducing all the old laws of thermodynamics, it demonstrated that thermodynamics is at bottom a statistical theory of molecules moving and interacting according to the laws of Newtonian physics. Thus it provided the last span in a bridge that Maxwell and Clausius had begun to build some years earlier between the island of thermodynamics and the mainland of classical mechanics. But Boltzmann's theory did even more. It described a wide range of important phenomena that lie outside the domain of classical thermodynamics, such phenomena as heat conduction, molecular diffusion, and the viscous decay of internal motions in a gas. The rates of all these processes could now be deduced from the properties of molecules and molecular interactions.

Yet Boltzmann's theory seemed to be built on sand, for his use of microcells to calculate the entropy blatantly contradicted a fundamental tenet of classical physics: that a freely moving particle has a continuous range of possible states, represented not by cells but by points. Boltzmann himself seems to have regarded the use of microcells as a harmless computational device, and many modern textbooks present it in that way. But microcells aren't really a computational device. Instead of facilitating a difficult calculation, as computational devices are supposed to do, they allowed Boltzmann to replace a legitimate calculation that gave a meaningless answer with an illegitimate calculation that gave what was demonstrably the right answer.

Had Boltzmann been gifted with second sight and an unnatural degree of self-confidence, he might have argued as follows: "My theory of entropy rests on a premise that contradicts a basic tenet of classical physics. My theory is clearly right. Therefore its premise must be right. So classical physics must be wrong." As we will see, an argument of this kind was actually made by Albert Einstein thirty years later in a closely related context. But Boltzmann's "computational device" wasn't legitimized until 1925—more than half a century after he first used it. Quantum physics represents the microstates of a particle by cells of finite and fixed volume in six-dimensional position–velocity space.

Step 2: Planck's Theory of Blackbody Radiation

An evacuated box whose walls are kept at a fixed temperature contains an intangible "gas"—light emitted by the heated walls. Like an ordinary gas, light carries energy and exerts pressure, both of which increase with the temperature of the walls. Experiments show that the light, like the gas, eventually settles into a state of equilibrium in which all its measurable properties are determined by the temperature of the walls. Light in this state is called *blackbody radiation.*

In the late nineteenth century, physicists began to study the *spectrum* of blackbody radiation and its dependence on temperature. (By the spectrum of blackbody radiation, I mean the curve that specifies how the intensity of the radiation varies with color or wavelength.) Theorists made some progress. They showed that if the spectrum is known for a single temperature, it can then be calculated for any other temperature by means of a simple rule. But in spite of strenuous efforts, they failed to predict the spectrum. Meanwhile, experimenters were trying to mea-

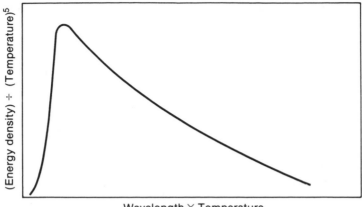

Wavelength × Temperature

FIGURE 5.5 Spectral distribution of blackbody radiation. "Energy density" means energy per unit volume and per unit wavelength; "temperature" means absolute temperature. If energy density divided by the fifth power of the temperature is plotted against the product of wavelength and temperature, a single curve represents the spectral distribution of radiant energy at any given temperature. Thus the wavelength at which the curve peaks is inversely proportional to the absolute temperature, and the area under the curve, which represents the radiant energy per unit volume, is proportional to the fourth power of the absolute temperature.

sure it. Early in the year 1900 they succeeded, obtaining the curve shown in Figure 5.5

Max Planck had been grappling with this theoretical problem for several years. He had also been keeping a close watch on the experiments. As soon as he heard about the new measurements, he began to look for a formula that would fit them and that would also satisfy the known theoretical constraints. He soon found one. But even knowing what he was convinced was the correct formula, he couldn't derive it theoretically. At this point Planck turned to Boltzmann's theory of randomness, to which he had previously paid little attention. He set out to adapt Boltzmann's way of calculating the most random distribution of molecular velocities to the problem of predicting the most random spectral distribution of light.

Planck considered a box filled with light and with a gas composed of rather peculiar molecules. Each molecule consisted of an electron vibrating inside an extended spherical cloud of positive electric charge. According to Maxwell's electromagnetic theory, such a system absorbs and emits light of a single wavelength. The *frequency* (reciprocal period) of the absorbed or emitted light waves is equal to the vibrating electron's frequency of vibration. Planck reasoned that in the most random state, the combined energy of the light and the gas would be split three ways. Part of the energy would be in the form of light; part of it would be in the form of kinetic energy associated with the motions of whole molecules; and part of it would be in the form of energy associated with the vibrations of electrons within individual molecules. The third share, vibrational energy, was the crucial one. If he could calculate how it was distributed among the molecules, he would

78

be able to deduce the spectrum of the light; for he knew that, in equilibrium, the rates at which molecules absorb and emit energy must be equal, and he knew how to calculate those rates for a vibrating electron. So everything hinged on calculating the distribution of vibrational energy among the molecules.

It was at this point that Planck turned to Boltzmann's earlier calculation of the most random distribution of molecular velocities. He found that the most random distribution of vibrational energy did indeed lead to his earlier formula for the spectrum of blackbody radiation, but only if he assumed that *the microstates of vibrational energy form a discrete aggregate.* More precisely, Planck had to postulate that the energy of a vibrating electron is made up of finite parcels of energy and that the size of an energy parcel (or quantum) is a fixed multiple of the vibrational frequency:

$$\text{energy quantum} = \text{fixed constant} \times \text{frequency}$$

Physicists call the fixed constant Planck's constant and represent it by the letter h.

In replacing the classical continuum of vibrational energies by a discrete aggregate, Planck was following in Boltzmann's footsteps. But in Boltzmann's calculation of the most random velocity distribution, the exact size of the microcells didn't matter. In Planck's derivation, the size of the energy packets—that is, the value of the constant h—was crucial. To make the predicted blackbody curve agree with experiment, Planck had to set the constant h equal to a specific number, whose value could be deduced from the curve's measured shape.

Planck himself found this very puzzling, as he tells us in reminiscences published forty-three years later:

> Now the theoretically most difficult problem arose of giving a physical meaning to this peculiar constant [h], whose introduction meant a break with classical theory that was much more fundamental than I had suspected at the outset. . . . Over a period of many years I tried again and again to fit [it] into the framework of classical physics.[1]

Step 3: The Discreteness of Light

Einstein, who was twenty-two years old when Planck's paper appeared, became convinced that Planck's constant and the discreteness that it symbolized would never fit into the framework of theoretical physics. In a paper entitled "On a Heuristic Point of View about the Creation and Conversion of Light," published in 1905, he put forward the bold hypothesis that monochromatic light of frequency f can be emitted or absorbed only in discrete packets, or quanta, of energy hf. (Monochromatic light is light of a single color or frequency. Newton discovered that a prism separates light into monochromatic rays, which can't be further split up. Monochromatic rays—or rather, their energy quanta—are thus the elementary constituents of light, in somewhat the same way that molecules are the elementary constituents of a gas.)

Einstein's hypothesis flouted conventional wisdom. Planck and nearly everyone else believed, and continued to believe for many years after the publication of Einstein's paper, that classical electromagnetic theory fully and correctly describes light itself, if not the interaction between light and matter. They assumed that the origin of the mysterious discreteness that Planck had been forced to introduce into his derivation lay elsewhere. Characteristically, Einstein met this plausible (but unfounded) belief head-on. His paper opens with the remark that physicists treat matter and light differently: they assume that matter is fundamentally discrete but that light is fundamentally continuous. Perhaps, he suggested, they are *both* discrete. True, the wave theory of light works very well and is probably here to stay, but

> optical observations refer to time averages and not to instantaneous values. . . .
> [So] it is quite conceivable that [the wave theory] will lead to contradictions when
> it is applied to the phenomena of the creation and conversion of light.[2]

What Einstein had in mind was that a beam of light might be made up of *spatially* distinct units, like a beam of particles. But this was just a picture accompanying his actual hypothesis: that monochromatic light of frequency f can gain or lose energy to matter only in packets of magnitude hf.

To justify this hypothesis, Einstein showed that the entropy of monochromatic light of sufficiently high frequency is given by a formula that coincides with Boltzmann's formula for the entropy of an ordinary gas if, in that formula, the number of particles is set equal to the energy of the light divided by Planck's energy quantum. In other words, the entropy of high-frequency light is the same as the entropy of gas whose molecules are Planck's energy quanta.

Einstein proposed three experimental tests of this hypothesis.

1. *Photoluminescence.* A photoluminescent substance absorbs light of one frequency and emits light of another frequency. Einstein's hypothesis predicts that the frequency of the emitted light should always be less than that of the absorbed light. For example, a photoluminescent substance could absorb ultraviolet light and emit yellow light but it couldn't, according to Einstein's hypothesis, absorb yellow light and emit violet light. Why? When I buy something with a dollar bill, the change is always less than a dollar. A photoluminescent substance that has absorbed a quantum of energy emits a quantum of smaller energy and hence lower frequency. The rest of the energy may be emitted at another frequency, as illustrated in Figure 5.6.

2. *Photoionization.* Physicists had noticed that an electrically nonconducting gas may become conducting after it has been irradiated with ultraviolet light, but not with visible or infrared light. According to Einstein's hypothesis, ultraviolet light consists of more energetic quanta than visible light. An ultraviolet quantum may transfer enough energy to an electron bound in a gas molecule to tear the electron loose (Figure 5.7). Unbound electrons (and positively charged ions) can move freely through a gas in response to an applied electric field, so an ionized gas conducts electricity.

3. *The photoelectric effect.* Light falling on the surface of some metals causes

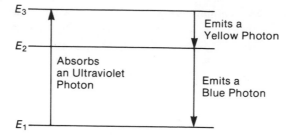

FIGURE 5.6 Photoluminescence. The molecules of a photoluminescent substance are normally in their state of lowest energy E_1. If a molecule is in the excited state of energy E_3, it can emit a visible photon by jumping to an intermediate state whose energy E_2 lies between E_1 and E_3. A second jump, back to the ground state, may also be accompanied by the emisison of a visible photon. To get to state E_3, however, the molecule must absorb an ultraviolet photon whose energy is at least as great as the energy difference between E_3 and E_1 divided by Planck's constant.

electrons to be ejected. Einstein's explanation of this phenomenon is similar to his explanation of photoionization. An electron bound in a metallic lattice (which we may think of as a large molecule with repetitive structure) may acquire enough energy to break free from the lattice if the lattice absorbs a sufficiently energetic light quantum. Just as is true for photoionization, the quality (that is, color) of the light is what counts. Weak ultraviolet light falling on a photosensitive metal causes electrons to be ejected, albeit slowly; red light, however intense, is completely ineffective.

All three predictions were consistent with the experimental data then available, but the experiments were not good enough to provide a crucial test. In both photoionization and the photoelectric effect, part of the absorbed quantum of energy is used to liberate an electron from an atom, a molecule, or a crystal lattice. The rest of the energy appears as the kinetic energy of the ejected electron. Thus electrons ejected by light of relatively high frequency should be moving faster than electrons ejected by light of lower frequency. Robert A. Millikan, who first accurately measured the electric charge of a single electron, set out to test this prediction for the photoelectric effect. In 1915, he summarized his results, and his conclusions about Einstein's hypothesis, in these words:

> Einstein's photoelectric equation . . . appears in every case to predict exactly the observed results. . . . Yet the semicorpuscular theory by which Einstein arrived at his equation seems at present wholly untenable.[3]

Not just untenable, but "wholly untenable." Abraham Pais has remarked that Millikan's attitude toward the hypothesis that light exchanges energy with matter only in discrete packets was not unrepresentative: "Rather, the physics community at large had received . . . the hypothesis with disbelief and with skepticism bordering on derision."[4]

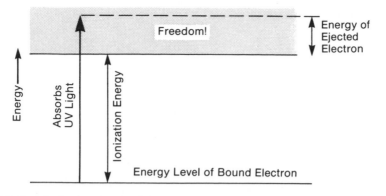

FIGURE 5.7 Photoionization. An ultraviolent quantum may transfer enough energy to an electron bound in a gas molecule to tear the electron loose. An ionizing photon's frequency must exceed the ionization energy divided by Planck's constant.

The 1905 paper has another interesting feature. Einstein visualized energy quanta as localized in space:

> *According to the assumption considered here, when a light ray starting from a point is propagated, the energy* is not continuously distributed over an ever-increasing volume but *consists of a finite number of energy quanta*, localized in space, which move without being divided and *which can be absorbed or emitted only as a whole.* (italics added)

This sentence summarizes Einstein's conclusion. But what his argument actually establishes—or at least makes highly plausible—is conveyed by the part of the sentence that I've italicized:

> [The energy carried by a light ray] consists of a finite number of energy quanta which can be absorbed or emitted only as a whole.

The nonitalicized parts of the original sentence convey a *picture* of energy quanta: they are localized in space and move without being divided. This may be the only picture one can form of energy quanta that is consistent with their testable properties. It doesn't follow that the picture is true, however. A modern physicist would say that energy quanta can't be pictured at all; certainly they can't be localized. Although Einstein continued to believe in the possibility of describing nature in a way that can be pictured, in his later papers on quantum physics he scrupulously refrained from drawing conclusions that went beyond the strict limits of his mathematical arguments.

Step 4: Energy Levels

As we saw, Planck derived his radiation formula by considering the interaction between blackbody radiation and a collection of tiny oscillators tuned to precise frequencies. In 1906 Einstein showed that Planck's theory requires these oscillators to behave in a way that contradicts classical theory. The amplitude of vibration of an oscillating electrically charged particle has a continuous range of possible values. The larger the amplitude, the greater the vibrational energy. Moreover, classical electromagnetic theory predicts that the energy of the oscillator decreases or increases continuously as it emits or absorbs light. Einstein showed that this picture can't be valid for the oscillators that figure in Planck's theory. A Planck oscillator must have a discrete set of possible energies: 0, *hf, 2hf, 3hf*, and so on, where, as before, *h* denotes Planck's constant and *f* denotes the frequency of the oscillator. Furthermore, when a Planck oscillator emits or absorbs an energy quantum, its energy must decrease or increase *discontinuously* by the amount *hf*. Although these conclusions were logically entailed by Planck's theory, Einstein's energy-quantum hypothesis, and the principle of conservation of energy, they blatantly contradicted classical electromagnetic theory, which says that an oscillating electric charge radiates energy continuously. But the worst was yet to come.

During the first decade of the twentieth century, most physicists subscribed to a model of the atom invented by J. J. Thomson, the discoverer of electrons. Thomson's atom has pointlike electrons embedded in a sphere of positive electricity, like blueberries in a blueberry muffin. Planck's oscillators are the simplest exemplars of this model. According to classical electromagnetic theory, stable configurations of this kind do exist, and an electron vibrating about its equilibrium position does emit light of a definite frequency, so Thomson's model had much to recommend it. True, it didn't behave as Einstein said an oscillator should behave, emitting energy in discontinuous bursts, but few people took Einstein's argument seriously.

In 1911, the bubble burst. Ernest Rutherford and two assistants, Hans Geiger and E. Marsden, discovered experimentally that real atoms don't in the least resemble Thomson's model. Instead, they are constructed like miniature solar systems, with the nucleus playing the part of the Sun and the electrons the part of the planets. But if such an atom obeyed Newton's and Maxwell's laws, it couldn't possibly emit or absorb radiation at only a discrete set of frequencies; it would emit and absorb a continuous band of frequencies.

Inspired by Rutherford's discovery and Einstein's deductions about energy quanta and the quantized energy states of Planck oscillators, Niels Bohr set out to construct a new atomic theory. He began by laying down two postulates:

> 1. The energy of an atom or a molecule has a discrete set of possible values. Every atom and molecule has its own set of possible energies, or *energy levels*.
> 2. Changes in the energy of an atom or a molecule take place discontinuously, in quantum jumps. The energy gained or lost in a quantum jump may be supplied or taken by another object (for example, an electron or another atom) or by light. In the second case, a single energy quantum is absorbed or emitted.

Both rules generalize deductions that Einstein had made earlier in specific contexts (Figures 5.6 and 5.7). Their importance lay not only in this fact, but also

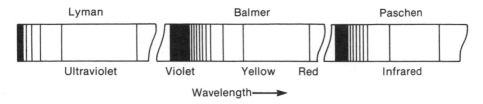

FIGURE 5.8 The spectrum of atomic hydrogen is a series of series. The first series, the Lyman series, is produced by transitions from the ground state to states of higher energy. The second series, the Balmer series, is produced by transitions from the first excited state to states of higher energy, and so on.

in the fact that they offered a new and fruitful starting point for the analysis of atomic and molecular spectra. Consider the spectrum of hydrogen (Figure 5.8). It consists of several distinct series of lines. The first series (the Lyman series) is in the ultraviolet part of the spectrum; the second (the Balmer series) is in the visible part of the spectrum; the third (the Paschen series) is in the infrared; and so on. With the help of Bohr's second rule, which connects the frequencies of spectral lines with differences between energy levels, we can construct an energy-level diagram (Figure 5.9) that represents all these series and predicts many others whose wavelengths fall outside the range accessible to laboratory spectroscopy. For example, quantum jumps between adjacent high-lying levels give rise to radio waves. Astronomers have actually observed this radio-frequency radiation, emitted by hydrogen atoms in interstellar space.

Bohr was not content with stating these rules. He set himself the task of reconciling Rutherford's planetary model of the hydrogen atom with the energy-level diagram shown in Figure 5.9. Consider, to begin with, what classical physics has to say about Rutherford's model. The force of attraction between the positively charged nucleus of a hydrogen atom and the negatively charged electron is qualitatively identical with the gravitational attraction between the Sun and a planet: both are proportional to the inverse square of the distance between the attracting bodies. Since the possible orbits of a planet are ellipses with the Sun at one focus, the possible orbits of the electron in a hydrogen atom should be ellipses with the nucleus at one focus. According to Newton's theory, the energy of a particle moving in such an orbit depends only on the size of the orbit's larger diameter and not on the ratio between the small and the large diameters. Thus infinitely many orbits correspond to a given value of the energy (those with a given major diameter and all possible values of the minor diameter), and the energy itself can have any negative value. (Positive values of the energy correspond to hyperbolic orbits.)

These predictions clash with the requirement that the hydrogen atom have a discrete set of possible states and energy levels. But Bohr, in 1913, was not willing to abandon Newton's theory outright. Einstein had shown that a Planck oscillator must have a discrete set of energy levels. Bohr asked himself: Is there an analogous rule that selects the possible orbits of an electron in a hydrogen atom from the continuously infinite set of orbits allowed by Newton's theory?[5]

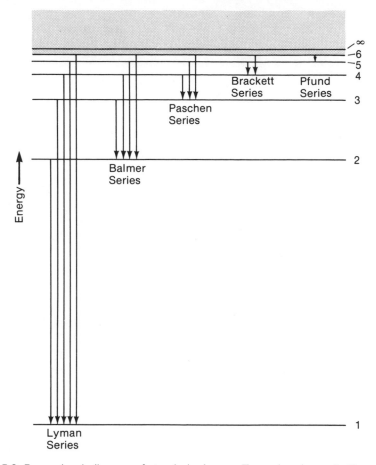

FIGURE 5.9 Energy-level diagram of atomic hydrogen. The series shown in Figure 5.8 are produced by the transitions illustrated here. A single sequence of energy levels serves to predict an infinite number of spectral sequences.

According to Newton's theory, the size and shape of an orbit are determined by two "constants of the motion"—quantities whose value doesn't change as the electron travels along the orbit. One of these is energy; the other (discovered by Johannes Kepler) is called *angular momentum.* An electron's angular momentum is equal to twice the product of its mass and the rate at which an arrow drawn from the nucleus to the electron sweeps out area (Figure 5.10). Angular momentum happens to have the same physical dimension (energy × time) as Planck's constant *h.* Bohr had the brilliant idea of postulating that the electron in a hydrogen atom must travel in a circular orbit and that its angular momentum must be a whole number of angular-momentum units, an angular-momentum unit being some suitably chosen multiple of Planck's constant (Figure 5.11). This idea worked like a charm. Combined with the Newtonian formula for the energy of an electron

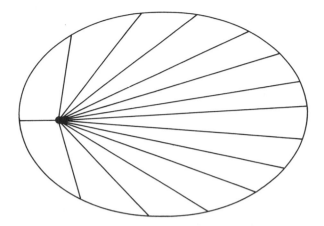

Figure 5.10 Kepler's law of areas (conservation of angular momentum). A planet orbiting the Sun sweeps out equal areas in equal times. Radii drawn from the Sun to the planet's position at a sequence of equally spaced moments in timebound sectors of equal area. As a consequence, the planet moves fastest when it is closest to the Sun.

moving in a circular orbit, it yields a formula that exactly reproduces the energy-level diagram illustrated in Figure 5.9, provided the unit of angular momentum is set equal to Planck's constant divided by 2π—the ratio of a circle's circumference to its radius. Factors like 2π and 4π (the ratio of the surface area of a sphere to the square of its radius) are always turning up in the laws of physics, probably because circles and spheres are such basic elements of the mathematical language in which the laws are expressed. So the appearance of the factor 2π lent an air of authenticity to the proceedings, like the appearance of a congressman at a clambake.

An even more impressive triumph followed shortly. The helium atom has two electrons, and its nucleus has twice the electric charge and four times the mass of the nucleus of a hydrogen atom. At very high temperatures (which spectroscopists can produce for short periods by using electric arcs and sparks), helium atoms part with one of their electrons. A singly ionized helium atom is just like a hydrogen atom except that its nucleus has twice the charge and four times the mass of the hydrogen nucleus, so Bohr could use his theory to predict energy levels of ionized helium. The predictions differed from measured values by less than one-thousandth of 1 percent. Rarely in the history of physics has such a spectacular success been achieved by such simple means.

Bohr's picture of the hydrogen atom resembles Aristarchus's picture of the Solar System. (Aristarchus postulated that the Earth and the five bright planets travel at constant speeds in coplanar circular orbits centered on the Sun.) The two theories are similar in other ways, too. Both rest on ad hoc assumptions; both account for complex phenomena in a surprisingly simple and accurate way; and both served as points of departure for theoretical developments that led eventually to much deeper theories.

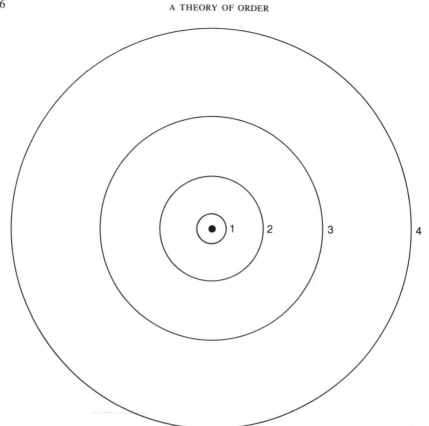

FIGURE 5.11 The first four Bohr orbits. The orbits are circles with radii proportional to n^2. Thus the four radii are in the ratios 1:4:9:16. The angular momentum of an electron in a Bohr orbit is an integral multiple of Planck's constant divided by 2π.

Step 5: Photons and Quantum Jumps

Having taken a few years off to develop the general theory of relativity, a task completed in 1915, Einstein once more turned his attention to quantum physics. While most of his colleagues were busy exploring the ramifications of Bohr's pioneering study of the hydrogen atom, Einstein characteristically turned his attention to more fundamental problems. His 1917 paper "On the Quantum Theory of Radiation" strengthened the foundation of quantum physics and permanently changed the way physicists view the interaction between light and matter.

In 1905 Einstein had speculated that light might behave as though it consists of discrete particles, and had produced convincing arguments—strongly confirmed by later studies—that the *energy* of monochromatic light of frequency f can increase or decrease only by whole energy quanta of magnitude hf. A classical particle has three characteristics: its interactions occur at definite points in space and definite moments in time; it carries a definite energy; and it carries a definite momentum. (The momentum of a slowly moving classical particle is the product of its mass and its velocity.) In classical physics the energy and momentum of a free particle are simply related. The kinetic energy of a slowly moving particle is equal to the square of its momentum divided by twice its mass, while according to special relativity, a particle of energy E moving at or close to the speed of light c has momentum E/c. Maxwell's theory shows that the same relation holds between the energy and the momentum carried by a beam of monochromatic light. So it was obvious in 1905 that energy quanta *in a parallel beam of light* must also carry momentum and that the energy of an energy quantum is c times its momentum, as it is for the beam as a whole. Einstein, however, refrained from stating this law until 1917. In the introduction to his 1917 paper he explains why the preceding argument is not airtight:

> Let us consider the emission [or light by a molecule] from the point of view of classical electrodynamics. If a body emits the energy E, it receives a recoil (momentum) E/c if all the radiation is emitted in the same direction. If, however, the emission takes place isotropically—for instance, in the form of spherical waves—no recoil at all occurs.

In 1905, Einstein hadn't been able to rule out the possibility that energy quanta are emitted isotropically. Twelve years later, he presented a beautiful argument showing that "we arrive at a consistent theory only if we assume each elementary [absorption or emission] to be completely directional." He considered a box filled with radiation and with a gas so dilute that its molecules interact only with the radiation and not with one another, and proved that recoils are needed to maintain the statistical distribution of molecular velocities appropriate to the equilibrium state. The proof relies on a statistical argument that Einstein had used in a famous paper on Brownian motion published in 1905, the year in which his papers on special relativity and energy quanta also appeared.

Einstein had now established that monochromatic light behaves as though it consists of particles (subsequently dubbed *photons*). Still, most physicists didn't accept this conclusion until six years later, when Arthur H. Compton showed experimentally that in interactions between light and electrons, both energy and momentum are exchanged exactly as in a collision between two material particles. Even then, in 1923, several influential physicists—most notably Niels Bohr—still refused to accept the photon hypothesis.[6]

Einstein's 1917 paper not only gave the photon hypothesis a secure theoretical foundation, but also contains what Einstein, in a letter to Michele Besso written in the summer of 1916, describes as "an astonishingly simple derivation of the Planck formula, I might even say *the* derivation. Everything quantal." As we saw earlier, Planck had derived his famous formula by considering the interaction between light and a collection of oscillators. These oscillators were hybrid objects, endowed with contradictory properties. They had discrete energy levels, but radiated and absorbed light in accordance with Maxwell's theory. Einstein put forward a new theory of radiation that made no appeal to Maxwell's theory. He accepted Bohr's postulate that an atom or a molecule has a discrete set of energy levels and emits or absorbs light only when it makes a jump from one level to another. He then postulated that such jumps are *random processes, governed by statistical laws*. Einstein was ambivalent about this postulate, which marks the entrance of chance on the stage of quantum physics:

> The weakness of the [present] theory lies, on the one hand, in the fact that it does not bring any nearer the connexion with the wave theory [of light] and, on the other hand, in the fact that it leaves the moment of occurrence and the spatial direction of the elementary processes [of absorption and emission] to "chance"; all the same, I have complete confidence in the reliability of the method used here.[7]

It took physicists another thirty years to explain fully the connection between quantum physics and the wave theory of light, but the second weakness that Einstein perceived in his theory—"that it leaves the moment of occurrence and the spatial direction of the elementary processes [of absorption and emission] to 'chance' "—became a permanent and central feature of quantum physics; chance shed its quotation marks. To the end of his life, however, Einstein resisted the view that chance is inherent in natural phenomena. We will see why in Chapter 6.

Step 6: Electrons as Waves

At this point, our path divides. We will follow in the footsteps of Louis de Broglie and Erwin Schroedinger. The other branch was cleared by Werner Heisenberg, Max Born, Pascual Jordan, Paul A. M. Dirac, and Wolfgang Pauli. They reached the summit several months before Schroedinger, but their path is steeper and less direct than the one we will follow, and hard to negotiate without the ropes and pitons of abstract mathematics.

The most characteristic property of a wave—the property that distinguishes it from a beam of particles—is its ability to exhibit *interference*. A sharp-edged screen in the path of a parallel beam of particles casts a sharp-edged shadow; in the path of a parallel monochromatic light wave, it produces a shadow edged with interference fringes (Figure 5.12a). Figure 5.12b illustrates how such fringes are produced when different parts of the same wave that have traveled along slightly unequal paths are reunited.

FIGURE 5.12 (a) An opaque screen in front of a parallel beam of monochromatic light doesn't cast a sharp shadow, because of interference. Instead, a series of closely spaced bright and dark fringes appears at the place where one would expect the edge to be. (b) Every point of the plane that contains the screen, except points on the screen itself, acts as a source of "secondary" spherical waves. These waves arrive at a given point on the photographic plate with different phases because they have traveled different distances. Interference between these secondary waves produces a gradual transition from shadow to light and a series of alternating bright and dark fringes at the edge of the shadow.

(a)

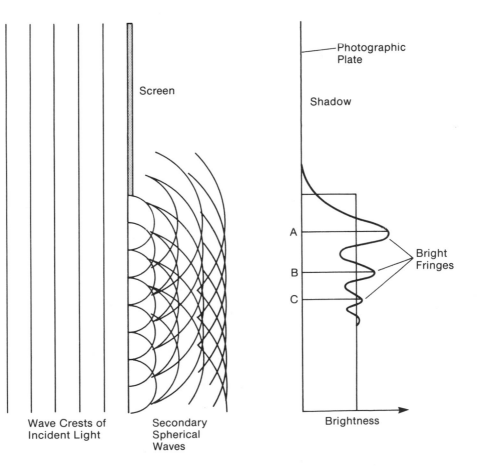

Screen

Photographic Plate

Shadow

A

B

C

Bright Fringes

Wave Crests of Incident Light

Secondary Spherical Waves

Brightness

(b)

Shortly after Compton's experiment confirmed that monochromatic light behaves as though it consisted of particles, Louis de Broglie conjectured that a beam of *particles* might, under appropriate conditions, behave like a *wave*. He predicted that a sharp-edged screen in the path of a beam of electrons would cast a shadow edged with interference fringes. Later experiments confirmed this prediction. In 1924, however, direct evidence of this kind didn't yet exist. Electron waves had to be invented before they could be observed.

De Broglie postulated that the frequency and wavelength of an electron beam are related to its energy and momentum by the same formulas that relate the frequency and wavelength of a light wave to the energy and momentum of its photons. Specifically,

$$\text{energy} = h \times \text{frequency} \qquad \text{momentum} = h \div \text{wavelength}$$

where, as usual, h stands for Planck's constant. This may not seem like such a remarkable step. If a light wave behaves, in certain respects, like a beam of particles, why shouldn't a beam of particles behave, in certain respects, like a wave? But there is more to de Broglie's hypothesis than that. How is the speed of the wave related to the speed of the associated particles? You may recall that the speed of a wave is equal to the product of its wavelength and its frequency. (Because the frequency of a vibration is the reciprocal of its period, the last statement is just another way of saying that during a time interval equal to its period, a wave travels a distance equal to its wavelength.) From the displayed equations, we can infer that the product of the frequency and the wavelength of a de Broglie wave, and hence its velocity, is equal to the energy of one of the particles in the beam divided by its momentum. This, however, is *not* the velocity of a material particle. The de Broglie wave moves much faster than the particles in the beam. But for the hypothesis to make sense, the wave must have the same speed as the particles. Or so it seems at first sight.

To resolve this apparent contradiction, we have to look more closely at the relation between electrons and de Broglie waves. By ''the velocity of an electron,'' we mean the distance it travels during a short interval of time, divided by that interval. So we can talk about an electron's velocity only when its position at any given moment has a well-defined value. But a monochromatic wave traveling in a definite direction fills all space. The wave crests of such a wave are infinitely extended parallel planes perpendicular to the direction in which the wave is moving. A monochromatic wave, therefore, can't represent a particle that occupies a tiny volume of space. To represent such a particle, we have to superimpose waves of slightly different wavelengths moving in slightly different directions. We can select these waves in such a way that they reinforce one another in a small region and interfere destructively outside this region, as illustrated in Figure 5.13. Such a collection of waves is called a *wave packet*. The velocity of a wave packet differs from that of its constituent waves, which move *through* the wave packet. If the relation between the speed of a wave and its wavelength is known, as it is for de Broglie waves (because Newton's and Einstein's theories tell us how an electron's energy and momentum are related), it is easy to calculate the velocity of a wave packet. By doing this calculation, de Broglie was able to show that the

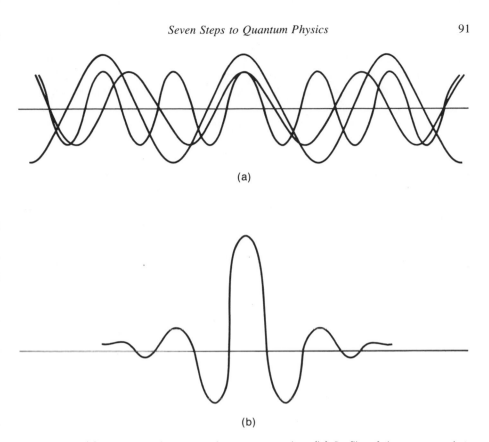

(a)

(b)

FIGURE 5.13 (a) Representative waves in a wave packet. (b) Profile of the wave packet formed by constructive and destructive interference of the waves that make up the packet.

velocity of a wave packet does indeed coincide with the speed of the particle the packet is supposed to represent. Thus de Broglie's hypothesis, fantastic as it seems at first sight, is internally consistent. This demonstration immediately convinced Einstein and a few others, among them Erwin Schroedinger, that de Broglie had made an important conceptual breakthrough.

The most famous consequence of de Broglie's hypothesis is the statement known as Heisenberg's uncertainty relation, which says that the position and momentum of a particle can't *simultaneously* be known with arbitrary precision. The more precisely we know the particle's position, the less precisely we can know its momentum, and vice versa. A wave packet like that shown in Figure 5.13 is made up of de Broglie waves with a certain range of wavelengths. These waves reinforce one another inside the wave packet and destroy each other outside. Let the wave packet have width W. Then it contains W/w de Broglie waves of wavelength w and W/w' waves of another wavelength w'. If the two waves are in phase (reinforce each other) at the center of the wave packet and out of phase (destroy each other) at its edges, these two numbers, W/w and W/w', must differ by a whole number. So they must differ by at least 1. Thus, the spread in the range of reciprocal wavelengths $1/w$ of the de Broglie waves that make up a wave packet, multiplied by the width of the wave packet, can't be less than 1. The narrower a wave packet, the greater the spread of the reciprocal wavelengths of its constituent waves; the smaller the spread of reciprocal wavelengths, the broader the wave packet.

Now recall that the momentum of the particle associated with a de Broglie wave is equal to Planck's constant times the wave's reciprocal wavelength. It follows that the width of a wave packet, multiplied by the spread in the momenta associated with its constituent waves, can't be less than Planck's constant. We conclude, finally, that the product of the uncertainties in the position and momentum of the particle whose physical state the wave packet represents can't be less than Planck's constant. This is Heisenberg's uncertainty relation.

We are now, at last, able to justify Boltzmann's assumption that the microstates of particles are represented by finite regions (microcells) of the six-dimensional position–velocity space rather than by points. To begin with, we may use momentum instead of velocity to define a particle's state of motion. Heisenberg's uncertainty relation tells us that a particle's momentum and its position can't both be specified with arbitrary precision. For one-dimensional wave packets, the product of the uncertainties in position and momentum can't be less than Planck's constant. In the two-dimensional position–momentum space of a particle confined to a straight line, the particle's microstates are represented by rectangles whose area (the product of the uncertainties in position and momentum) is equal to Planck's constant. Analogously, in the six-dimensional position–momentum space of a particle that can roam freely in three dimensions, the particle's microstates are represented by cells whose volume is equal to the cube of Planck's constant. The uncertainty principle doesn't restrict the *shape* of a microcell, only its volume.

Step 7: Stationary Atomic States and Standing Waves

Schroedinger reasoned that if a free electron is represented by a packet of de Broglie waves, the possible state of an electron bound in a hydrogen atom should be represented by standing de Broglie waves with definite frequencies, exactly analogous to the harmonic modes of a vibrating string or drumhead. To test this idea he needed a way to calculate the frequencies of such harmonic modes of vibration.

Most of the waves that physicists had studied were vibrations of a material medium—a string, drumhead, a column of air, or a body of water. Light waves were a recent exception to this rule. From the time of Newton until 1905, physicists had imagined them to be vibrations in a subtle, all-pervading medium, the "luminiferous ether." Einstein's special theory of relativity dispensed with the ether, and eventually physicists stopped believing in it. The standing waves that Schroedinger was trying to describe were presumably of this immaterial kind— vibrations without a vibrating medium.

Water waves, sound waves, and light waves are different species of the same mathematical genus. Each species is governed by a "wave equation" that differs slightly from the wave equations governing other species. The wave equations of classical physics are themselves consequences of more fundamental laws. For example, the wave equation governing light is a consequence of Maxwell's laws of electromagnetism. Schroedinger, however, couldn't derive the wave equation he was seeking from more fundamental laws because they didn't yet exist. He arrived at it by another route, which I will now try to explain.

All wave equations have a family resemblance, and each is connected by a somewhat complicated but well-understood mathematical transformation to a formula that relates the frequency of the wave to its wavelength:

wave equation ↔ relation between frequency and wavelength

The Einstein–de Broglie relations enable us to translate frequency into energy and wavelength into momentum, so we need a formula that relates the *energy* of an electron bound in a hydrogen atom to its *momentum*. Schroedinger, like Bohr, took this formula from classical physics. In fact, Schroedinger and Bohr used the same classical formula, but they used it differently. Bohr used it to express the energy of an electron in a circular orbit; Schroedinger used it to derive a relation between frequency and wavelength for de Broglie waves in a hydrogen atom. By subjecting the resulting equation to the mathematical transformation symbolized by the displayed relation, he derived the famous wave equation that bears his name. Suitably generalized, Schroedinger's wave equation still forms the basis for most calculations in atomic, molecular, and solid-state physics.

Schroedinger himself carried out the first such calculation. He showed that the original equation has standing-wave solutions for only a discrete set of frequencies. These frequencies, translated into energies, coincide exactly with those given by Bohr's theory.

Schroedinger's theory also enabled him to make predictions that lay beyond the reach of Bohr's theory. For example, spectroscopists had discovered that when hydrogen atoms are placed in an electric field, some of their energy levels split into clusters of closely spaced levels. The presence of an electric field adds an extra term to the energy of an electron, which translates into an extra term in the wave equation. Schroedinger was able to solve the modified wave equation and to predict the observed splitting of the energy levels.

*   *   *

The ascent had been slow and arduous. Rarely could those who led it see beyond the next handhold or foothold. Now broad new vistas opened up on every side, and physicists rushed to stake out their claims. Old sciences like chemistry were taken over and reorganized from top to bottom; new sciences like nuclear physics took root and flourished like weeds. At the same time, the language of quantum physics gradually became more powerful and more abstract. Schroedinger's standing waves came to be seen as one of many equivalent ways of describing the physical state of a bound particle, all of them "representations" of a more abstract mathematical object, the "state vector." Schroedinger himself proved that "wave mechanics"—the theory based on his wave equation and its generalizations—was mathematically equivalent to "matrix mechanics"—the more abstract theory that had been developed slightly earlier by Heisenberg, Born, Jordan, and Dirac. Physicists discovered that electrons and other particles have internal states that can be described mathematically but can't be represented spatially, and hence can't be pictured.

Familiarity and use gradually lent the abstract language of quantum physics an intuitive quality of its own. But now a new difficulty arose. As physicists became more fluent in the new language, they became less certain about what it meant. What was the relation between quantum physics and reality? Quantum physics clearly wasn't a *picture* of physical reality. What, then, was it? Einstein, Bohr, Heisenberg, and Schroedinger attached great importance to this question, which they answered in different ways. Einstein's views were the most divergent. Until the end of his life, he remained convinced that quantum mechanics couldn't be a *complete* description of physical reality. During Einstein's lifetime, few of his colleagues shared this view. In a letter to Schroedinger in December 1950, Einstein wrote:

> You are the only contemporary physicist, besides Laue, who sees that one cannot get around the assumption of reality—if only one is honest. Most of them simply do not see what sort of risky game they are playing with reality—reality as something independent of what is experimentally established. They somehow believe that the quantum theory provides a description of reality, and even a *complete* description. . . .[8]

Many of Einstein's colleagues believed him to be the victim of philosophical prejudices inherited from the prequantum era—an ironic judgment about the person who, more than anyone else, had been responsible for freeing physics from the view of physical reality underlying classical physics and who had been among the first to recognize and applaud the conceptual breakthroughs of de Broglie and Schroedinger. Since Einstein's death, however, many physicists have come to recognize that Einstein's unwillingness to accept quantum physics as a complete description of physical reality was rooted in more than just prejudice. In Chapter 6 we will see why.

6

Alice in Quantumland

Alice A graduate student of philosophy
Zeno A quantum physicist and student of quantal paradoxes

ALICE: Your account of the steps leading from Boltzmann to Schroedinger seems logical enough, although I must confess I haven't understood every detail.

ZENO: The process itself wasn't logical, of course. I've made it seem so by leaving out all the failed efforts and ignoring all the trails that led nowhere. I also left out many successful efforts that in retrospect seem less important than the ones I mentioned, although they seemed very important at the time.

ALICE: I can well believe it. But that isn't what I find so confusing about quantum physics. I know I must seem very stupid, but I'm sure it would all be ever so much clearer if you could answer just one question for me. What do quantum physicists actually mean when they say that Schroedinger's standing waves *represent* the discrete states of a hydrogen atom? I know what a standing wave on a taut string represents: it represents the string's displacement from its resting position. I think I know what a sound wave represents: alternate compressions and rarefactions of the air in which the wave propagates. I even have an idea, although a less clear one, of what a light wave represents: an oscillating pattern of electric and magnetic forces. But what do Schroedinger's waves represent?

ZENO: That's not the easiest question you could have picked. It troubled Schroedinger, and it continues to trouble quantum physicists to this day. I'm not sure I can answer it to your satisfaction, but I'll try.

The first point to notice is how the *amplitude* of a wave is related to its *intensity*. The amplitude of a vibration is proportional to its width—the distance traversed by the back-and-forth motions at a given point. Intensity is the quantity that measures how much energy a wave carries. The intensity of a wave is proportional to the *square* of its amplitude.

The brightness at any point in an interference pattern (Figure 5.12) is proportional to the intensity of light at that point. But light is made up of photons. So the intensity of monochromatic light at any point in the interference pattern also measures the rate at which photons are arriving there. Now remember that a parallel beam of electrons shows the same kind of interference effects as a parallel beam of monochromatic light. So the intensity of a de Broglie wave—the square of its amplitude—must measure the space density of electrons.

ALICE: That makes it a *little* clearer, but I'm afraid I'm still puzzled. Let's talk about the electron in a hydrogen atom. Isn't it true that according to quantum physics, the standing de Broglie wave packet that represents the electron in its state of lowest energy is spread out over the whole atom?

ZENO: Yes. In fact, if you remember, the wave profile looks something like the curve illustrated in Figure 6.1. So the wave packet has no real edge, although its intensity is very small beyond a certain distance from the nucleus, which corresponds to the radius that a chemist, for example, would assign to the atom in its ground state.

ALICE: Does that mean that the electron itself is spread out over the whole atom, and beyond it? In other words, does the intensity of the wave packet measure the density of negative electric charge in the atom?

ZENO: I believe that Schroedinger himself initially interpreted his wave packets in the way you just described, but that interpretation can't be right. Consider a free electron. Its wave packet is made up of de Broglie waves moving in slightly different directions with slightly different speeds. Because of these slight differences in speed and directions, the wave packet of a free electron—unlike that of an electron bound in an atom or a molecule—gradually disperses. It comes to occupy an ever-increasing volume of space. But experiments show that free electrons scatter light as though they were tiny spheres of negative electricity a million-millionth of a centimeter in radius. The radius of a hydrogen atom is 10,000 times greater.

You may have been present at one of these experiments without being aware of it. During a total eclipse of the Sun, you can see a faint glow surrounding the black disk of the Moon. Most of the glow is sunlight scattered into our line of sight by free electrons in the Sun's outer atmosphere, just as the blue light of the sky when the Sun *isn't* being eclipsed is sunlight scattered into our line of sight by air molecules. Comparisons between measurements of the glow and calcula-

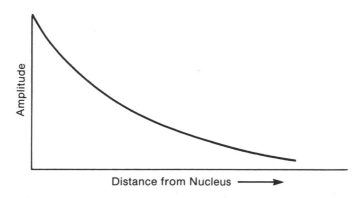

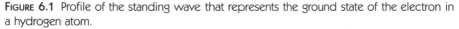

FIGURE 6.1 Profile of the standing wave that represents the ground state of the electron in a hydrogen atom.

tions based on electromagnetic theory have shown that the electons responsible for the glow do indeed behave like tiny spheres.

ALICE: So electrons really are pointlike objects. Yet the state of an electron in a hydrogen atom is represented by a wave packet that fills the whole atom, and even spills over a little. I was prepared to believe Bohr's picture of electrons moving in circular orbits, but I don't see how an electron—or its physical state (whatever *that* means)—can be spread over a three-dimensional region and yet remain a pointlike object.

ZENO: You put the paradox very neatly. It was resolved by Max Born in 1926, the same year Schroedinger published his theory. I think it will help you to understand Born's resolution of the paradox if I describe its context.

Born, remember, was one of the inventors of the earlier, more abstract form of quantum mechanics (which he himself usually referred to as "Heisenberg's theory"). That theory had been used to make successful predictions of energy levels and of the intensities of spectral lines. Heisenberg himself had carried out the first such calculation, predicting the energy levels of a Planck oscillator, and a few months later Wolfgang Pauli succeeded in calculating the energy levels of the hydrogen atom. But no one had been able to find a way of using the theory to describe the quantum jumps themselves. Some physicists believed that the theory was fundamentally incomplete. Born, however, "impressed with the closed character of the logical nature of quantum mechanics, came to the presumption that this theory is complete and that the problem of transitions [that is, quantum jumps between energy levels] must be contained in it."[1]

Born had another, closely related reason for this "presumption." It had been known for some time that quantum jumps occur not only in interactions between atoms and light, but also in interactions between atoms and other particles—electrons, protons, helium nuclei (alpha particles), and other atoms. In 1914, for example, James Franck and Gustav Hertz had carried out a famous experiment showing that electrons fired at silver atoms cause quantum jumps to take place in the atoms

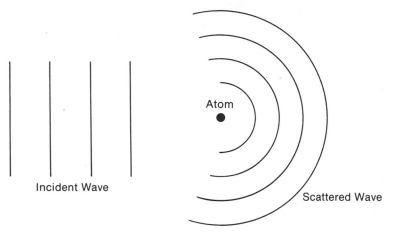

FIGURE 6.2 Scattering of a de Broglie wave by an atom. The incident wave, representing a collimated beam of electrons, is a plane wave. The scattered wave is a spherical wave centered on the atom.

if the energies of individual electrons are greater than the energy difference between the levels connected by the jump. But surely, Born reasoned, if quantum mechanics could describe the outcomes of interactions between particles *within* atoms, it also ought to be able to describe the outcomes of collision processes, as in the experiment of Franck and Hertz. Hence quantum mechanics should be capable of describing quantum jumps. No one, however, could figure out how the theory of Heisenberg, Born, Jordan, and Dirac could be applied to this problem.

When Schroedinger's theory appeared, Born quickly realized that he could use it to describe collisions between atoms or between atoms and particles. Consider, for example, a collision between a hydrogen atom that is initially at rest and an electron that is initially traveling toward it in a straight line. Although the interaction between the electron and the atom may be very complicated, it lasts for only a short time, while the electron is within a few atomic radii of the atom. Before and after the interaction, the electron is a free particle and is represented by a de Broglie wave. Born recognized that the problem of describing such a collision, within the framework of Schroedinger's theory, was mathematically similar to one he was already familiar with:

> Indeed one has nothing more than a "diffraction problem" in which an incoming plane wave is refracted or scattered at an atom. In place of the boundary conditions one uses in optics to describe the diffraction diaphragm, one has here the potential energy of interaction between the atom and the electron.[2]

The outgoing, or "scattered," wave consists of spherical de Broglie waves traveling radially outward from the scattering atom (Figure 6.2).

After deriving an approximate formula for the amplitudes of these scattered de Broglie waves, Born asked, "What is the physical interpretation of this formula?"

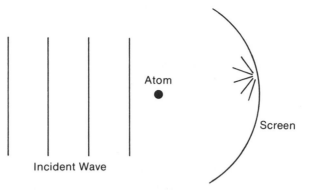

FIGURE 6.3 Observing individual scattering events. The outcome of a single scattering event is unpredictable, but in a long series the fraction of scatterings in a given direction is equal to the squared amplitude of the corresponding de Broglie wave.

"Only one interpretation is possible," he replied: the squared amplitude of a scattered de Broglie wave must represent the *probability* that an electron will be scattered in that direction.[3]

ALICE: Why is this the only possible interpretation?

ZENO: Born doesn't say, but he probably had an argument like the following in mind. Suppose we consider a beam of electrons, all moving at the same speed in the same direction, falling on a small cloud of identical atoms, all at rest. (In fact, that's the way such experiments are always done—not with individual electrons and single scatterers, but with beams of electrons and clouds of scatterers.) Assume that each electron is scattered by a single atom, so that the mathematical description of the process is the same as before. The squared amplitude of a scattered de Broglie wave then represents the *fraction* of scattered electrons coming off in that direction. Now suppose we reduce the intensity of the incident beam until only one scattering event at a time is being recorded. Scattering events might be recorded by scintillation screens (which emit a flash of light when struck by a high-speed electron) that are placed as shown in Figure 6.3. We would then observe a *random sequence* of scatterings in many directions; but if the experiment went on long enough, the fraction of scatterings in any particular direction would be given by the squared amplitude of the corresponding de Broglie wave. That is what de Broglie meant by saying that the squared amplitude of a de Broglie wave gives the probability of scattering in the corresponding direction.

Born recognized that if an extended wave is to represent a localized outcome, such as the impact of an electron at a point on a scintillation screen, it can represent only the *probability* of that event—that is, the frequency with which that outcome would occur in an infinite sequence of identical experiments.

Born must also have been thinking of Einstein's 1917 paper on the quantum theory of radiation. You will recall how Einstein postulated that there is a definite *probability* that an atom irradiated by monochromatic light of the right frequency will absorb or emit a photon during a given interval of time, but whether a particular atom will do so can't be known. Similarly, the probability that a photon in a well-collimated beam will be scattered in a particular direction by an atom in its path is proportional to the intensity of the scattered light in that direction.

ALICE: I also recall that Einstein had misgivings about the role of chance in his theory. Did Born share these misgivings?

ZENO: Not at all. Born addresses the question explicitly in his short preliminary paper (it runs to less than four pages) on atomic collisions. He points out that Schroedinger's theory provides a definite description of atomic collisions, but not a deterministic description. It doesn't predict definite outcomes, but only the probabilities of possible outcomes. Then he continues:

> Here the whole problem of determinism comes up. From the standpoint of . . . quantum mechanics there is no quantity that in any individual case causally determines the outcome of a collision; nor have experiments so far given us any reason to believe that there are any inner properties of the atom that condition a definite outcome. Ought we to hope that such properties remain to be discovered? Or ought we to believe that the agreement between theory and experiment (about the impossibility of prescribing conditions that determine a definite outcome) is . . . founded on the nonexistence of such conditions? I myself am inclined to give up determinism in the world of atoms. But that is a philosophical question for which physical arguments alone are not decisive.[4]

History has vindicated Born's point of view. Moreover, Born's interpretation of the squared amplitude of a de Broglie wave as a probability turned out to be a special case of a much more general rule that tells us what is predictable in the quantum world and what we have to calculate to make a prediction.

ALICE: Could you make that a little more explicit?

ZENO: I'll try. In classical physics, we typically ask questions of the form: What values will such and such quantities have at time t if they, or some other set of quantities, have such and such values at time t_0? For example: Where will the planets be at midnight on 31 December 1999, given their present positions and velocities? In quantum physics, we ask instead: What are the *probabilities* that such and such quantities will have such and such values at time t, given that these, or some other, quantities have such and such values at time t_0? For example, we might ask: What is the probability that the electron in a hydrogen atom will be found within a certain distance of a certain point when the atom is in its ground state? Or we might ask: What is the probability that an atom known to be in an excited state (a state other than the ground state) at time t_0 will be found in the ground state at a later time t? Or possibly: What is the probability that an electron that approaches an atom from a given direction will be scattered through a certain angle?

Each of these questions refers to two quantum states, called the *initial state* and the *final state*. The initial state is the state about which information is given;

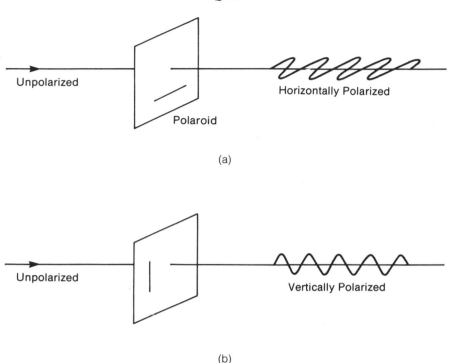

FIGURE 6.4 (a) A sheet of Polaroid marked with a horizontal arrow transmits only horizontally polarized light. (b) When the Polaroid is rotated through 90°, it transmits only verticlaly polarized light.

the final state is the state about which information is requested. The questions all have the form: *What is the probability of observing a certain final state, given a certain initial state?* And the answers are all given by the following rule: *The desired probability is the square of the amplitude of the final state relative to the initial state.*

ALICE: All that seems very abstract. Could you give me a concrete example, please?

ZENO: Let's look at the way quantum physics describes polarized light. I happen to have in my pocket some small pieces of Polaroid, the stuff sunglasses are made of. Light that has passed through a sheet of Polaroid is said to be polarized. Its *state of polarization* is represented by a line perpendicular to the direction of the light. The orientation of that line is determined by the crystalline structure of the Polaroid. I've marked it on each of these pieces. When I rotate the Polaroid, the polarization of the transmitted light changes in a corresponding way (Figure 6.4).

When polarized light—light that has passed through one Polaroid—passes through a second Polaroid, it becomes weaker, unless the lines marked on the two Polaroids happen to be parallel. If the lines are perpendicular, none of the light transmitted by the first Polaroid gets through the second Polaroid. In intermediate cases,

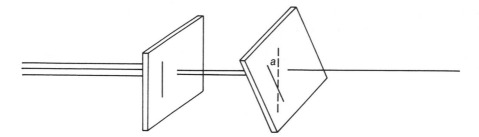

FIGURE **6.5** Verticlaly polarized light is partially transmitted by a Polaroid whose polarization axis makes an angle *a* less than 90° with the vertical direction. The fraction of photons
transmitted is the square of the cosine of the angle *a.*

an intermediate fraction of the light gets through (Figure 6.5). Those are the facts
we need. Now let's see how they translate into the language of quantum physics.

Because light consists of photons, we must assign each photon a state of polarization. For photons that have passed through a piece of Polaroid, this state is
represented by the line that represents the polarization of the light—that is, a line
parallel to the line marked on the Polaroid. We've just seen that only a fraction
of the photons that pass the first Polaroid pass the second as well, unless the
Polaroids are so oriented that the lines marked on them are parallel. So a single
photon emerging from the first Polaroid has a certain *probability* (equal to the
fraction I just mentioned) of passing through the second Polaroid. According to
the rule I mentioned earlier, that probability is the square of a relative amplitude.
This relative amplitude is easy to calculate. It is just the cosine of the angle
between the two lines. Thus the probability that a photon will get through both
Polaroids is the square of the cosine of the angle between the lines marked on
them (Figure 6.6).

ALICE: This example does make it clear why probabilities have to enter into the
description of what happens to individual photons. But there's something about
the description that I find confusing. Did I understand correctly that the first Polaroid passes photons in a particular polarization state, and the second Polaroid
passes photons in another particular polarization state?

ZENO: Yes. A sheet of Polaroid is a filter that passes light of a particular polarization the way a piece of colored glass passes light of a particular color.

ALICE: But none of the photons that reach the second Polaroid is actually in the
polarization state appropriate to that filter; they're all in the state appropriate to
the first filter. So why do any of them get through?

ZENO: A good question. The answer is that, although every photon that passes
through the first Polaroid is definitely and completely in the first polarization state,
it's also *partly* in the second polarization state.

ALICE: Curiouser and curiouser.

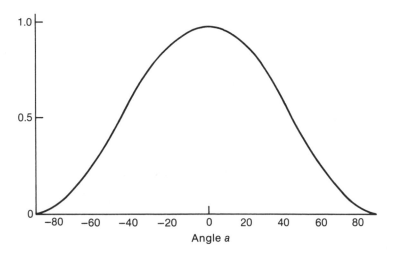

FIGURE 6.6 Probability that a vertically polarized photon (a photon that has passed the first Polaroid) will get through the second Polaroid, whose axis makes an angle *a* with the vertical direction. The probability is 1 if the axes of the two Polaroids coincide, zero if they are perpendicular.

ZENO: Still, perhaps we can make some sense of it. Consider our old friend, the vibrating violin string. You will recall that a violin string vibrates in several modes at once—the fundamental and its overtones—and that the amplitude of the vibration is the algebraic sum, or resultant, of the amplitudes of these simple vibrations. We could say that the vibrating violin string is partly in each of the states represented by the fundamental vibration and its overtones. Analogously, we may say that when a photon is in a definite polarization state it is partly in other definite polarization states. This statement is illustrated in Figure 6.7. Here polarization states are represented by arrows parallel to the direction of polarization. The statement "When a photon is in polarization state *A*, it is partly in each of the polarization states *B* and *C*" is represented in the diagram by the fact that the arrow representing *A* is the "sum" of the arrows representing *B* and *C*. (To form the "sum" of two arrows, join the end of one to the tip of the other. The "sum" is the arrow joining the free end to the free tip.) The length of an arrow represents the relative amplitude of the corresponding polarization state. If arrow *A* has unit length, the amplitude of *B* relative to *A* is the cosine of the angle between the two arrows.

ALICE: It seems very odd that a photon should be able to be in two mutually exclusive states at the same time. Isn't there a less provocative way of translating the mathematics into English?

ZENO: It *is* odd, I admit. But it's not just a peculiarity of the language that quantum physicists speak. It's an essential feature of the world they're trying to describe. Consider the benzene molecule. You probably learned in high-school chemistry that it has a hexagonal shape, with carbon atoms at the vertices and a

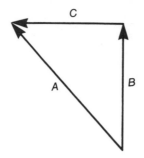

FIGURE 6.7 Diagrammatic representation of the statement "When a photon is in polarization state A, it is partly in each of the polarization states B and C." Arrow A has unit length and is parallel to the polarization axis marked on a Polaroid that passes photons in state A. Arrows B and C are any two arrows whose lengths are less than the length of arrow A and that can be joined in the manner illustrated. The polarization axes of states B and C are parallel to arrows B and C, respectively. The probability of finding a photon known to be in state A in state B (or C) is the square of the length of arrow B (or C).

hydrogen atom attached to each carbon (Figure 6.8). The short connecting lines in this diagram represent pairs of shared electrons. Each carbon atom shares one pair of electrons with a hydrogen atom, one pair with one of its neighboring carbon atoms, and two pairs with its other neighboring carbon atom. Now, there's another way of drawing this molecule (Figure 6.9). The only difference is in the placement of the short connecting lines.

ALICE: Which way is right?

ZENO: Neither. Quantum theory and experiment agree that the benzene molecule must be simultaneously in *both* states. A correct—or more nearly correct—diagram looks something like Figure 6.10.

ALICE: How strange it must feel to be a benzene molecule! I'm sure *I* wouldn't know how to be in two different states at the same time.

FIGURE 6.8 A possible state of the benzene molecule. Each line segment represents a *covalent bond*, formed by a pair of shared electrons. Each carbon atom in the ring shares two electrons with one carbon neighbor, four with the other carbon neighbor, and two with a hydrogen atom.

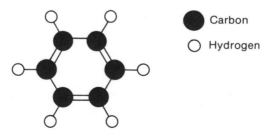

● Carbon

○ Hydrogen

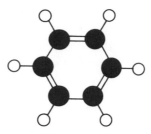

FIGURE 6.9 Another possible state of the benzene molecule. This ring is identical with the ring represented in Figure 6.8 except that it is rotated through 60°.

ZENO: Exactly. Macroscopic objects like you and me can't be in two mutually exclusive states at the same time; quantal objects can be, and often are. That's a very significant difference between the world of experience and the quantum world. And that difference is at the root of a paradox that has tantalized quantum physicists for well over half a century. Quantum physics seems to contain classical physics as a limiting case . . .

ALICE: Why do you say ''seems to''? Surely physicists know by now whether it does or doesn't.

ZENO: From a strictly mathematical point of view, there's no question that the laws of classical physics are limiting cases of the laws of quantum physics in just the same way that Newton's theory of gravitation is a limiting case of Einstein's. Most textbooks on quantum physics take this straightforward view of the relationship. Some physicists, however, maintain that quantum physics presupposes classical physics. Quantum physics, they say, is a theory about the possible outcomes of measurements. Every measurement, however, has a macroscopic component—a pointer reading or something similar—and relies for its interpretation on classical physics.

ALICE: That seems a very strong argument. But I interrupted your train of thought. Please go on.

ZENO: Let's for the moment accept the ''naïve'' view that quantum physics is valid in the macroscopic, as well as the microscopic, domain. If that is so, one

FIGURE 6.10 According to quantum theory, the actual state of a benzene molecule is approximated by a superposition of the states represented in Figures 6.8 and 6.9. (The symbols representing carbon and hydrogen atoms have been omitted.)

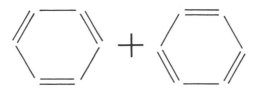

can devise situations in which a *macroscopic* object is in two mutually exclusive states at the same time. Such situations are actually very common. They arise whenever measurements are carried out to determine the quantum state of an atom or another microscopic object. Schroedinger invented what he called a ridiculous example to illustrate this point:

> A cat is placed in a steel chamber, together with the following hellish contraption. . . . A Geiger counter contains a tiny amount of radioactive substance, so tiny that within an hour one of the atoms may decay, but it is equally probable that none will decay. If one decays the counter will . . . activate a little hammer which will break a container of cyanide.[5]

What is the state of the cat-plus-hellish-contraption at the end of an hour? According to quantum physics, the whole system is half in the state "cat alive, bottle intact" and half in the state "cat dead, bottle broken." The state of the cat is exactly analogous to the state of our benzene molecule (Figure 6.11).

ALICE: But that conclusion surely can't be right. Either the bottle will be intact and the cat still alive, or else the poor cat will be dead.

ZENO: I agree. The conclusion that the whole system is half in one state and half in the other does contradict everyday experience. It also contradicts classical physics.

ALICE: But you said a moment ago that quantum physics contains classical physics as a limiting case. I don't see how that can be true if quantum physics leads to conclusions that contradict classical physics. Wouldn't it be more logical to consider the paradox—I would call it a contradiction—as evidence that classical physics *isn't* a limiting case of quantum physics?

ZENO: That, in effect, is the "orthodox view." According to this view, which appears in most textbooks and is held by most of my colleagues, a physical system can change in two distinct ways. The changes that occur in isolated systems—systems not interacting with any measuring apparatus—are governed by the laws of quantum physics. They are *continuous* and *predictable*.

ALICE: Hold on, Zeno! You say that systems governed by the laws of quantum physics change in a continuous manner. I thought the quantum world was fundamentally discrete.

ZENO: It is. The energy levels of an atom or a molecule make up a discrete aggregate, and when an atom or a molecule is in a state of definite energy, it has a definite structure. But, as we have been discussing, an atom or a molecule—and even, it would seem, Schroedinger's cat—can be in two or more discrete states at the same time. What changes continuously and predictably are the relative contributions of these discrete states to the system's actual state. For example, a radioactive nucleus begins in its undecayed state. After a little while, it is partly in its decayed state. After a long time—that is, a time much longer than its half-life—the nucleus is almost, but not quite, certain to be in its decayed state. The

FIGURE 6.11 The state of the cat in Schroedinger's thought experiment, according to quantum theory.

transition from the initial undecayed state to the final almost-certainly-decayed state is both smooth and predictable. Think of a violinist playing an open string and making her bow gradually get closer to the bridge. You hear the quality of the sound change: it becomes thinner, less rich in overtones. The overtones themselves don't change; they gradually disappear.

ALICE: Like the Cheshire cat. And the cat's grin, which remains behind, is like the set of overtones or pure harmonic vibrations. But I interrupted you.

ZENO: I was saying that, according to the orthodox view of quantum physics, a system can change in two completely different ways. While it isn't being observed, it evolves continuously and predictably. But if we carry out a measurement to discover which of its discrete states the system is actually in, its state changes *discontinuously* and *unpredictably*. Schroedinger's thought experiment illustrates the second kind of change. At the end of the experiment, the cat is certainly in one of its two possible states. Which one is a matter of chance.

ALICE: I'm puzzled. Are you saying that if no measurement takes place, a quantal system evolves smoothly and predictably? And that coupling the system to a macroscopic measuring device causes the state of the whole system to change discontinuously and unpredictably?

ZENO: Exactly.

ALICE: But surely the decay of a radioactive atom is both discontinuous and unpredictable whether or not we observe it. Isn't the behavior of the macroscopic measuring apparatus discontinuous and unpredictable just *because* it's coupled to a microscopic process with these properties? Your remarks seem to imply that the opposite is true—that the discontinuity and unpredictability stem from the macroscopic measuring device.

ZENO: A very acute objection! An orthodox quantum physicist would meet it by denying your premise that the decay of a radioactive atom is discontinuous and unpredictable whether or not we observe it. As long as we don't observe a radio-active atom—or, more precisely, as long as the atom doesn't interact with a de-vice that could tell an observer whether a decay has occurred—we can't describe its history in those terms. We can't say, for example, that at time t it hasn't yet decayed, but at a later time t' it has. What we *can* say is that a radioactive atom is partly in the decayed state and partly in the undecayed state. As time goes on, the decayed component of this *composite state* predominates more and more.

ALICE: That sounds a little mystical to me.

ZENO: Perhaps a bit of mathematical language will demystify it.

ALICE: You remind me of someone I met once in a dream. She offered me a dry biscuit to quench my thirst. But I see you've uncapped your fountain pen, so go ahead.

ZENO: I promise that it won't be *very* dry. Let S stand for the state of a radioac-tive atom at some arbitrary time; let I stand for the initial undecayed state; and let F stand for the final decayed state. Then S is expressed by the formula

$$S = aI + bF$$

where a and b are numbers whose squares add up to unity:

$$a^2 + b^2 = 1$$

Initially, a is equal to 1, so b is equal to zero. As time goes on, a gradually decreases toward zero (Figure 6.12). At the same time, b gradually increases toward 1. The quantity b^2 is the probability that the atom will be found to have decayed. This probability is zero initially and gradually approaches unity.

If there were only one kind of process in the world—if the laws of quantum physics applied universally, without restriction—all processes, including macro-scopic ones, would be smooth and predictable. But the world isn't like that. Ra-dioactive atoms do decay, and the particles they emit leave definite tracks in cloud chambers and bubble chambers.

ALICE: Well, the mathematics does demystify things a little. But the "second kind of process" still seems mysterious to me. Do measurements cause the laws of quantum physics to be suspended? If so, why? And what laws take their place?

ZENO: Now your questions are really getting close to the bone. There have been many attempts to answer those questions, none of them, in my opinion, wholly satisfactory. One of the most interesting—and also the most speculative—seeks to relate the dualistic view of quantal processes I've been describing to a more fun-damental kind of dualism, the dualism of mind and matter.

ALICE: Surely *that* old philosophical chestnut can't have anything to do with the interpretation of quantum physics?

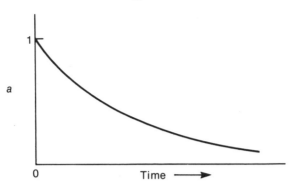

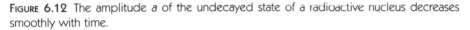

FIGURE 6.12 The amplitude *a* of the undecayed state of a radioactive nucleus decreases smoothly with time.

ZENO: Don't be too sure, Alice. Suppose I observe a radioactive atom with an apparatus that will allow me to detect a decay event if it occurs, and suppose that at the end of one minute the probability that I will have observed such an event is ½. If quantum physics correctly described the state of my brain, that state would be given by a formula like the one I wrote down a moment ago:

$$S = aY + bN$$

Here *S* represents the state of my brain; *Y* represents the state YES (a decay has occurred); *N* represents the state NO; and the numbers *a* and *b* are both equal to $1/\sqrt{2}$. (Remember that the probability that the system—in this case, my brain—will be found in the state *Y* is a^2—in this case, one-half.) But my brain is certainly in no such composite state. Either I have seen a record of a decay event or I haven't. My brain is *either* in state *Y* or in state *N*, not in both at once.

ALICE: That's just Schroedinger's thought experiment, with your brain playing the part of the cat. I don't see that that changes anything very much.

ZENO: Not yet. The next point, though, is that there is no *logical* reason to assume that the recording device—the device I consult to decide whether a decay has occurred—is *not* in this odd composite state.

ALICE: Are you saying that, although your brain is certainly in the YES state *or* the NO state but not in both, a recording device could be in both at the same time?

ZENO: Precisely. At least, that's what Eugene Wigner, who is responsible for the view I'm describing, has suggested.

ALICE: But what's so special about brains? If the laws of quantum physics apply to recording devices, why don't they apply to brains?

ZENO: Because, Wigner says, "the quantum description of objects is influenced by impressions entering my consciousness." The soul acts on the brain and resolves the ambiguity that would characterize the state of an inanimate measuring device.

ALICE: Surely modern physical scientists don't talk about souls acting on brains?

ZENO: Let me quote from an article by Professor Wigner published in 1962:

> Until not many years ago, the "existence" of a mind or soul would have been passionately denied by most physical scientists. The brilliant successes of mechanistic and, more generally, macroscopic physics and of chemistry overshadowed the obvious fact that thoughts, desires, and emotions are not made of matter, and it was nearly universally accepted among physical scientists that there is nothing besides matter. The epitome of this belief was the conviction that if we knew the positions and velocities of all the atoms at one instant of time, we could compute the fate of the universe for all future. . . .
>
> There are several reasons for the return, on the part of most physical scientists, to the spirit of Descartes's "*Cogito ergo sum*," which recognizes the thought, that is, the mind, as primary. First, the brilliant successes of mechanics . . . were . . . recognized as partial successes, relating to a narrow range of phenomena, all in the macroscopic domain. When the province of physical theory was extended to encompass microscopic phenomena, through the creation of quantum mechanics, the concept of consciousness came to the fore again: it was not possible to formulate the laws of quantum mechanics in a fully consistent way without reference to consciousness.[6]

At this point, Wigner quotes Heisenberg:

> The laws of nature which we formulate mathematically in quantum theory deal no longer with the particles themselves but with our knowledge of the elementary particles.[7]

Of course, not all physicists share these views. Some of us still cling to the belief that there is an objective reality and that it is knowable through the natural sciences.

I must admit, though, that Wigner's resolution of the paradox is perfectly logical. After all, the only direct evidence we have that quantum physics doesn't have unlimited validity comes from introspection. When I say, "This recorder is not in the composite state predicted by quantum physics; it is in a specific macroscopic state," what I really mean is, "I *perceive* that this recorder is in a specific macroscopic state."

ALICE: True, but if both of us look at the device, we receive the same impression. How does Wigner's suggestion account for that? I can't believe that the first observer to come along resolves the ambiguity for all subsequent observers. And what about Schroedinger's poor cat? Does it remain in its limbo until someone happens to glance at it?

ZENO: I don't know how Professor Wigner would answer those questions, but let me suggest a possible answer. I recently came across a set of lectures given by

Erwin Schroedinger in 1956.[8] In one of them he argues that "the multiplicity [of minds] is only apparent, in truth there is only one mind." Schroedinger points out that this notion is commonplace in Eastern religious thought and that it may be found in mystical writings from many places and periods, even though in the West it "has little appeal, it is unpalatable, it is dubbed fantastic, unscientific." If you suppose there is only one mind, you can attribute the consistent resolution of composite quantal states to interaction with that mind.

ALICE: I'm afraid that sounds very mystical to me.

ZENO: Of course it is, but only marginally more so than a belief in mind as something that acts on matter but is at the same time distinct from matter. It seems to me that if you're willing to accept *that* belief—which, incidentally, draws no support from modern scientific studies of the relation between mind and brain— you may as well follow Schroedinger and go the whole way.

ALICE: Is there no alternative?

ZENO: There is one, but it has a price. It requires us to give up the time-honored idea that the task of physics is to describe "physical reality." The main task of any physical theory is to predict the outcomes of possible experiments and observations. Perhaps we must admit that quantum physics does that, and no more— that it is merely a computational device for predicting the outcomes of possible measurements. More specifically, it helps us answer questions of the following form: If certain measurements have had certain outcomes, what are the probabilities that certain other measurements will have certain other outcomes?

If we accept this austere view of quantum physics, questions about the states of physical systems and how they change don't even arise, because none of the mathematical quantities that figure in the theory actually represents the state of a physical system. The equation

$$S = aY + bN$$

which I wrote down earlier, still holds, but S no longer represents the state of a physical system. It is simply part of a scheme for calculating the probabilities of experimental outcomes.

I admit that that is a revolutionary view of physics.

ALICE: It's revolutionary in a very literal sense. The turn of the wheel seems to have brought quantum physicists back to ideas that prevailed before Copernicus.

ZENO: How so?

ALICE: Listen to this.

> It is the job of the astronomer to use painstaking and skilled observation in gathering together the history of the celestial movements, and then—since he cannot by any line of reasoning reach the true causes of these movements—to think up or construct whatever causes or hypotheses he pleases such that, by the assump-

tion of these causes, those same movements can be calculated from the principles of geometry for the past and for the future too. . . . And if [mathematical astronomy] constructs and thinks up causes—and it has certainly thought up a good many—nevertheless it does not think them up in order to persuade anyone of their truth but only in order that they may provide a correct basis for calculation.[9]

That's from Osiander's notorious preface to Copernicus's book *On the Revolution of the Heavenly Spheres,* which I just happen to have with me. Change a few words here and there, and you have the view of quantum physics that you've just described. Some philosophers call it instrumentalism.

ZENO: I, for one, don't find instrumentalism that congenial, and I suspect that many of my colleagues who maintain that quantum physics is nothing more than a computational device hold that view only on Sundays. They attend the instrumentalist church, but deep down they are unreconstructed realists. They may pay lip service to the view that "the function of quantum mechanics is not to describe some 'reality,' whatever that means, but only to furnish statistical correlations between subsequent observations,"[10] but I doubt whether many of them think of themselves as mere calculators of statistical correlations. Far from believing that theories are made for experiments, the theoretical physicists that I know believe the chief function of experiments is to assist in the formulation of theories, and that these theories furnish an increasingly truthful description of an underlying physical reality.

But I haven't yet touched on what is perhaps the deepest difficulty assocciated with this view. It was raised in a paper published in 1935 by Einstein, Boris Podolsky, and Nathan Rosen. (That paper, incidentally, motivated the essay in which Schroedinger describes his grisly experiment.) Einstein, Podolsky, and Rosen set out to demonstrate that quantum physics doesn't provide a complete description of physical reality. To that end, they devised a thought experiment. Let me describe, in a schematic and oversimplified way, a similar experiment that was actually carried out by several experimental groups some thirty-five years later (Figure 6.13). A light source emits a pair of photons in different directions. With the help of Polaroids located at some distance from the source, we measure the polarization of each photon. Quantum theory makes the following predictions:

1. Neither photon is in a definite polarization state *before* its polarization is measured. Thus the outcome of a single polarization measurement, on either photon, is unpredictable. (The theory does, however, tell us the probabilities of the possible outcomes.)

2. Although the polarization states of the two photons are individually unpredictable, they are *correlated.* As soon as we have measured the polarization of one photon, we can predict the polarization of the other photon. For example, theory might predict that the two correlation states are mutually perpendicular. Then if we measure the polarization of the first photon and find that it is horizontal, we can be sure that a measurement of the polarization of the second photon will show that it is vertical.

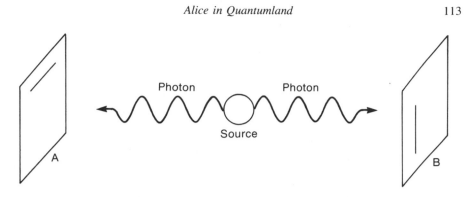

FIGURE 6.13 Schematic representation of the two-photon experiment. A light source emits a pair of photons in opposite directions. Quantum theory doesn't predict the polarization state of either photon, but it does predict that if the outcome of a polarization measurement on one photon turns out to be *H*, then the outcome of a simultaneous polarization measurement on the other photon will be *V*, and vice versa. Thus a polarization measurement of A determines the outcome of a simultaneous polarization measurement at B, even though the outcome at B was indeterminate before the measurement at A took place and even though news about the outcome of that measurement could not have reached B in time to influence the measurement there.

Experiments fully confirm these predictions.

ALICE: The first prediction seems to be consistent with the orthodox view. Before an actual polarization measurement, a photon isn't in a definite polarization state. The act of measurement causes it to jump into one of its possible polarization states.

ZENO: Right. The difficulty arises from the second prediction. Suppose I measure the polarization of the first photon, causing it to jump into the state that corresponds to the outcome of my measurement. Until that moment, the outcome of a polarization measurement on the second photon was equally unpredictable. The second photon was not in a definite polarization state. But when the first photon makes its jump, then, at that very moment, the second photon also jumps into a definite polarization state.

ALICE: I think I'm missing something. You said, didn't you, that the polarizations of the two photons are measured at well-separated places?

ZENO: Yes.

ALICE: So there's no possibility that the first polarization measurement can physically influence the second photon or the outcome of a measurement on the second photon?

ZENO: That's right.

ALICE: Then your account of the experiment doesn't make sense to me. If the measurement carried out on the first photon can't affect the second photon or its measurement, how can it cause a change in the second photon's physical state? If no signal can be transmitted to the second photon, how can it know that a measurement has been carried out on the first photon?

ZENO: Einstein had a sardonic answer to that question: by telepathy. It seemed to him that the orthodox view of measurement logically entailed a belief in the instantaneous, nonphysical communication between distant physical objects—that is, telepathy.[11]

ALICE: Another difficulty with your account occurs to me, Zeno. You said that the state of the second photon changes *at the very moment* when the polarization of the first photon is measured. But according to Einstein's special theory of relativity, if two events appear to be simultaneous to one observer, there must be other observers for whom the second event occurs *earlier* than the first. For such an observer, the outcome of the second measurement would be determined even before she made the first measurement. That makes no sense at all, even if you believe in telepathy.

ZENO: A good point! It emphasizes that the relation between the two measurements can't be a causal relation.

ALICE: What did Einstein conclude from the thought experiment?

ZENO: Einstein considered it unreasonable to suppose that the physical state of a photon can be influenced by an event with which it can have no causal connection. Given that the experiment happens in the way I've described it—and it does—the only logical alternative is to suppose that the second photon didn't suddenly jump into its measured polarization state, but *was there all along*. Since the situations of the two photons are quite symmetrical, the same would have to be true of the first photon. In short, the polarization of a photon before a polarization measurement is not *indeterminate*, according to Einstein; it is merely *unknown*.

ALICE: Wouldn't Einstein's view also extricate us from our uncomfortable position vis-à-vis Schroedinger's cat?

ZENO: Indeed it would. A cat in a state we have not been able to predict is a very different animal from a cat in an indeterminate state.

ALICE: I take it, though, that quantum physicists didn't welcome Einstein's modest proposal with open arms.

ZENO: Far from it. Although Schroedinger and a few others continued to be troubled by the paradoxes we've been discussing, Einstein stood virtually alone in his conviction that the photons in the two-photon experiment and the cat in the cat-in-the-box experiment are in definite but unknown physical states.

ALICE: Why was Einstein's view so unpopular? It sounds perfectly reasonable to me.

ZENO: Any native speaker of the language of quantum mechanics—as, of course, Einstein wasn't; he merely helped invent it—would have great difficulty imagining how that language could be embedded in a classical, deterministic language. The quantal language is more general than the classical language, which it contains as a kind of local dialect, somewhat in the way the language of fractions contains the language of whole numbers. That, I think, was what kept theoretical physicists from taking Einstein's view of quantum physics more seriously: they didn't see how it could be expressed by an adequate mathematical theory. Einstein himself tried to construct such a theory, but never published an account of his efforts. We now know that the kind of theory Einstein was trying to construct doesn't exist.

ALICE: How can that possibly be known?

ZENO: Through a simple but highly ingenious argument published by John S. Bell in 1964. The argument runs as follows.

Let's suppose that Einstein is right and that in the experiment we've been discussing, the polarization states of photons, before any measurement has taken place, are not indeterminate, as quantum mechanics says they are, but merely unknown. Let's also assume, as experimental evidence strongly suggests, that quantum mechanics' statistical predictions are correct. Then quantum mechanics must be an incomplete version of a more complete theory in which the state of a photon is specified by additional variables (called *hidden variables*) whose values determine the precise outcome of any polarization measurement. On this hypothesis, we are unable to predict the outcome of the first polarization measurement because we don't know the values of the hidden variables that specify that outcome.

Einstein's hypothesis implies that the hidden variables are related to the standard variables of quantum theory, such as polarization, in much the same way that the positions and velocities of molecules are related to the macroscopic properties of a gas, such as temperature and density. (Einstein himself used this analogy.) And just as Maxwell and Boltzmann assumed that the velocities of the (then unobserved) molecules of a gas have a definite probability distribution, so we may assume that the values of the hidden variables have a definite probability distribution. This assumption—obvious only in retrospect—is the key step in Bell's argument.

ALICE: I have a vague idea of what you mean by the phrase "probability distribution," but your analogy troubles me slightly. We can say that a certain fraction of the molecules of a gas have velocities between such and such limits. That fraction is the probability that the velocity of an *individual* molecule will be found to lie between those limits, and the set of all such probabilities defines the "probability distribution." Does the probability distribution of hidden variables have that kind of concrete interpretation?

ZENO: Yes. Imagine that the experiment we've been discussing has been repeated a very large number of times. The hidden variables that determine the outcome of the first polarization measurement will have different values in different repetitions of the experiment, but each set of values occurs in a definite fraction of the repetitions. That fraction is, by definition, the probability that the hidden variables will take on that set of values. It is also the probability of the outcome specified by those values.

ALICE: So the probability distributions of the hidden variables determine the statistical properties of the outcomes of polarization measurements.

ZENO: Precisely. Bell now asked: Is it possible to devise probability distributions for the hidden variables that will ensure that the outcome of polarization measurements have the statistical properties predicted by quantum mechanics?

ALICE: Surely it must be possible. After all, you've imposed no restrictions on how many or what kind of hidden variables there may be.

ZENO: Bell proved that it *isn't* possible. He proved that a theory that invokes hidden variables to account for the uncertain outcomes of polarization measurements and that assumes that the orientation of one Polaroid doesn't affect the outcome of a remote measurement made with the help of a second Polaroid necessarily makes some statistical predictions that differ from those of quantum mechanics. He not only showed that such predictions must exist, but actually found some. The simplest predictions of this kind involve correlations between measurements that refer to three separate directions of polarization. The experiments that Bell envisaged would make it possible to test Einstein's conjecture that quantum mechanics is related to a deeper deterministic theory in the way that macroscopic descriptions are related to molecular descriptions.

ALICE: I suppose that experimenters rose to the challenge?

ZENO: Indeed they did. But decisive experiments proved surprisingly difficult to mount. The first such experiment was not carried out until 1972. Its results supported quantum mechanics and refuted Einstein's conjecture. Subsequent experiments have strengthened this conclusion. We can now be confident that quantal indeterminacy has its roots in something deeper than human ignorance.

ALICE: But in what, Zeno? Remember the dilemma that hidden variables seemed to offer a way of avoiding: if we accept the orthodox view of quantum mechanics, we must believe that measuring the polarization of one photon "causes" a simultaneous change in the polarization state of a second, remote photon. (But the "causation" involved is of a kind unknown to physics.) The alternative view, instrumentalism, holds that quantum mechanics doesn't describe "physical reality" (an empty phrase for those who hold this view) but merely provides a basis for calculating the probabilities of experimental outcomes.

ZENO: No one gave more thought to these questions than Niels Bohr. Bohr held that we may not in general regard phenomena as existing independently of the means by which we record them. In *classical* physics, we can pretend that the phenomena and our observations of them are distinct because we can acquire information about a classical system without appreciably altering its physical state. This isn't the case when we deal with atomic and subatomic systems. For example, to measure the position of an electron in a hydrogen atom, we would have to let the electron interact with a photon whose wavelength was small compared with the radius of the atom. Such an interaction, however, would knock the electron out of the atom. For this reason, Bohr maintained, the "objective mode" of description, which makes a sharp distinction between the object and the observer-plus-measuring-apparatus, must lead to paradoxes. Bohr's prescription for avoiding these paradoxes was to use

> the word *phenomenon* to refer only to observations obtained under circumstances whose description includes an account of the whole experimental arrangement.
> . . . . [E]very atomic phenomenon is closed in the sense that its observation is based on registrations obtained by means of suitable amplification devices with irreversible functioning such as, for example, permanent marks on a photographic plate, caused by the penetration of electrons into the emulsion. . . . . [T]he quantum-mechanical formalism permits well-defined applications referring only to such closed phenomena.[12]

Bohr's philosophy of measurement offers a rationale for the instrumentalist view; but many physicists who accept the instrumentalist view don't subscribe to Bohr's philosophy. That's why I haven't discussed it separately. Besides, I have to admit I've never been able to persuade myself that I really understand it. Neither, for that matter, could Einstein, who once remarked that despite "much effort," he had never been able to achieve a "sharp formulation" of the central notion in Bohr's philosophy, the principle of complementarity.

On occasion, Einstein could be even more scathing. In a letter to Schroedinger, dated 9 August 1939, he refers to Bohr as a

> mystic, who forbids, as being unscientific, an inquiry about something that exists independently of whether or not it is observed, i.e. the question whether or not the cat is alive at a particular instant before an observation is made.[13]

And in an earlier letter to Schroedinger (31 May 1928), he writes:

> The Heisenberg–Bohr tranquilizing philosophy—or religion?—is so delicately contrived that, for the time being, it provides a gentle pillow for the true believer from which he cannot very easily be aroused. So let him lie there.[14]

Of course, Einstein's colleagues—apart from Schroedinger—were just as unsympathetic to *his* philosophical views.

ALICE: Surely, though, all the issues we've been talking about have now been settled. That's the difference between philosophy and physics, isn't it? We philosophers are still debating issues raised by Democritus and Socrates; you physicists solve your problems and go on to new ones.

ZENO: Quantum physics has indeed made enormous strides during the past half-century. Yet, strangely enough, the issues we've been discussing aren't yet settled. True, Einstein's conjecture that the outcome of every measurement is in principle predictable has been refuted; I think everyone now agrees that quantum physics isn't incomplete in the way he imagined. But that makes the two puzzles we've discussed more puzzling, if anything. None of the solutions we've discussed seems to me completely convincing. I find the orthodox view that measurements cause unpredictable changes in a system's physical state unsatisfactory for the same reasons that Einstein did. I'm not yet ready to believe in telepathy. Nor, for that matter, in Mind as a separate kind of reality that acts on physical reality. Yet I can't accept the instrumentalist view, which denies that the notion of physical reality has any meaning.

ALICE: You haven't answered my original question, Zeno. I still don't know exactly what physicists mean when they say that the states of a hydrogen atom are *represented* by standing wave packets, but I'm glad to learn that I'm in such distinguished company.

7

The Strong
Cosmological Principle
and Quantum Physics

At first blush, it may seem absurd to suggest that there is a direct connection between quantum physics and cosmology. Yet there are two clues that point in this direction.

The first has to do with chance. According to quantum physics, some experiments and observations have inherently unpredictable outcomes. Observations of the radioactive decay of individual radioactive nuclei and measurements of the polarization states of individual photons are the examples we have discussed most. As we saw in Chapter 6, recent experiments have all but ruled out the possibility that this kind of unpredictability (which I will refer to as *quantal indeterminacy*) arises from our ignorance of more fundamental, fully deterministic laws. In Chapter 3 we discussed another kind of indeterminacy, which I will refer to as *cosmological indeterminacy*. The Strong Cosmological Principle implies that a complete description of the Universe lacks a vast quantity of microscopic information. Are there really two distinct kinds of objective indeterminacy in the world? Or are quantal and cosmological indeterminacy somehow related?

The second clue comes from the quantal paradoxes. These have to do with the outcomes of measurements carried out on quantal systems. (The cat in Schroedinger's thought experiment is cast in the unhappy role of a measuring device.) Measurement also figures prominently in the interpretation of quantum theory. In the orthodox interpretation, measurements cause discontinuous and unpredictable changes in the physical states of quantal systems; the instrumental interpretation views quantum physics as a computational device for making predictions about the outcomes of measurements. Now, we will see that the critical property of measurements is *irreversibility*. Every measurement necessarily produces a record, hence an irreversible change in the macroscopic world. But according to the argument sketched in Chapter 3, irreversibility is linked to the Strong Cosmological Principle. So, once again, quantal and cosmological indeterminacy seem to be linked.

In this chapter, I will develop an interpretation of quantum physics that unites quantal and cosmological indeterminacy and resolves the quantal paradoxes. This interpretation combines three elements: Einstein's idea that quantum theory describes assemblies rather than individual systems; the Strong Cosmological Principle; and the theory of irreversible processes sketched in Chapter 3.

Einstein's Interpretation of Quantum Physics

Einstein believed that quantum physics correctly describes the statistical properties of assemblies of physical systems but doesn't describe the physical states of individual systems. He also believed that there exists a classical, deterministic description of the microscopic world. We now know that the second opinion is almost certainly mistaken. Bell's theorem shows that if a classical description of microscopic reality exists, some of its predictions must differ measurably from those of quantum physics; and relevant measurements support quantum physics. But what about the opinion that quantum theory describes assemblies rather than individual systems?

If this opinion could be justified, it would resolve both of the paradoxes that we discussed in Chapter 6. Consider Schroedinger's thought experiment. If we assume that the laws of quantum physics apply to individual objects, large and small, we find that the box containing the cat and the hellish contraption ends up in a composite state, with the cat partly dead and partly alive. But if, following Einstein, we regard quantum physics as a theory that describes infinite assemblies of systems, we reach the sensible conclusion that an assembly of initially identical boxes containing identical live cats and undecayed radioactive nuclei evolves into an assembly in which a fraction of the boxes contains dead cats and decayed nuclei, while the remaining fraction contains live cats and undecayed nuclei. Again, Einstein's interpretation of the two-photon experiment avoids the unpalatable conclusion that measuring the polarization of a photon at detector A somehow causes the photon at detector B to jump into a correlated polarization state. It leads instead to the sensible conclusion that in a certain fraction of the members of an assembly of two-photon emitters, the photon at A is horizontally polarized and the photon at B is vertically polarized, while in the remaining fraction, the photon at A is vertically polarized and the photon at B is horizontally polarized.

Some people have suggested that Einstein disliked quantum theory because he was convinced that "God doesn't play dice." Certainly, he had misgivings about the role of chance. Yet, as we have seen, it was Einstein who invented the first quantal theory of the interaction between matter and radiation, a theory in which chance plays a central role. I think Einstein's distrust of quantum theory had more to do with the paradoxes we have been discussing than with his dislike of chance. The aspect of quantum theory that Einstein criticized most frequently and in most detail was not its reliance on chance but its claim to describe individual physical systems.

The Strong Cosmological Principle implies that individual systems don't figure in a complete description of physical reality. Hence it is incompatible with the claim that quantum theory describes individual systems. Yet quantum theory makes correct statistical predictions. Can we find an interpretation of quantum theory that doesn't alter the theory's statistical predictions but is consistent with the Strong Cosmological Principle?

An Assembly of Polarization Measurements

We discussed polarization states of photons in Chapter 6. Recall that a sheet of Polaroid marked with a horizontal line is transparent to horizontally polarized photons (photons in state H) and opaque to vertically polarized photons (photons in state V), while a Polaroid marked with a vertical line is transparent to photons in state V and opaque to photons in state H. Both Polaroids are partially transparent to photons whose polarization state is neither H nor V. A single photon whose polarization state is neither H nor V has some probability p between zero and 1 of passing through the Polaroid marked with a horizontal line, and it has probability $1-p$ of passing through the Polaroid marked with a vertical line. Such a photon is partly in both polarization states. Its polarization state S is represented by the formula

$$S = aH + bV \qquad (1)$$

where $a^2 = p$ and $b^2 = 1 - p$.

Now consider an infinite, unbounded assembly of photons, all in the same polarization state S. We may represent the physical state of the assembly by a row of Ss

$$. . . SSSSS . . . \qquad (2)$$

with S given by formula (1). Each S in row (2) refers to one of the photons in the assembly. Instead of saying, "This photon (that is, the photon whose polarization state is measured by a particular detector) is in polarization state S," we could equally well say, "This photon belongs to an infinite assembly of photons, all of which are in the state S." The two statements mean the same thing, but the second statement expresses the meaning in a way that is more obviously consistent with the Strong Cosmological Principle.

Now suppose that the polarization state of each photon has been measured. In principle, we can perform such a measurement with a crystal of Iceland spar (calcite). When you look at an object through a calcite crystal, you see two images. As illustrated in Figure 7.1, the crystal splits a beam of unpolarized light into two linearly polarized beams with perpendicular axes of polarization, which

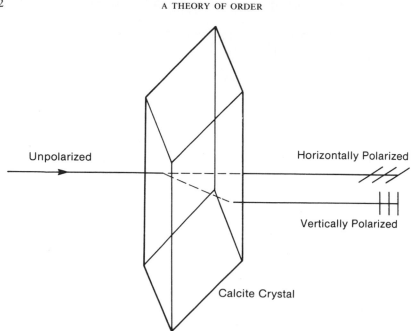

FIGURE 7.1 A properly oriented calcite crystal splits a beam of unpolarized light into two beams, one horizontally polarized, the other vertically polarized.

we may identify with H and V. What happens to a single photon? If its initial polarization is parallel to H, the photon will emerge in the upper beam. If it is parallel to V, the photon will emerge in the lower beam. If the polarization of the incident photon is neither horizontal nor vertical, it will emerge in one beam or the other, but we can't predict whether it will take the high road or the low road.

The polarization states of an assembly of emergent photons are represented by an endless row of Hs and Vs

$$\ldots HHVHVVVH \ldots \qquad\qquad (3)$$

in which the fraction p of the places is filled by Hs, and the remaining fraction $1-p$ is filled by Vs. In all other respects, the distribution of Hs and Vs is random. Some of the photons in this assembly have polarization H, some have polarization V, but we can't predict the measured polarization of an individual photon.

Row (2) represents the physical state of an assembly of photons in identical polarization states *before* a polarization measurement has taken place. Row (3) represents the physical state of the assembly *after* a measurement has taken place. Are these states the same?

To answer this question, let's consider the kinds of information needed to specify the two states. Each entry in row (3) represents the outcome of an experiment whose two possible outcomes, H and V, have probabilities p and $1-p$, respectively. Otherwise, the Hs and Vs are randomly distributed. The structure of

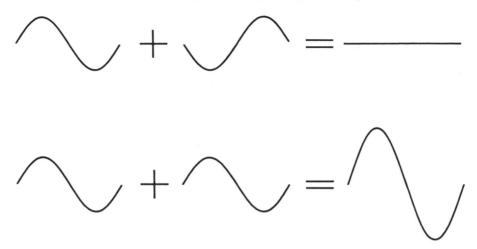

FIGURE 7.2 Destructive and constructive interference. Two identical waves of opposite phase cancel when added. Two waves with the same phase combine to make a wave with double the original amplitude.

the row is therefore completely determined by the single number p, whose value we can estimate by counting the Hs in a large sample of the row. To specify row (2), we have to specify the coefficients a and b. But a is the square root of p, and b is the square root of $1-p$. So it may seem at first sight that the number p specifies the structure of row (3) as well as the structure of row (2).

But we have overlooked a seemingly trivial point. The square root of p can be negative as well as positive; so can the square root of $1-p$. If a is a square root of p, then so is $-a$; and if b is a square root of $1-p$, then so is $-b$. So we need more information to construct row (2), which represents the premeasurement state of the assembly, than to construct row (3), which represents the postmeasurement state. What does the extra information mean?

Recall that physical states are represented in quantum physics by vibrations. Now, when we add two identical vibrations, they may cancel each other or they may reinforce each other, as illustrated in Figure 7.2. The vibrations reinforce each other if they are *in phase*—that is, if their crests coincide; they cancel each other if they are *out of phase*—that is, if the crests of one coincide with the troughs of the other. If we represent identical vibrations that are *in* phase by the same arrow, so that their ''sum'' is an arrow twice that length, then we should represent identical vibrations that are *out* of phase by arrows of equal length pointing in opposite directions. We may denote the arrow that cancels the arrow H by $-H$. The rule for ''adding'' arrows tells us that the sum of these arrows is indeed an arrow of zero length, which we may denote by 0.

A Polaroid sheet doesn't discriminate between vibrations along the same axis, such as H and $-H$ or V and $-V$. (That's why we used lines rather than arrows to indicate the polarization state of photons that a Polaroid sheet passes.) But, as

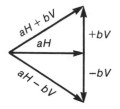

FIGURE 7.3 The polarization states $aH + bV$ and $aH - bV$ define different polarization axes. Only the *relative* sign of the two coefficients matters. Changing the sign of both coefficients changes the direction of the arrow but not the polarization axis.

Figure 7.3 shows, the expressions $aH + bV$ and $aH - bV$ define different axes of linear polarization. If we know only the squares of a and b, we can't tell which of these polarization states the photons were in before the measurement.

To summarize the argument up to this point:

1. Row (2) represents the premeasurement state of an assembly of photons, all in the polarization state S.
2. Row (3) represents the postmeasurement state of the assembly.
3. Some of the information needed to specify the premeasurement state of the assembly is not present in the postmeasurement state.

Somehow the measurement process transforms an assembly of photons in state (2) into an assembly of photons in state (3). Our next task is to understand how this comes about. To do this, we must take a closer look at the measurement process itself.

What Is a Measurement?

All measurements have two key properties. First, every measurement brings about a correlation between the states of two physical systems: the target of the measurement and a device that records the outcome of the measurement—a pointer, a magnetic tape, a photographic plate, or the like. A thermometer measures the temperature of the air in a room because the height of the mercury is correlated with the average kinetic energy of the air molecules.

The second key property of every measurement is *irreversibility*. When we measure the temperature of the air, heat flows between the thermometer and its surroundings until their temperatures are equal. The reverse process—a spontaneous flow of heat that would cause the thermometer and the surrounding air to assume different temperatures—never occurs. It would violate the second law of thermodynamics. Chemical reactions also generate entropy and hence never spontaneously reverse themselves.

Measurements of classical systems and measurements of quantal systems share these properties, but they also differ in important ways. Measurements carried out on macroscopic systems can be nonintrusive, in the sense that they don't signifi-

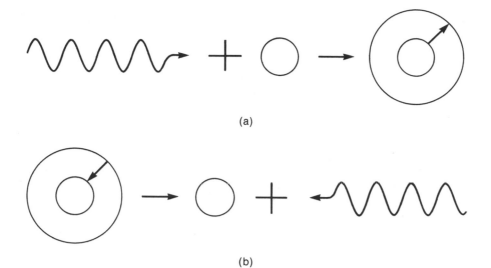

(a)

(b)

FIGURE 7.4 The absorption of a photon by an atom is a reversible process. (a) An incident photon is absorbed, and the atom that absorbs it jumps to an excited state. (b) An atom in the excited state drops to a state of lower energy, emitting a photon.

cantly disturb the target system. We don't noticeably alter the temperature of air in a room by measuring it, even if the thermometer has just come out of a refrigerator, because the heat capacity of the air is so much greater than the heat capacity of the thermometer. By contrast, measurements carried out on quantal systems are often intrusive. For example, to measure the position of an electron in a hydrogen atom, we must use at least one photon, and the wavelength of that photon must be smaller than the diameter of the hydrogen atom. The energy of such a photon is at least 1,000 times larger than the energy needed to disrupt the atom.

There is another important difference between classical and quantal measurements. The fact that classical measurements are irreversible doesn't distinguish them from other macroscopic processes. *All* real macroscopic changes are irreversible; reversible macroscopic change is an unattainable ideal. But all purely quantal processes are reversible, because the laws that govern them don't distinguish between the two directions of time. Consider, for example, the emission of a photon by an atom in an excited state. If the photon's direction were reversed, it would be reabsorbed, leaving the atom in its original excited state (Figure 7.4). It is true that some quantal processes are easier to reverse than others. Reversing the decay of a free neutron (which spontaneously disintegrates into three particles: a proton, an electron, and an antineutrino) can't be done in the laboratory. Nevertheless, the reverse process doesn't violate the laws of quantum physics. There are even physical conditions—those prevailing in the early Universe, for example—under which the direct and reverse processes occur at comparable rates. Quantal *measurements,* however, are irreversible processes. The measuring device is a macroscopic system, and its interaction with the target system makes the measurement process irreversible.

An Assembly of Quantal Systems Coupled to Measuring Devices

To understand what happens during a measurement, we have to consider the state of the measuring device as well as the state of the target quantal system. An ideal measurement causes states of the measuring device to become correlated with states of the target system. Let h and v denote *microstates* of the polarization-measuring device correlated with the polarization states H and V, respectively, and let hH and vV denote joint states of the photon and the polarization-measuring device. Suppose a photon is in the polarization state S ($=aH+bV$) before a measurement has taken place. Quantum theory tells us that after the measurement has taken place, the joint state of the photon and its polarization-measuring device is $ahH+bvV$. That is, the structure of the joint state of the photon and the measuring device mirrors the structure of the photon's state before a measurement has taken place.

We haven't yet used the fact that measurement is an irreversible process. In Chapter 3, I argued that systems that display irreversible behavior are in definite *macrostates* but not in definite *microstates*. More precisely, information that would specify the microstate of such a system is not merely unknown to human observers but (by virtue of the Strong Cosmological Principle) objectively absent. In the last paragraph, h and v are *microstates* of the polarization-measuring device. But we can't specify these microstates; we can specify only the macrostates to which they belong. Let h', h'', and so on denote other microstates belonging to the same macrostate as the microstate h; and let v', v'', and so on denote other microstates belonging to the same macrostate as the microstate v. After a polarization measurement has taken place, the state of the assembly of photons and coupled polarization-measuring devices is represented by the row

$$\ldots (ahH+bvV)\ (ah'H+bv'V)\ (ah''H+bv''V)\ \ldots \tag{4}$$

whose entries are drawn randomly from a huge collection of joint microstates.

We come now to the crux of the argument. The initial microstates of the measuring device make up a random assembly, and any correlation between the microstates h and v, h' and v', and so on that might arise during the course of a measurement is destroyed by random external disturbances, as we discussed in Chapter 3. It follows that terms like $ahH+b(-v)V$ occur just as frequently in row (4) as terms like $aH+bV$. Consequently, row (4) represents the same physical state as the row

$$\ldots (ahH-bvV)\ (ah'H-bv'V)\ (ah''H-bv''V)\ \ldots \tag{5}$$

We can construct row (5) by changing b to $-b$ in every term of row (4). Yet, as we have just seen, the two rows represent the same physical state, because the microstates h and v have random signs.[1] Because rows (4) and (5) don't, despite their appearance, contain information about the relative sign of the coefficients a and b, they represent the same physical state as the row

$$. . . (hH) (h'H) (v''V) (h''H) (v'V) . . . \qquad (6)$$

in which a fraction a^2 of the entries contain Hs and the remaining fraction b^2 contains Vs. Thus polarization measurements carried out on an assembly of photons each of which is initially in the composite state $aH + bV$ cause the assembly of photons and coupled polarization-measuring devices to evolve into an assembly in which a fraction a^2 of the photons is in state H (with measuring devices in microstates of type h) and the remaining fraction b^2 is in state V (with measuring devices in microstates of type v).

This is the conclusion we have been working toward. As in the orthodox and instrumental interpretations of quantum theory, measurements have definite but unpredictable outcomes. But we haven't had to postulate that measurements cause discontinuous and unpredictable changes in the physical states of the measuring device and the target system. Instead, we have seen that an assembly that represents what can (in principle) be known about the premeasurement states of a quantal system and a macroscopic measuring device evolves during the measurement into an assembly of quantal systems and coupled measuring devices, each in a microstate that corresponds to one of the possible outcomes of the measurement.

The orthodox and instrumental interpretations both *postulate* that the probability of finding a photon in the polarization state H is a^2 if the photon was in state $aH + bV$ before the measurement. In the present interpretation, this rule follows from more fundamental postulates, including the Strong Cosmological Principle.[2]

The present interpretation of quantum theory, like Einstein's, easily resolves the two quantal paradoxes discussed in Chapter 6. Thus in Schroedinger's thought experiment, we may let H and V denote the two possible states of the cat, and h and v corresponding microstates of the "measuring device" to which the cat is coupled. In the final state (or in any intermediate state), a certain fraction of the cats in the assembly are alive and the cyanide bottles are intact; in the remaining fraction, the bottle is broken and the cat is dead.

In the two-photon experiment, we may let H and V denote the two possible outcomes of a *pair* of polarization measurements, carried out at well-separated stations A and B. For each member of the assembly, the measurements carried out at the two stations have definite and correlated outcomes. A polarization measurement at A doesn't cause the measured photon's polarization state to jump into a definite polarization state, as in the orthodox interpretation; it merely tells us whether that photon belongs to a member of the assembly for which the polarization measurement at A has the outcome H or to a member of the assembly for which the polarization measurement has the outcome V. And knowing that the outcome at A is H, say, we also know that the outcome at B must be V.

But the present interpretation of quantum theory differs from Einstein's in an important way. Einstein believed that quantum theory applies to assemblies rather than to individual systems because individual systems are governed by as-yet-undiscovered deterministic laws. I have argued that quantum theory applies to assemblies rather than to individual systems because a complete description of physical reality doesn't refer to individual systems but only to assemblies. The smallest fragment of the Universe we can meaningfully describe is an assembly. If the members of the assembly are in identical microstates, there is no harm in treating them as individuals. But if they are quantal systems coupled to (macroscopic) measuring devices, we run into paradoxes like those we have discussed when we assume that quantum theory applies to them directly as individuals.

Do We Exist in Multiple Copies?

Are the assemblies we have been discussing "real"? Does the Strong Cosmological Principle imply that somewhere in the Universe there is a star very much like the Sun; and orbiting that star, a planet very much like the Earth; and on that planet, a person very much like you, the reader, reading a book very much like this one? Of course, such near-replicas of the Earth and its inhabitants would be very thinly distributed in space. Although I haven't made a serious estimate, I am confident that the nearest one would lie well beyond the most distant galaxy we could observe, even with infinitely sensitive instruments. Even so, the idea is unsettling, however familiar it may be to readers of science fiction. Must we accept it if we accept the Strong Cosmological Principle?

I think not. The Strong Cosmological Principle doesn't prescribe the contents of the Universe; on the contrary, it drastically limits the predictive scope of physical laws. In the Universe of Laplace and Einstein, there is no gap between what exists and what can be predicted. A Laplacian Intelligence can know the state of the Universe at any given moment in time. With the help of this knowledge, it can predict all past and future states. The Strong Cosmological Principle implies that such an Intelligence can't exist. What can be known and predicted are statistical properties only. Statistical predictions, however, do not prescribe all the properties of infinite collections. For example, suppose that a certain outcome has zero probability of being realized. Does that mean it doesn't occur? Not at all. The probability of an outcome is the fraction of times it occurs in an infinite set of "trials." Any outcome that occurs a finite number of times has zero probability.

Are there possible outcomes—outcomes permitted by the laws of physics—that have zero probability? I don't know; this is a question that needs further thought. But if, as I suspect, there are, there could be a considerable gap between what is real and what is predictable.

The Many-Worlds Interpretation of Quantum Theory

The interpretation of quantum theory discussed in this chapter resembles in some respects the "many-worlds" interpretation proposed by Hugh Everett in 1957.[3] Everett, in a Ph.D. thesis supervised by John Wheeler, suggested that every measurement or measurement-like process causes the Universe to split into a vast number of "parallel universes," in each of which one possible outcome of the measurement is realized. In one set of universes, Schroedinger's cat lives; in another, it dies. Quantum theory, according to this interpretation, doesn't describe individual physical systems, as in the orthodox and instrumental interpretations; nor does it describe assemblies of physical systems, as in the interpretation based on the Strong Cosmological Principle. It describes a multitude of universes, each of which splits at every moment into a multitude of parallel universes. All these universes are equally real, but only the one we happen to be in is real to us; all the others are completely inaccessible to us.

According to the many-worlds interpretation, the probability that a measurement has a given outcome is equal to the fraction of the parallel universes in which that outcome occurs. Since probabilities are real numbers that can assume any value between zero and one, the set of parallel universes must be infinite. Every measurement or measurement-like process in every universe therefore creates an infinity of new parallel universes.

The many-worlds interpretation shares two attractive features of the interpretation based on the Strong Cosmological Principle. It avoids the paradoxes that result from the conventional assumption that quantum theory describes individual systems. And it predicts, instead of merely positing, the basic rule mentioned earlier for calculating the probabilities of experimental outcomes. For example, it predicts that when someone (in any universe) measures the polarization of a photon initially in state $aH + bV$, that universe splits into an infinite collection of parallel universes, in which the fraction a^2 is in state H and the fraction b^2 is in state V.

But the many-worlds interpretation also has a serious flaw (apart from metaphysical extravagance). Like the orthodox interpretation, it takes as a given the distinction between reversible quantal processes and irreversible measurement-like processes. According to the orthodox interpretation, measurements and measurement-like processes cause unpredictable and discontinuous changes in the state of a physical system; according to the many-worlds interpretation, they cause the universe in which they occur to split into an infinity of parallel universes. Now, we have seen that measurements are irreversible processes that create correlated states of the measuring device and the target system. The many-worlds interpretation makes irreversibility a primitive, and hence inexplicable, feature of the world (or rather, of an infinity of infinities of parallel worlds). The orthodox interpretation faces the same problem, and indeed several people have tried to construct a theory that would explain *why* measurements cause unpredictable and discontinuous changes in the states of quantal systems. So far, none of these attempts has succeeded, but an explanation is at least conceivable. By contrast, the splitting of a universe into parallel universes isn't susceptible to scientific explanation. If we accept the many-worlds interpretation, we must renounce any attempt to *understand* the role of irreversibility. This seems to me a fatal flaw.

✱ ✱ ✱

In this chapter, I have argued that Einstein was right in his belief that there is only one kind of chance in the world. According to the theory I have described, the chance that figures in a description of radioactive decay and the chance that figures in a complete statistical description of the Universe are the same actor in different costumes. But this kind of chance, contrary to Einstein's belief, is *irre-*

ducible. In other words, it is a consequence not of human ignorance or of our meager capacity to store and process information, but of a fundamental symmetry in the Universe, the equivalence of all positions in space and of all directions at each point in space. By virtue of this postulated symmetry, an Intelligence of the kind envisioned by Laplace and Einstein can't exist—not because there is too much information in the world but because there is too little, and because, as we will see in the following chapters, the future is not entirely predictable.

The interpretation of quantum physics outlined in this chapter also accords with Einstein's belief in the objectivity of physics. Since the mid-1920s, when the basic laws of quantum physics were formulated, quantum physicists, philosophers of science, and (especially) popularizers have repeatedly asserted that quantum physics undermines the old scientific belief in a physical reality that exists independently of human observers. Quantum physics has taught us (they say) that reality is created, at least in part, by interactions with observing and measuring devices—or even, perhaps, by interactions with human minds. John Wheeler puts this point very clearly:

> [N]ature at the quantum level is not a machine that goes its inexorable way. Instead what answer we get depends on the question we put, the experiment we arrange, the registering device we choose. We are inescapably involved in bringing about that which appears to be happening.[4]

Some physicists, among them Werner Heisenberg, have gone still further, maintaining that quantum physics doesn't describe external reality at all, but only our knowledge of it. The degree to which physicists stress (and welcome) the subjective or "participatory" aspect of the quantal world picture varies widely, but as long as measurement remains a primitive notion of the theory, some degree of subjectivity is unavoidable. And measurement must remain a primitive notion as long as we adhere to the traditional view that quantum physics describes individuals systems.

If, however, quantum physics describes assemblies of identical systems obeying the Strong Cosmological Principle, as I have proposed in this chapter, it doesn't have to be supplemented by ad hoc postulates about measurement. Formulated in this way, the theory *predicts* that measurements have definite but unpredictable outcomes, and that the probability of any given outcome is given by the usual rule.

Unlike earlier interpretations of quantum theory, the interpretation that I propose doesn't presuppose classical physics. Instead, classical physics appears as a true limiting case of quantum theory. Although quantum theory's description of physical reality clashes with some of our intuitive physical ideas, its account of physical reality isn't, according to the present interpretation, less objective than that of classical theory. Nevertheless, the present interpretation retains—and to some degree even explains—the most innovative aspect of the quantal worldview: the irreducible character of chance. To Einstein's objection that "God doesn't play dice," it suggests the reply, "He must play dice because He doesn't play favorites."

II

ASPECTS
OF
TIMEBOUND
ORDER

8

Cosmic Evolution:
The Standard Model

Boltzmann's Hypothesis for the Origin of Order

The second law of thermodynamics says that all natural processes generate entropy. Entropy is a measure of disorder. To Rudolf Clausius, discoverer of the Second Law, and Ludwig Boltzmann, inventor of the statistical theory of entropy, the cosmological implication of these statements seemed obvious: the Universe must be running down, its supply of order continuously diminishing.

Indeed, if the Universe was infinitely old, as nearly all scientists from Newton onward had assumed, it must already have run down—attained its state of maximum entropy. Why, then, do we see so much order around us? Boltzmann offered an ingenious answer that can be summed up in a single word: *fluctuations*.

Randomness and fluctuations are two sides of a coin. Consider a random sequence of coin tosses. If the sequence is long enough, we are likely to find, somewhere in it, a sub-sequence of 10 heads. In an infinite random sequence of coin tosses, we are *certain* to find sub-sequences of 10—or 100 or 1 million—heads if we look long enough. Again, imagine an infinitely long, random string of lower- and upper-case letters of the English alphabet, marks of punctuation, and spaces. Because any given sequence of characters has a finite probability of occurring, it *must* occur—and, indeed, infinitely often—in such a string. Thus the string contains every piece of English prose and poetry that has ever been written or that ever will be written, each in a vast number of slightly differing versions. It also includes English translations, of varying degrees of merit, of all literary works in foreign languages, along with working drafts of all these translations.[1] Of course, sequences that are meaningful in *any* language are exceedingly infrequent in the string. Meaningful phrases are oases of order in a Sahara of disorder. To the student of probability theory, they are *statistical fluctuations*. Boltzmann speculated that our corner of the Universe is such a fluctuation, its orderliness a consequence of our parochial point of view. From a cosmic perspective, our world is just a fantastically improbable fragment of perfect chaos.

The theory of probability tells us that fluctuations continuously form and dissolve. And because there is no preferred direction of time in a Universe wholly lacking in order, any given fluctuation is as likely to grow as to decay. The second law of thermodynamics tells us that the fluctuation we inhabit is decaying. Boltzmann speculated that the inhabitants of what *we* would describe as a growing fluctuation would nevertheless perceive their world as sinking into, rather than emerging from, chaos. Their time, therefore, would run in the opposite direction from ours.

> This . . . seems to me to be the only way [Boltzmann wrote] in which one can understand the second law—the heat death of each single world—without a unidirectional change of the entire universe from a definite initial state to a final state.[2]

Astronomical Evidence

Although Boltzmann's hypothesis was widely accepted in its day, advances in our understanding of the astronomical Universe have made it untenable. These advances were spearheaded by observational studies carried out during the 1920s and 1930s by Edwin P. Hubble with the 60- and 100-inch telescopes at the Mount Wilson Observatory in southern California. Hubble drew two main conclusions about the astronomical Universe, both of which have been strongly confirmed by subsequent studies.

First, he concluded that apart from local irregularities, the spatial distribution of galaxies is statistically uniform.[3] As we saw in Chapter 3, modern astronomical observations strongly support this hypothesis. Our corner of the Universe is not a highly improbable fluctuation but, as Hubble said, a fair sample.

Hubble's second major finding was even more spectacular. Studies of systematic shifts in the wavelengths of lines in the spectra of distant galaxies led him to conclude that the astronomical Universe is not static but is undergoing a uniform expansion. Distant galaxies are receding from us at rates proportional to their distances. It is easy to see that if this relation between recession speed and distance holds for one observer, then it holds for every observer (Figure 8.1). Thus the cosmic expansion looks the same to every observer.

Because the galaxies are moving farther apart, they must have been closer together in the past. Einstein's theory of gravitation tells us that a uniform, unbounded distribution of mass must either contract or expand in accordance with

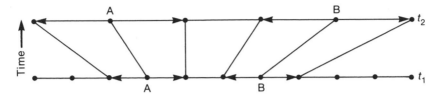

FIGURE 8.1 Uniform expansion of an infinite row of equally spaced points. At time t_2, the distance between adjacent points is twice as great as at time t_1. Notice that any point (A or B, for example) is at the center of a uniform expansion. At any given moment, the relative velocity of any two points is a fixed multiple of their separation.

Hubble's law (recession speed proportional to distance), and predicts that the rate of expansion (or contraction) decreases with decreasing mass density. According to present estimates, the cosmic expansion began between 10,000 and 15,000 million years ago.

The start of the cosmic expansion was also the beginning of the history of the Universe; according to Einstein's theory, there was no earlier moment. This may seem odd. It is hard to think of time as having a beginning. We are tempted to ask: What happened *before* the Universe began to expand? Einstein's theories have taught us, however, that our intuitive ideas about space and time are not to be trusted beyond the limited contexts in which they evolved. We should think of space and time as elements in theories whose ultimate justification lies in their ability to represent experience. Einstein's theory of gravitation tells us that space is expanding and that time has a beginning. These are strange ideas, but they are internally coherent and they agree with experience.

The inference that the Universe is 10 to 15 billion years old is supported by two independent methods of estimating astronomical ages: radioactive dating and estimates of stellar ages based on the theory of stellar evolution.

Radioactive Dating

Certain rocks—meteorites, Earth rocks, Moon rocks—contain undecayed radioactive elements along with their decay products. All radioactive elements decay according to the same law: half of any sample of a given radioactive element decays after a fixed period of time, called the half-life. Different elements have different half-lives, but the half-life doesn't depend on the size of the sample. This implies that the *logarithm* of the undecayed fraction of a given sample decreases at a constant rate, as illustrated in Figure 8.2. With the help of this law, geologists have been able to deduce the ages of rocks from measurements of the relative abundances of radioactive elements and their decay products trapped within the rocks. The oldest terrestrial rocks turn out to be 3.8 thousand million years old; Moon rocks and meteorites are about 4.6 thousand million years old. The last figure is usually taken to be the age of the Solar System.

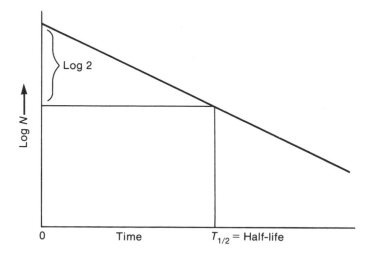

FIGURE 8.2 The law of radioactive decay. The logarithm of the number N of radioactive nuclei decreases at a constant rate. This implies that the fraction of undecayed nuclei that decays during any given period doesn't depend on the number present to begin with. When the logarithm of the number of undecayed nuclei decreases by log 2 (about 0.3), the number itself is halved. The corresponding time interval is called the half-life.

Stellar Ages

"Look! The remains of a fire. And it's still warm! They can't have gone far." Astrophysicists use this reasoning, familiar to readers of Westerns and children's literature, to estimate the ages of stars. Heat and light are forms of energy, so a body that gives off heat and light must have a finite lifetime. If we can identify the source of the Sun's energy and estimate the size of the energy supply, we can estimate its age from the known rate at which it is losing energy in the form of sunlight.

The Sun and most stars consist mainly of hydrogen. At the temperatures and densities that prevail near the centers of stars, hydrogen fuses into helium. This is the principal source of stellar energy. It is a very efficient process, converting seven-tenths of 1 percent of the mass of the fuel, hydrogen nuclei, into energy. (Oil and gas heating are a million or so times less efficient.) Even so, when the Sun was formed, it had only enough fuel to shine for about 10,000 million years before turning into a red giant. As we have just seen, age estimates based on radioactive dating indicate that the Sun is somewhat less than 5,000 million years old, so its life as a normal star is now about half over. Many other stars—perhaps the bulk of them—have already burned out. The oldest stars we can still see were probably formed soon after the beginning of the cosmic expansion. Their ages are currently estimated at between 12,000 and 18,000 million years.

Is the Universe Running Down?

To sum up, modern astronomical observations, interpreted in the light of well-established theories, suggest that the Universe is from 10,000 to 15,000 million years old and that its structure is very complex and orderly. What are we to conclude from this? The Second Law seems to tell us unequivocally that the Universe is running down. There would seem to be no escape from the conclusion that the Universe had even more order, in some form or other, when it began. As Eddington wrote in 1929:

> There is no doubt that the scheme of physics as it has stood for the last three-quarters of a century postulates a date at which either the entities of the universe were created in a state of high organization, or pre-existing entities were endowed with that organization which they have been squandering ever since. Moreover, this organization is admittedly the antithesis of chance. It is something which could not occur fortuitously.[4]

In the same vein, Richard Feynman wrote in 1967:

> [T]he success of all those sciences [based on the Second Law] indicates that the world did not come from a fluctuation, but came from a condition which was more separated, more organized, in the past than at the present time.[5]

This view of cosmic evolution is still held by nearly all physical scientists. Although based on a scientific argument, it has often been pressed into the service of theology. The quotation from Eddington continues:

> This has long been used as an argument against a too aggressive materialism. It has been quoted as scientific proof of the intervention of the Creator at a time not infinitely remote from today.

Yet, Eddington adds, we should not draw hasty conclusions:

> Scientists and theologians alike must regard as somewhat crude the naive theological doctrine which (suitably disguised) is at present to be found in every textbook of thermodynamics, namely that some billions [that is, millions of millions] of years ago God wound up the material universe and has left it to chance ever since. This should be regarded as the working-hypothesis of thermodynamics rather than its declaration of faith. It is one of those conclusions from which we can see no logical escape—only it suffers from the drawback that it is incredible. As a scientist I simply do not believe that the present order of things started off with a bang; unscientifically I feel equally unwilling to accept the implied discontinuity in the divine nature. But I can make no suggestion to evade the deadlock.

Eddington's scientific intuition that "the present order of things" could not have "started off with a bang" was well grounded. The preceding argument, which seems to prove that the world "came from a condition which was more separated, more organized, in the past than at the present time," is in fact fallacious. Before describing the fallacy, however, let me dispose of an equally fallacious counter-argument that has achieved some popularity in philosophical circles. Stripped of excess verbiage, it runs as follows: the Second Law has been formulated and tested for closed systems. The Universe is not a closed system. Hence the Second Law doesn't apply to it.

The Second Law is not merely a law about closed systems, however. If it were, it wouldn't be much use, because closed systems aren't found in nature. Nor is the Second Law merely a law that applies to "ideal" closed systems, in the way that the mathematical theory of frictionless fluids, for example, applies to ideal fluids. The Second Law is a statement about *processes*. It says that all natural macroscopic processes generate entropy. Of course, this assertion hasn't been verified for all conceivable processes, and as we saw in Chapter 3, there *are* situations—for the most part rather artificial ones—in which entropy, as it is normally defined, does increase, at least for a while. Despite these *caveats,* we may be confident that the total rate of entropy production within any region large enough to be considered a fair sample of the Universe is indeed positive, and has always been positive. That is how Clausius's assertion that "the entropy of the Universe always increases" should be—and was probably intended to be—interpreted.

Randomness and Order

It doesn't follow, however, that the Universe was initially more orderly than it is now. Earlier, we defined order as absence of disorder. Thus order is present whenever the maximum possible randomness consistent with given constraints exceeds the actual randomness. The order present is measured by the size of the gap. In symbols, if R stands for randomness (given by Boltzmann's rule: the randomness of a macrostate is the logarithm of the number of its microscopic realizations, or complexions) and R_{max} stands for the greatest possible value of R under given conditions, their difference, $R_{max} - R$, which I will call I (for *information*), is a measure of order:

$$I = R_{max} - R$$

In thermodynamic contexts, we may identify the randomness of appropriately defined macroscopic states with entropy, but, as we discussed in Chapter 2, the notions of randomness and information apply in wider contexts. According to the Second Law, macroscopic processes generate entropy and thus increase the value of R. Hence macroscopic processes destroy macroscopic information, represented by I. But information is *generated* by any process that increases the quantity R_{max}, the greatest possible randomness available to a given system under specified conditions. Do such processes exist?

Consider the system illustrated in Figure 8.3: a large sphere, empty except for a much smaller gas-filled balloon at its center. Initially the air in the balloon is in equilibrium. Its entropy is accordingly as large as it can be under the circum-

<small>Figure 8.3</small> A gas-filled balloon at the center of an otherwise empty sphere.

stances, and the system contains no order. Now imagine that the gas filling the balloon is an ideal gas, platonium, whose molecules don't interact. Suddenly the balloon bursts, and molecules stream out in all directions. They do not, however, interact. Each molecule moves with the velocity it had at the moment when the balloon burst—at least until it reaches the inner surface of the large sphere. Just before the first molecule arrives there, the velocities of the molecules are distributed as shown in Figure 8.4: all but the slowest molecules are moving almost directly away from the center and have sorted themselves into concentric shells according to their initial speeds, with the fastest molecules most distant from the center. This is clearly an orderly state of affairs. The motions of the molecules are far from random, and their spatial distribution is far from uniform (the average density of molecules is highest at the center of the sphere and lowest at the surface). What has happened?

The randomness of the gas hasn't changed, but its largest allowed value has increased, because of the greatly increased volume accessible to each molecule. (The average energy of the molecules remains the same during the expansion of the gas, so the accessible volume of "velocity space" doesn't change.) As the gas expands into the newly accessible region beyond its original boundary, a gap opens up between its greatest allowable randomness and its actual randomness. This gap represents order, which manifests itself in the nonuniform spatial distribution of the molecules and the orderly arrangement of their velocities.

As the molecules rebound off the inner boundary of their spherical container, their distribution in space and velocity gradually becomes more random. The randomness of the gas increases toward its maximum allowable value, and eventually a new equilibrium condition is reached in which the two are equal. The changes in R, R_{max}, and I are shown graphically in Figure 8.5.

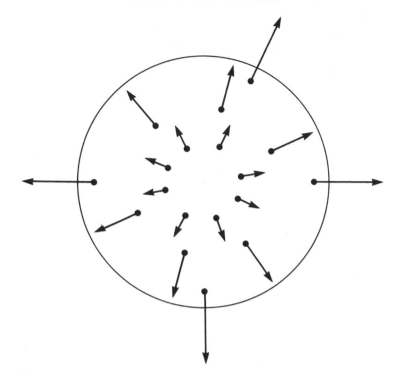

FIGURE 8.4 An orderly distribution of molecular velocities evolves from an initially random distribution. All but the slowest molecules are moving radially outward. The molecules farthest from the center have the largest velocites.

Suppose that the balloon, instead of bursting, expanded very slowly until it fitted snugly inside the large sphere. Then R and R_{max} would increase at the same rate; no gap would open between them, and order would not be generated. We can also imagine situations intermediate between the two extreme cases, in which randomness and order are generated simultaneously as the volume of state space accessible to the gas increases. All these possibilities are illustrated in Figure 8.5, from which we see that two conditions must be met for order to be generated. The maximum randomness R_{max} must increase, and it must increase faster than the randomness R itself.

The ultimate cause of order in the Universe is the cosmic expansion. (Boltzmann was correct in his conclusion that in a static universe—in his day, no one had imagined expanding or contracting space—order can only decay.) The processes by which the cosmic expansion generates order are subtler than the one we have just considered, but the underlying principles are similar. The cosmic expansion gives rise to two distinct kinds of order: chemical order and structural order. Chemical order manifests itself in the preponderance of hydrogen in the Universe; structural order, in the highly nonuniform spatial distribution of matter.[6] The processes that generate chemical order are well understood, although specific details

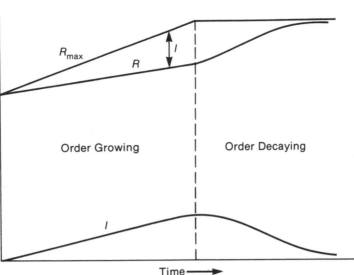

FIGURE 8.5 Growth and decay of order in the distribution of molecular velocities. As the molecules move into the empty region between the balloon and the spherical boundary, their distribution becomes more orderly. The randomness R of the molecular distribution grows more slowly than R_{max}, its maximum possible value. After the first molecules reach the spherical boundary (dashed line), R_{max} remains constant while R increases more rapidly. Hence the information I, the difference between R_{max} and R, decreases toward zero. Notice that the randomness continues to increase as long as it remains less than its maximum possible value.

are controversial. Those that generate structural order are poorly understood, and everything about them is controversial. So let's begin with chemical order.

Chemical Order

Why is the fact that the Universe consists mostly of hydrogen an aspect of order? Recall once more that order is absence of randomness. We have already discussed several kinds of randomness: in the spatial distribution of molecules, in the distribution of molecular velocities, and in the distribution of photon energies. Chemical order consists in the absence of chemical randomness.

To pin down this idea, consider a mixture of hydrogen and oxygen in an isolated vessel. Under ordinary conditions, both hydrogen and oxygen are composed of molecules, each containing two identical atoms. And under ordinary conditions, the two kinds of molecules simply mix without reacting chemically. But of course there are other ways in which the atomic building blocks might be arranged. For example, some of the hydrogen and oxygen might be in atomic form. Some of it might be combined in water molecules. And there are other, more complicated, possibilities. For the sake of simplicity, let's suppose that only hydrogen molecules, oxygen molecules, and water molecules are present and that there are twice as many hydrogen molecules as oxygen molecules. We may now ask: Given the numbers of both kinds of building blocks (hydrogen and oxygen atoms), the size of the vessel, and the temperature of the gas when (say) no water molecules are present, what ratio of water molecules to hydrogen molecules maximizes the randomness of the mixture?

One might at first suppose that the state in which the total number of molecules was greatest would be the most random. It is true that increasing the number of particles increases the number of ways in which a collection of particles can be distributed among the microcells of a box, because each particle contributes a factor equal to the number of microcells. But when two hydrogen molecules and an oxygen molecule combine to form two water molecules (thereby decreasing the number of molecules from three to two), a certain quantity of energy is released, and the temperature of the gas accordingly increases. *This, in turn, increases the range of molecular velocities.* Thus the formation of water molecules decreases the spatial contribution to the randomness but increases the "kinetic" contribution (that is, the contribution associated with the distribution of molecules in velocity space). Because randomness and order are complementary, it follows that the formation of water molecules increases the spatial order of the collection of molecules but decreases its kinetic order.

Which effect predominates depends on the temperature and on the concentrations of the various molecules. Low temperatures and high densities promote togetherness; high temperatures and low densities, separateness. At high temperatures, the most random states of our collection of hydrogen and oxygen atoms have a low proportion of water molecules; at low temperatures, a high proportion. At still lower temperatures, the most random state is one in which the molecules themselves are bound in a crystal (ice). When a supercooled liquid crystallizes, there is a net *decrease* of order. The increased kinetic disorder of the molecules, which vibrate about their mean positions in the crystal lattice, more than compensates for the increased spatial order.

Chemical reactions, like all other spontaneous processes, tend to increase randomness. When an isolated system is in its state of maximum randomness, neither its chemical composition nor any other statistical property can undergo a spontaneous change, because any spontaneous process would generate randomness, which is impossible. States of maximum randomness are thus states of equilibrium.

In principle, the converse is also true: equilibrium states are states of maximum randomness. But in practice, a system far from *chemical* equilibrium may remain in that condition for an indefinite period. For example, hydrogen and oxygen don't react chemically at ordinary temperatures and pressures, even though the formation of water molecules would increase the randomness of a mixture of the two gases. Analogously, a piece of dry wood is not in chemical equilibrium with the surrounding air, but it doesn't begin to burn until its temperature is raised to a certain critical value.

We are now in a position to understand the origin of chemical order. The early Universe was populated not by atoms and molecules, but by neutrons, protons, electrons, neutrinos, photons, and more exotic particles. But the principles governing combinations and rearrangements of these building blocks in elementary-particle reactions are the same as those governing combinations and rearrangements of atoms in chemical reactions today. A typical reaction is described by the formula

$$p + e \leftrightarrow n + \nu$$

Here p stands for a proton, e for an electron, n for a neutron, and ν for a neutrino. The double-headed arrow indicates that the reaction can go either way: a proton and an electron may come together and form a neutron and a neutrino, or a neutron and a neutrino may come together and form a proton and an electron.

In equilibrium, the "forward" and "backward" reactions go on at the same rate, so the ratio between the number densities of protons and neutrons doesn't change. Suppose that at some moment this ratio has the value appropriate to equilibrium at the prevailing temperature and density of the cosmic medium. As the Universe continues to expand, its density and temperature diminish, and the value of the proton–neutron ratio appropriate to equilibrium changes. Its *actual* value also changes, because the rates of the forward and backward reactions are no longer equal, and it changes in the direction that tends to restore equilibrium (the condition of maximum randomness). But whether equilibrium is actually restored depends on whether the reactions occur fast enough to keep up with the changing conditions. If they do not, a gap opens up between the maximum randomness (corresponding to chemical equilibrium) and the actual randomness; chemical order is generated.

The average rate at which a proton collides with electrons is proportional to the number of electrons per unit volume, which, in turn, is proportional to the mass density of the cosmic medium. Similarly, the average rate at which a neutron collides with neutrinos is proportional to the mass density. It follows from Einstein's theory of gravitation, though, that the cosmic medium expands at a rate proportional to the square root of the mass density. Thus the rates of the reactions that seek to maintain equilibrium are more sensitive to changes in the mass density than the rate of the cosmic expansion, which tends to destroy equilibrium.

Reaction rates depend on temperature as well as density. In general, they increase with increasing temperature. Because the temperature of the cosmic medium decreases as the Universe expands, the two effects of the cosmic expansion—decreasing temperature and decreasing density—reinforce each other.

It follows that the rates of equilibrium-maintaining reactions must have exceeded the rate of cosmic expansion early in the cosmic expansion. Eventually, however, the rate of any given equilibrium-maintaining reaction must become smaller than the rate of cosmic expansion, as illustrated in Figure 8.6. The curve representing the reaction rate is steeper than the curve representing the expansion rate. The two curves cross at a certain epoch. Before that epoch, the forward and backward reactions occur fast enough to keep the proton–neutron ratio close to its equilibrium value at the prevailing temperature and density; after that epoch, the same reactions no longer play a significant role in the chemical economy.

A similar argument applies to every reaction involving two or more particles. Any such reaction maintains equilibrium concentrations of the particles that are created or destroyed by the reaction until its rate becomes comparable with that

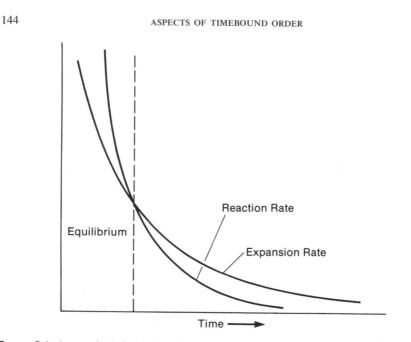

FIGURE 8.6 As we look back into the past, we see the rates of equilibrium-maintaining reactions increasing faster than the rate of cosmic expansion. For any given reaction, there is a critical epoch (dashed line). At earlier times, the reaction is fast enough to maintain equilibrium ratios of the particles that figure in the reaction; at later times, it isn't. Chemical order is created when the rate of an equilibrium-maintaining reaction falls below the rate of cosmic expansion.

of the cosmic expansion, and then ceases to be effective. It follows that *chemical equilibrium prevailed during the initial stages of the cosmic expansion, and only then. Chemical order was not present at the outset; it was created by the cosmic expansion.*

These considerations show, incidentally, that the onset of the cosmic expansion was not a "big bang" but an exceedingly gentle—although also exceedingly rapid—decompression. A bang, or explosion, creates large and sudden deviations from chemical, thermal, and pressure equilibrium; the cosmic medium began to expand in a state of chemical, thermal, and pressure equilibrium, and gradually deviated from chemical equilibrium as the expansion slowed down.

Chemical order resides in deviations from chemical equilibrium. This dry and technical definition may give the impression that chemical order has little to do with the kinds of natural order we see around us—the order of a living planet, with its seas and rivers, its mountains and plains, its myriad forms of life. But this is a false impression. Earth is a living planet precisely because chemical order was created 10 billion or so years ago, when the Universe was very young. Without sunlight, there could have been no life and none of the changes that life has wrought in the natural landscape. Sunlight is a by-product of the burning of hydrogen in the core of the Sun. And hydrogen is present in such abundance in the present-day Universe because the nuclear reactions that, in the early Universe, would have converted hydrogen into helium and heavier elements could not keep pace with the cosmic expansion.

The present preponderance of hydrogen is a clue to the physical conditions that prevailed in the early Universe—but a clue whose interpretation is ambiguous. Before discussing it, let's consider the second main type of order created by the cosmic expansion.

Structural Order

Why is the distribution of matter in the Universe so extravagantly clumpy? In the beginning, the cosmic medium was certainly gaseous and very nearly—if not perfectly—uniform. Why hasn't it remained uniform?

Clearly, the present lack of uniformity must have something to do with gravity. The cosmic medium is clumpy over an enormous range of scales, from the small solid bodies of the Solar System to the great superclusters of galaxies, and on all these scales gravity is the cohesive force, the glue that keeps astronomical systems from flying apart. This suggests that a tendency to form clumps is a "natural" property of an expanding, initially uniform gas filling all space, in somewhat the same way that clotting is a natural property of blood.

Newton believed that a uniform *static* medium would break up:

> [It] seems to me that if the matter of our sun and planets and all the matter of the universe were evenly scattered throughout all the heavens, and every particle had an innate gravity toward all the rest, and the whole space throughout which this matter was scattered was but finite, the matter on the outside of this space would, by its gravity, tend toward all the matter on the inside and, by consequence, fall down into the middle of the whole space and there compose one great spherical mass. But if the matter was evenly disposed throughout an infinite space, it could never convene into one mass; but some of it would convene into one mass and some into another, so as to make an infinite number of great masses, scattered at great distances from one to another throughout all that infinite space. And thus might the sun and fixed stars be formed, supposing the matter were of a lucid nature.[7]

Newton never submitted this idea to rigorous analysis, perhaps because his theory of gravitation doesn't support an unambiguous, self-consistent description of a uniform, unbounded medium. Einstein's theory does support such a description, and it predicts that a uniform, unbounded medium can't be static; it must either expand or contract. The question therefore arises whether an initally uniform, unbounded, expanding gas would tend to break up.

If the gas was *perfectly* uniform to begin with, it would remain uniform. Perfect uniformity, however, is an unattainable ideal. In a still room the number of air molecules per cubic centimeter fluctuates from place to place and from moment to moment. These random fluctuations obey well-understood statistical laws, among them the "square-root-of-N law," which says that if the average number of molecules in a sample is N, then the random fluctuations of this number are likely to be near the square root of N. Thus if the average number of molecules in a sample is nine, the actual number is likely to be between six and twelve. A cubic centimeter of air at room temperature and pressure contains around 10^{19} molecules, so the actual number is likely to deviate from the average by the square root of 10^{19}, or about 3×10^9. The corresponding *fractional* fluctuation—the fluctuation divided by the average value—is therefore very small, around 3×10^{-10}. To take another example, the Sun contains about 10^{57} hydrogen atoms. Random fluctuations containing this number of atoms would have been overdense by only three parts in 10^{29}. And they would have grown much too slowly to produce any detectable nonuniformity in the available time. Thus *gravitational attraction alone cannot account for the clumpy structure of the astronomical universe.*

And yet gravitation must have been the chief actor in shaping the hierarchy of self-gravitating systems. No other force could have given rise to structure on such large scales or on such a wide range of scales.

The structure of crystals, molecules, atoms, and atomic nuclei is, in a sense, inherent in the laws of quantum physics. We can predict the structure of ice or of molecular hydrogen or of atomic hydrogen without knowing or assuming anything about the way these substances actually form in nature. The laws of quantum physics enable us to *predict* at what temperatures and pressures water will crystallize and molecular hydrogen will dissociate into its constituent atoms. The preceding argument tells us that astronomical structure isn't inherent in the law of gravitation. The formation of astronomical structure is a *historical* process: we cannot hope to understand it until we have understood the circumstances in which it occurred. Knowing that gravitation played the central role is not enough; we have to know the setting and the supporting actors.

The Cosmic Microwave Background

Two radically different scenarios for cosmic evolution have been proposed. According to one, the world began in fire; according to the other, in ice. Let us begin with the "hot" scenario, which is the one favored by most cosmologists. It is often called "the standard model," although, as we will see, there have been many versions of it, each of which has run into serious difficulties.

All versions of the standard model take as their starting point a hypothesis that was put forward to explain the *cosmic background radiation,* a feature of the astronomical Universe that was discovered in 1965. This background radiation appears equally bright in all directions. It is like a thin wash spread evenly over the whole sky, an almost perfectly uniform background. The only well-established deviations from uniformity are accounted for by the Earth's motion relative to this sea of radiation: the background is brighter (by about one-tenth of 1 percent) in the direction of the Earth's motion, fainter by the same amount in the opposite direction. (A source of light looks brighter when we move toward it because the relative motion increases both the rate at which we receive photons and the frequency of each photon.)

Because there is no observational or theoretical reason to suppose that the

Earth or any of the larger systems to which it belongs occupies a special place in the Universe, we may reasonably assume that this sea of radiation would appear equally bright in all directions to any other observer, apart from the small distortion caused by that observer's motion relative to the sea. It follows that the sea must also be perfectly *uniform:* the number density of photons, or the density of radiant energy, must be the same at every point. For if it varied from point to point, some observers would be able to detect *directional* variations in brightness, contrary to our supposition.

The background radiation has a second remarkable property. Its spectrum is nearly that of blackbody radiation, with a temperature of 2.7 degrees on the Kelvin scale—that is, 2.7 degrees above absolute zero. For this temperature, the blackbody spectrum (Chapter 5) peaks at a wavelength of about 1 millimeter.

The uniformity, isotropy, and spectrum of the background, taken together, imply that it closely resembles blackbody radiation. This conclusion is more momentous than it may seem to be at first sight, because blackbody radiation is produced only under very special conditions—for example, within opaque boxes whose walls are maintained at a constant temperature and deep inside stars, where the physical conditions mimic those inside a black box. Sunlight is very far from blackbody radiation. Its spectrum resembles (although not closely) that of blackbody radiation at a temperature of 6,000 degrees Kelvin, the temperature of the Sun's visible surface. But the energy density of sunlight at the Earth's surface is far lower than it would be inside a box whose walls were maintained at 6,000 K. Now, the present-day Universe is just as transparent to the background radiation as it is to ordinary light. We are not living in the equivalent of an opaque box or inside an opaque gas. This means that the background could not have acquired its distinctive blackbody characteristics under present conditions. *The background radiation must be a relic of an earlier period of cosmic history, when the Universe was far denser and more opaque.*

Earlier we discussed the competition between chemical reactions, which tend to make the combinations of atomic and subatomic building blocks as random as possible, and the cosmic expansion, which continuously alters the conditions (temperature and density) that define the most random arrangement of building blocks. We saw that chemical equilibrium (in which this arrangement is as random as possible) persists as long as the rates of the reactions that maintain it exceed the rate of cosmic expansion. Analogous considerations apply to the processes that randomize radiation. Radiation is randomized by its interaction with matter. Sunlight, for example, is absorbed by the land and the oceans and is reemitted as infrared radiation, whose photons have only one-twentieth of the energy, on average, of sunlight photons. Blackbody radiation is radiation that has been thoroughly randomized through its interaction with matter. Absorption and emission processes randomize the frequency distribution of photons belonging to the cosmic background and adjust the energy density of the radiation to the prevailing temperature. Complete randomization can occur only if the rates of these processes exceed the rate of cosmic expansion, so that the average "age" of a photon is short compared with the time required for the temperature of the cosmic medium to change by an appreciable fraction of itself.

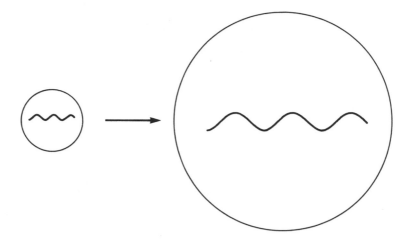

FIGURE 8.7 The wavelengths of photons in an expanding space expand at the same rate as the space itself, as though they were printed on the surface of an expanding balloon.

When the rate of a chemical reaction falls below the expansion rate, the relative abundances of the reactants are frozen in. Here the analogy between chemical reactions and the absorption and emission processes that randomize the radiation background ends. When the cosmic radiation ceases to interact with matter, its spectrum is not frozen in. As space expands, the wavelength of every photon expands with it, as though the wave were printed on the surface of an expanding balloon (Figure 8.7). (This is the cause of the cosmological redshift. When we observe a distant galaxy, we capture photons that have been traveling for a long time, and whose wavelengths have been expanding all the while.) How does the cosmic expansion alter the spectrum of the radiation background after the radiation has ceased to interact with matter?

Curiously enough, blackbody radiation remains blackbody radiation as the Universe expands, but its *temperature* steadily diminishes, according to a simple law: when the radius of a sphere that contains a fixed quantity of mass increases by certain factor, the temperature of the radiation decreases by the same factor. When the Universe was 1,000 times as dense as it is today, so that a sphere containing a fixed quantity of mass had one-tenth of its present radius, the temperature of the radiation background was 10 times its present temperature—27 K instead of 2.7 K. Now if we look back far enough, we must come to a time when the Universe was dense enough for the background radiation to have been thoroughly randomized, just as the radiation deep inside stars is thoroughly randomized. Later the Universe became transparent, but the background radiation kept its blackbody spectrum, its temperature declining as the Universe continued to expand. The cosmic radiation background can therefore be understood as a relic of an earlier, much hotter and denser period in the history of the Universe. How much earlier?

The Fireball Hypothesis

The standard model proceeds from the assumption that the cosmic radiation background was already present at the earliest times at which our present physical laws apply. What was the early Universe like, according to the standard model?[8] Consider an epoch when the density of matter was 10^{39} times its present value. Between then and now, the radius of a sphere containing a fixed number of protons has expanded by a factor of 10^{13} (the cube root of 10^{39}). According to our previous discussion, the temperature of the background radiation must have diminished by the same factor. It must therefore have been about 3×10^{13} K. At this temperature, the energy of an average photon is somewhat larger than the energy of an average proton, including the part of the proton's energy represented by its mass. But photons outnumber protons by at least 100 million to one, and that ratio hasn't changed because no new photons or new protons have been created. It follows that in the epoch we are considering, only a minute fraction of the mass density in the Universe was in the form of matter; the rest was in the form of radiation. At still earlier times, matter accounted for an even smaller fraction of the mass density. According to the standard model, then, radiation dominated the early Universe, and matter was an energetically insignificant impurity. In this context, how can we account for the clumpiness of the present-day Universe?

It isn't easy. A uniform sea of radiation is even less hospitable than a gas to the growth of random density fluctuations. There is no escape from the conclusion that density fluctuations large enough to give rise to the present clumpiness must have been present before our present physical laws began to apply. What were these density fluctuations like?

There are two extreme possibilities. We can postulate a *nonuniform* distribution of material particles embedded in a *uniform* sea of radiation, like raisins unevenly distributed in an expanding bread dough. Or we can postulate that the ratio of matter to radiation had the same value everywhere, but the density of radiant energy was not uniform. The first logical possibility seems to lead to a theoretical dead end. So far, no one has been able to invent a hypothesis that would plausibly account for fluctuations in matter density unaccompanied by fluctuations in the matrix of radiation in which the matter is embedded. The second possibility is more promising theoretically but has unpleasant consequences.

Just as air molecules tend to diffuse out of overdense regions of air, restoring uniformity, so photons tend to diffuse out of overdense regions in the cosmic sea of radiation. As these overdense regions flatten out, so do the embedded fluctuations in the density of material particles.

This process continues until the temperature of the cosmic medium has dropped to about 4,000 degrees Kelvin. Then the protons and electrons, which account for the bulk of the matter, combine to form hydrogen atoms. Because hydrogen atoms have no net electric charge, they can move freely through the sea of photons. (At earlier times, electrons and protons are prevented from moving relative to the radiation by a kind of electromagnetic friction.) So after the temperature of the cosmic medium has dropped to 4,000 degrees, overdense regions in the cosmic distribution of gas are at last free to contract under their enhanced gravitational self-attraction.

By then, however, the diffusion of photons has flattened out all overdense regions below a certain size. Calculations show that overdense regions whose masses are less than 10^{14} or 10^{15} times the mass of the Sun cannot survive. Thus the smallest clumps that can form, in the standard model, are as massive as the largest known self-gravitating astronomical systems—rich galaxy clusters and superclusters.

But now a new difficulty arises. If density fluctuations in the photon sea on these very large scales had *not* been smoothed out by diffusion (assuming they existed in the first place), they would have given rise to small variations over the sky in the brightness of the radiation background. Parts of the sky where the photon sea had been overdense would appear a little brighter than average. Recent calculations indicate that fluctuations large enough to produce large-scale clumpiness would also have produced variations in the brightness of the radiation background *slightly larger* than the present upper limit set by observation. This difficulty has not yet been resolved.

Suppose, however, that a resolution can be found. Two new difficulties then present themselves. The first concerns the nature of the predicted large-scale clumpiness. Theoretical arguments indicate that the overdense regions should separate and collapse in the form of large flattened gas clouds—"pancakes." But this prediction is difficult to reconcile with the large-scale structure of the Universe as revealed by recent observational work. It appears that, on the largest scales, galaxies and galaxy clusters are not clustered in pancakes but are concentrated in the interstices of enormous "bubbles"—roughly spherical regions that are significantly underdense (Figure 8.8).

The second difficulty concerns the origin of structure on scales extending from galaxies and galaxy clusters to stars. The standard model predicts that the first self-gravitating systems were structureless gas clouds of 10^{14} to 10^{15} solar masses. Can such clouds be expected to fragment hierarchically into galaxies, star clusters, and stars? This is a controversial question. Rich galaxy clusters, elliptical galaxies, the massive envelopes of spiral galaxies, and globular star clusters are all spheroidal systems. Theoretical arguments indicate that an initially structureless spheroidal gas cloud cannot fragment into smaller self-gravitating units, and this conclusion is supported by numerical simulations. The technical literature abounds in references to fragmentation, however, and many astrophysicists believe that there must be a way in which it could happen.

Let us suppose that these two difficulties can be resolved. The standard model then faces a pair of still more basic questions: What determines the present value of the photon–proton ratio? And what causes the primordial density fluctuations that the standard model has to postulate in order to account (if possible) for clumpiness?

The answers to both questions (if they have answers) must be sought in the *prehistory* of the Universe—the part of the early history in which our present physical laws break down and must be replaced by more fundamental laws. Because we don't yet know what these more fundamental laws are, cosmic prehis-

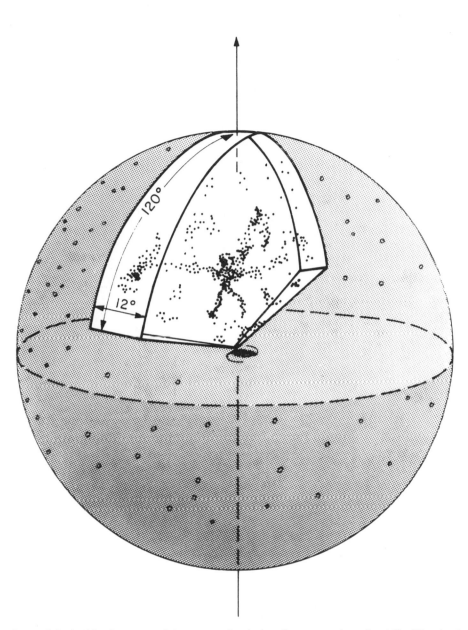

FIGURE **8.8.** In this diagram, points represent galaxies. Our own galaxy, the Milky Way, is at the center of the sphere. The estimated distances of the galaxies are proportional to their redshifts (the displacements of their spectral lines toward longer wavelengths). The figure shows the spatial distribution of galaxies in a 120° "slice" with an opening angle of 12°. The farthest galaxies are at distances of about 500 million light-years. On scales of tens of millions of light-years, galaxies and galaxy clusters appear to be concentrated in the interstices of enormous "bubbles"—roughly spherical regions that are significantly underdense. Notice the remarkably sharp edges of some of the large bubbles. (Smithsonian Astrophysical Observatory Illustration, drawn by John Hamwey, Cambridge, Mass., 1988. Reprinted by permission)

tory is considerably more speculative than the rest of cosmology. But it is also very exciting to theorists. The very early universe, inaccesible though it is, offers the best arena for applying and testing new theories of fundamental particles and their interactions. And it is here that the standard model splits into many branches, for there are many such theories and many ways in which they can be used to construct scenarios for the very early Universe. All this theoretical activity has, in important ways, been fruitful as well as exciting. But so far, it has produced more questions than answers.

9

Gravitational Clustering and Structural Order

The Fireball Hypothesis and Astronomical Observations

According to the standard cosmological model, the world was already quite complex when the present Constitution—the present set of physical laws—took effect. Supporters of this model, with its unexplained ratio of light to matter and its unexplained density fluctuations, hope to show that the primordial fireball evolved from a simple initial state and that, at the same time, the Constitution itself evolved from a simple primeval law. It is indisputably a grand vision, and it has inspired some of the most imaginative and abstruse theories in the history of physics.

To reconcile this vision with astronomical observations, supporters of the standard model have been forced to invent increasingly strange hypotheses. Consider, for example, this question: What is the world made of? In more formal language: What kinds of particles account for the bulk of the mass density of the Universe?

There is indirect but highly convincing evidence that between 90 and 99 percent of the mass in the Universe emits light so feebly as to be invisible. The evidence comes mainly from measurements of the motions of glowing clouds of atomic hydrogen in spiral galaxies and of the motions of galaxies in galaxy clusters. Just as we can infer the mass of the Earth from the speed of the Moon in its orbit and the diameter of the orbit, and the mass of the Sun from the speed of the Earth in its orbit and the diameter of the orbit, so we can infer the mass of a spiral galaxy from the measured speeds of orbiting clouds of atomic hydrogen, and the mass of a galaxy cluster from the measured speeds of its member galaxies.

Such estimates of mass, based on Newton's theory of gravitation, typically show that only 10 percent of the mass of a spiral galaxy like our own Milky Way and between 1 and 10 percent of the mass of a large galaxy cluster are sending out enough light to be recorded by our telescopes. Should we be surprised by this finding?

Perhaps not. It would be more surprising if visible stars contained most of the mass in the Universe. If a self-gravitating cloud of hydrogen is to collapse into a star, its mass must be at least one-tenth or so the mass of the Sun. If it is to collapse into a star whose lifetime is comparable with or greater than the age of the Universe—a star we are likely to observe before it burns out—its mass can't be much greater than the mass of the Sun. There is no known reason why most of the mass of the Universe should have congregated in self-gravitating clouds that meet these restrictive conditions. Certainly, it wouldn't be surprising if a large fraction of the mass had failed to condense or had condensed into substellar chunks of matter like the planets and their satellites or had condensed into massive stars that have long since burned out. In short, it is plausible that 99 percent of the mass of the Universe either never got a chance to shine or has already had its chance.

This conclusion draws support from observational and theoretical arguments that I will not pursue here. It conflicts, however, with an important prediction of the standard cosmological model. The early history of the Universe, according to the standard model, contained a brief period during which nuclear reactions created and destroyed deuterium, helium, and other light elements. These reactions are very well understood. By simulating them on computers, physicists can accurately predict what fraction of the matter was left in the form of hydrogen, what fraction in the form of deuterium, and what fraction in the form of helium. The predicted "primordial abundances" depend on the ratio between the number density of photons in the photon sea and the number density of nucleons (protons and neutrons, free or bound in atomic nuclei). This ratio has the same value now as it did during the period of nucleosynthesis. So if we knew the primordial abundance of helium, we could predict the ratio between the number densities of photons and nucleons.

Astronomers currently estimate the primordial abundance of helium as about 23 percent by mass, although there is some evidence that it might be 20 percent or even lower. The value 23 percent leads to a predicted photon–nucleon ratio of 10^{10}. Lower values of the primordial helium abundance lead to higher values of the photon–nucleon ratio. The cosmic number density of photons in the photon sea is about 300. Thus the predicted cosmic number density of nucleons is about 3×10^{-8}. The corresponding mass density of ordinary matter is about 5×10^{-32} grams per cubic centimeter. *This is very close to the mass density of luminous matter—matter in stars that are still shining.* (If the primordial helium abundance was as low as 20 percent, the predicted mass density of ordinary matter would be significantly *smaller* than the currently estimated mass density of luminous matter.) So the standard model implies that, as far as ordinary matter is concerned, what we see is what there is: nearly all the matter that could have formed stars did in fact condense into stars not much more massive than the Sun.

That is one side of the coin, and it is disconcerting enough. The other side of the coin is even more disconcerting: the building blocks of ordinary matter contribute less than 1 percent of the mass of the Universe. What contributes the remaining 99 percent?

Until recently, the answer favored by most theorists was *massive neutrinos*. Neutrinos aren't building blocks of ordinary matter, but in an important sense they are ordinary particles. Thus a free neutron decays into a proton, an electron, and an antineutrino (the neutrino's antiparticle). Neutrinos belong to the same "family" as electrons and the u and d quarks that make up protons and neutrons. The (rest) mass of the neutrino isn't known. Theorists used to assume that it was exactly zero, like the mass of a photon, but many now believe it will turn out to be finite. In the absence of strong theoretical or experimental constraints, cosmologists have felt free to speculate. Can the neutrino be *assigned* a mass that will enable it to play its appointed role?

At first, the answer seemed to be yes. For several years, most cosmologists were confident that the world consists almost entirely of massive neutrinos. But careful theoretical studies of how a universe dominated by massive neutrinos would evolve have now made that hypothesis untenable. Neutrinos may indeed have finite mass, but they aren't the building blocks of the Universe.

What is the alternative? Supporters of the standard model are left with the conclusion that 99 percent or more of the mass in the Universe resides in particles that don't figure in any well-established physical theory and whose properties we can only speculate about. I am happy to say that my colleagues haven't declined the invitation implicit in this situation. They have produced a long and imaginative list of candidate particles and an impressive technical literature in which the merits of the candidates are explored and compared.[1]

There is, of course, another possibility. The fireball hypothesis could be wrong. True, it accounts splendidly for the cosmic blackbody radiation. But so far it has failed to explain an equally important feature of the Universe: clumpiness. On scales ranging from planets and their satellites to giant clusters of galaxies, matter has congregated in nested self-gravitating systems. Many theorists believe that clumpiness is not one but many phenomena. They assume that the physical process that gave rise to galaxies, for example, was quite different from the process or processes that gave rise to stars. Of course, if the fireball hypothesis is correct, this *must* be the case: stars must (somehow) have condensed in collapsing gas clouds, while galaxies must have condensed from the expanding cosmic medium. But let's look at the evidence.

Gravitation, unlike the forces that hold atoms and atomic nuclei together, doesn't have a preferred interaction distance. A hydrogen atom and the proton at its center have definite sizes, but an isolated planetary system, star cluster, or galaxy could in principle have any size whatever. A planetary system could be as big as a star cluster; a star cluster could be as small as the Solar System.

The absence of a fixed interaction distance is also responsible for the fact that there is no preferred strength for gravitational binding, as there are preferred strengths for chemical and nuclear binding. (As a measure of the strength of binding, we may take the energy per unit mass needed to disperse the components of a bound

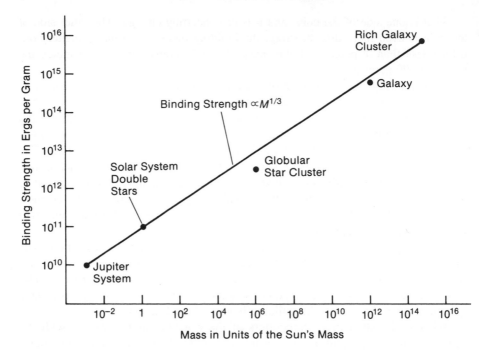

FIGURE 9.1 The binding strengths (binding energies per unit mass) of self-gravitating systems are approximately proportional to the cube root of their masses. This relation holds over eighteen decades of mass, from the Jupiter and Saturn systems to rich clusters of galaxies.

system.) The law of gravitation mandates no particular binding strength for a system of given mass. Yet astronomical systems of comparable mass do have comparable sizes and binding strengths. And when we compare the binding strengths of astronomical systems of different masses, a striking regularity emerges (Figure 9.1). The binding strength increases with mass according to a simple rule: it is proportional to the cube root of the mass. For example, the masses of double stars, globular star clusters, and galaxies are in the ratios $1:10^6:10^{12}$, and their binding strengths are in the ratios $1:10^2:10^4$. This rule holds over a mass range of more than twelve powers of ten.

This finding comports with the view that self-gravitating systems of different sizes have been formed by a single process—that the structural order of the astronomical Universe has a single, unified explanation.

The binding strengths of self-gravitating systems have another noteworthy property. *Their magnitudes are similar to the magnitudes of chemical binding strengths.* This suggests that chemical binding is somehow implicated in the origin of astronomical systems. Of course, this suggestion makes no sense in the framework of the standard model, because the unconventional particles that dominate the Universe, according to this model, are strangers to chemistry.

Early History of the Cold Universe

It isn't hard to understand why the standard model has so much difficulty accounting for the clumpiness of the astronomical Universe. The high temperatures that
prevail in the primordial fireball are very inhospitable to the formation and pres-
ervation of self-gravitating systems. Let's consider the hypothesis diametrically
opposed to the fireball hypothesis: that the Universe began to expand from a
macroscopically uniform state whose temperature was at or close to absolute zero.
This hypothesis implies that whatever processes occurred during the first fraction
of a microsecond, before the present Constitution was in place, didn't give rise to
macroscopic density variations or release large quantities of energy.

At densities much greater than the density of water, the cosmic medium would
have been gaseous, in spite of its low temperature. This is a consequence of
Heisenberg's uncertainty relations (Chapter 5), which imply that at zero temperature, the average momentum of an electron or a proton is approximately equal to
Planck's constant divided by the average distance between particles. The energy
per unit mass of a uniform medium is made up of two contributions, one from the
motions of its particles, the other from their interactions. The first contribution,
the kinetic energy, is positive; the second, the interaction energy, is negative. In
solids and liquids, the negative contribution predominates, and the sum of the two
contributions is negative: it takes energy to overcome the internal cohesion of a
solid or liquid. In gases, the kinetic energy exceeds the interaction energy, and
the total internal energy, the sum of the kinetic and interaction energies, is positive. Chemical cohesion originates in the electrostatic attraction between oppositely charged particles, and the electrostatic energy is inversely proportional to
the average distance between particles. Kinetic energy, however, is proportional
to the square of momentum and is thus, by virtue of Heisenberg's relations, inversely proportional to the *square* of the average distance between particles. So at
high densities, the kinetic contribution predominates and the medium is gaseous,
even though its temperature is as low as it can be.

Figure 9.2 shows how the kinetic energy, the interaction energy, and their
sum vary as the zero-temperature gas expands and the average distance between
particles increases. At the point marked A, the energy becomes negative. At the
point marked B, the energy has its largest negative value. Between A and B, the
medium liquefies and then freezes. At B, the medium is solid and its pressure is
zero. At slightly higher densities, corresponding to points to the left of B, the
solid medium is under pressure; at slightly lower densities, corresponding to points
to the right of B, it is under tension.

The curve beyond B is dotted because it doesn't correspond to what actually
happens when the medium expands past B. Instead of continuing to expand uniformly, a solid medium under tension would break up into fragments. What occurs at B is a kind of *phase transition,* analogous to the transition between a vapor
and a liquid or a liquid and a solid. Here the transition is between a uniform solid
and a "gas" composed of solid fragments.

The solid medium fragments in the way that requires the least expenditure of
energy. This implies that the fragments are as large as possible. For the larger a

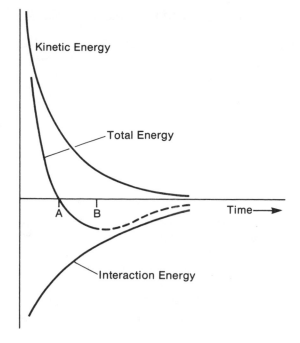

FIGURE 9.2 Early history of the cold Universe. At early times (before A), the cosmic medium is a dense gas with positive internal energy. As the Universe expands, the kinetic contribution to the internal energy decreases, as does the contribution from electrostatic interactions between charged particles. But the kinetic contribution decreases faster than the (negative) interaction contribution. Between points A and B, the medium liquefies and then freezes. At B, the medium is solid and has zero pressure. Beyond point B, the medium is under tension. Shortly after B, the medium breaks up into fragments.

fragment, the smaller the ratio between its surface area and its volume, and hence the less work per unit mass that has to be done at its surface to separate it from the surrounding medium. If a fragment is too large, however, its internal cohesion isn't strong enough to restrain its continued expansion, and it breaks up into smaller pieces. The largest possible fragments are those whose cohesion, combined with self-gravitation, is just strong enough to keep them from flying apart. Such fragments are held together equally by gravity and chemical cohesion. If, as seems likely, the particles of the early Universe were made of metallic hydrogen, their masses would have been about ten times that of the Earth.

Gravitational Clustering

The cosmic medium is now in an interesting condition. It is an expanding gas whose particles are planet-size chunks of solid hydrogen. The temperature of this gas, immediately after it has formed, is close to absolute zero, like the solid medium from which it was born. And most important of all, the dominant interaction between the particles of the gas is gravitation.

In ordinary gases, gravitational interactions between the molecules are of negligible importance, because gravitation is inherently a weak force. The electrostatic attraction between an electron and a proton exceeds their gravitational attraction by a factor of 10^{39}, a number comparable with the mass ratio between the Sun and a grain of sand. But gravitation, unlike the other three forces, is *cumulative*. An electron in an ionized gas feels the electrical attraction only of neighboring positively charged ions; ions beyond a certain small radius might as well not be there as far as their effect on the electron is concerned. A proton and a neutron attract each other even more strongly than a proton and an electron, but only when they are within 10^{-12} centimeter of each other. The interaction between a neutrino and a neutron turns on at an even shorter distance. But a particle at the surface of the Earth is attracted by the whole Earth, not just by the particles nearest to it. It is true that gravitational attraction diminishes with increasing distance from its source, but the distant particles—those beyond 100 kilometers, say—more than make up for their smaller individual attractions by their greater numbers. In a gas whose "particles" are of planetary mass, gravitation overwhelms all other forces.

We can gain insight into the process of gravitational clumping in a cold, expanding gas of gravitating particles by considering a somewhat analogous process: the liquefaction of an ordinary gas. Imagine a cylinder equipped with a movable piston and filled with, say, nitrogen gas. If the cylinder was thermally insulated, the gas would become hotter when it was compressed. Suppose, however, that the cylinder is in thermal contact with a heat bath maintained at a constant temperature, so that the temperature of the gas remains constant as its volume is gradually decreased. Now suppose that the gas is very slowly and gently compressed. If the initial temperature is low enough, there will come a point beyond which any further compression, however carefully executed, causes a finite fraction of the gas to liquefy.

The course of events just described is represented by the curve labeled *a* in Figure 9.3. This curve is a plot of the pressure in the cylinder against the volume of the gas as it is compressed at constant temperature. The curve terminates in the point *A,* at which the gas becomes unstable. Now imagine that the same experiment is repeated with the thermostat on the heat bath set at different temperatures. For temperatures close to that corresponding to curve *a,* we get qualitatively similar curves, such as *a'*. The points at which these curves terminate lie on the curve labeled *k*.

If the initial temperature is too high, as for the curve labeled *b,* the gas remains stable no matter how much it is compressed. The curve labeled *c* represents a watershed. Lower settings of the thermostat yield curves that look like *a;* higher settings, curves that look like *b.* Curve *c* itself belongs to neither category. Its key property is that it contains the point marked *K,* the *critical point.* This point represents a physical state that is neither gaseous nor liquid but in some sense both. When nitrogen, or any other substance, approaches this state, its appearance changes dramatically. It becomes milky and iridescent, like an opal.

Critical-point opalescence, as the phenomenon is called, signals a kind of identity

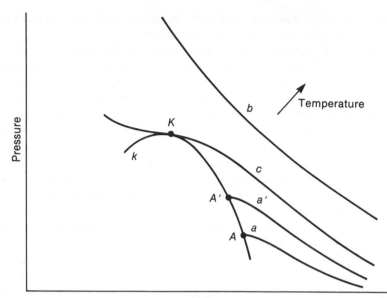

FIGURE 9.3 Curves a, a', b, and c represent the variation of pressure with volume at constant temperature in an ordinary gas, for a series of increasing temperatures. Curves a and a' terminate at points A. and A', respectively, where their tangents become horizontal. These points lie along a curve k whose highest point is K. If the gas is compressed at the (constant) temperature corresponding to curve a, it becomes absolutely unstable at point A, where a finite fraction of the gas liquefies. If the gas is compressed at the higher temperature corresponding to curve b, it remains gaseous at all pressures. Point K is a critical point, at which the gas exhibits macroscopic density fluctuations on a wide range of scales.

crisis. The substance doesn't know whether it is a vapor or a liquid. Macroscopic density fluctuations large enough to deflect light-rays, and thus produce the vapor's milky appearance, continually form and dissolve on all scales up to the size of the container, as if the vapor were uncertain whether to become a liquid or remain a gas. But the overdense regions never actually turn into liquid drops.

In both the gas and the liquid phases, the spatial distribution of the molecules is completely disordered. The two phases differ only in the extent to which the molecules cohere. Because intermolecular forces have a finite and fixed range, the liquid phase can't exist when the average distance between molecules exceeds a certain value (the value that corresponds to the density at the critical point). What is the analogue of the liquid phase for a gravitating gas, a gas in which gravitation is the dominant interaction?

As we discussed earlier, gravitational attraction has no fixed scale. Hence the analogue of the liquid phase in a gravitating gas must exist at all densities (if it exists at all). Moreover, gravitational cohesion, unlike chemical cohesion, is not produced by attractions between neighboring particles, but increases with the number of particles involved. Finally, gravitational cohesion, unlike chemical cohesion, requires large-scale density variations. In a uniform gravitating gas, the attractions exerted on a given particle by distant particles cancel one another out. The forces exerted by nearby particles fluctuate in magnitude and direction and don't give rise to internal cohesion. Conversely, the mutual gravitational attraction of particles belonging to an overdense region containing a large number of particles slows down its rate of expansion. The internal cohesion of such a region increases with the contrast between its density and the average density of the gas. If, then, we take internal cohesion to be the defining characteristic of the liquid phase, we may conclude that (1) in a gravitating gas the liquid phase is clumpy, and (2) there is no sharp line of demarcation between the gas and the liquid phases.

We saw earlier that significant density variations don't arise spontaneously in an initially hot, uniform gas. This conclusion holds for a gas of gravitating particles as well as for an ordinary gas. Let's now see what happens to the cold, expanding "gas" whose "particles" are newly formed planet-size chunks of solid hydrogen.

Every "particle" will be at rest (relative to the local standard of rest in the expanding space) immediately after it is formed. If the "particles" were identical spheres situated at the lattice points of a perfect cubic grid, they would remain at rest, because the gravitational attractions exerted on every "particle" by its neighbors would exactly cancel. In reality, the "particles" must have somewhat different masses, and they must be distributed somewhat irregularly. As a result, the gravitational attractions exerted on a given "particle" by its neighbors cannot cancel exactly, so the particle must begin to move.

To understand the subsequent evolution of our cold, expanding "gas," we have to focus attention on its *internal energy*. The internal energy of a gas is the part of its energy associated with the *random* motions of the particles and with their interactions. The first contribution, the kinetic energy of random particle motions, is positive. The second contribution results from the particles' mutual gravitational interaction. It is negative. It represents the energy per unit mass that would be needed to disperse the particles in a representative sample of the gas if all the particles were at rest. In an ideal gas, there are no interactions between the particles, and the internal energy is equal to the kinetic energy of the particles' random motions. In a real gas, the attractive interactions make a relatively small, negative contribution to the internal energy. Near the critical point, however, the kinetic and interaction contributions to the internal energy are comparable in magnitude. In our cold, expanding "gas," the internal energy is initially *negative*, and that—as we will now see—makes all the difference.

As long as the average pressure in the cosmic medium is positive, its internal energy must decrease as the medium expands, just as the internal energy of gas in an insulated cylinder fitted with a piston decreases as the piston is withdrawn. (In the case of the cylinder, the energy lost by the gas reappears as work done on the piston. In the expanding cosmic medium, the energy simply disappears. Neither energy nor momentum is conserved in expanding space. Freely moving material particles slow down; photons lose energy and momentum as their frequency decreases.) If the internal energy of the cosmic medium is positive to begin with,

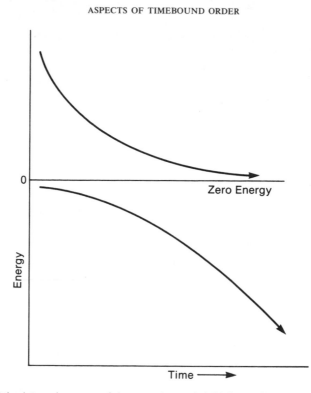

FIGURE 9.4 If the internal energy of the cosmic gas is initially positive, it decreases toward zero as the Universe expands (upper curve). If it is initially negative, it becomes increasingly more negative as the Universe expands (lower curve).

as in the fireball scenario we considered in Chapter 8, it decreases toward zero. But if the internal energy is negative to begin with, as it is in the present scenario, and the average pressure is positive, as it proves to be, then *the internal energy must assume larger and larger negative values as the Universe expands.* These conclusions are illustrated in Figure 9.4.

Now we have seen that negative contributions to the internal energy come from gravitational interactions whose magnitude is determined by the scale and magnitude of macroscopic density variations—that is, gravitational clumping. It follows that gravitational clumping must steadily increase in an expanding cosmic medium whose internal energy is initially negative and whose average pressure is positive. Conversely, the internal energy of the cosmic medium could never have become negative if it had been positive to begin with. I think this is the root of the difficulties that people have encountered in their (up until now fruitless) efforts to understand how gravitational clumping could have arisen in the initially hot universe.

Let's now try to see in a little more detail how and why gravitational clumping occurs in the cold, expanding "gas" whose "particles" are planet-size chunks of solid hydrogen. Three processes are going on at once.

First, "particles" are being accelerated by gravitational forces arising from macroscopic density variations in the medium. If space were not expanding, the resulting motions would neither create nor destroy internal energy, but merely redistribute it between the kinetic and the interaction components.

Second, the expansion of space tends to slow down "particles." (A skater gliding on a frictionless expanding ice rink would continuously slow down relative to the ice immediately beneath her skates.) Hence the expansion of space tends to reduce the kinetic component of the internal energy.

Finally, because the average distance between "particles" is continuously increasing, the expansion of space tends to reduce the magnitude of the (negative) gravitational component of the internal energy.

The combined effect of these three processes depends on the relative magnitudes of the kinetic and interaction components of the internal energy. In particular, if the first of the three processes were to keep the ratio between these magnitudes at a fixed value greater than ½ but less than 1, the magnitude of the internal energy would continuously increase, as illustrated in Figure 9.4. The condition that the kinetic energy be greater than half the magnitude of the gravitational energy implies that the average pressure is positive; the condition that the kinetic energy be less than the magnitude of the gravitational energy implies that the internal energy is negative.

As we have seen, the expansion of space tends to reduce the magnitudes of both the kinetic and the gravitational energies. But it tends to reduce the kinetic energy faster than the gravitational energy (in fact, at twice the rate). Hence it tends to reduce the *ratio* between the magnitudes of the kinetic and gravitational components. Calculation shows, however, that when this ratio falls below ⅔, the "gas" becomes violently unstable against the growth of density fluctuations on all scales. The motions resulting from this instability cause the ratio to increase. If it were to increase above ⅔, the instability would be suppressed. But then the expansion of space would once again drive the ratio below ⅔ and reestablish the instability. The ratio therefore remains slightly below ⅔. The opposing tendencies of the expansion and the instability are then in balance. This mildly unstable state is analogous to the state of an ordinary gas at a point like *A* or *A'* in Figure 9.3, where a certain fraction of the gas turns into a liquid.

Because their ratio maintains a fixed value between ½ and 1, both the kinetic energy and the magnitude of the gravitational energy increase steadily as the cosmic medium expands. The magnitude of the (negative) internal energy increases at the same rate, as shown in Figure 9.4. Thus the cosmic medium must become continually clumpier as it expands. How does this happen?

Shortly after the cold gravitating "gas" has been formed, it becomes clumpy on a scale somewhat greater than the average separation of the "particles" (planet-size chunks of solid hydrogen). The overdense clumps expand more slowly than the medium as a whole. Because they have negative energy, they soon separate as autonomous self-gravitating *clusters*. Assume for the sake of simplicity that all these clusters form at the same time. The state of the cosmic medium is now qualitatively similar to its state at the beginning of the process: it is a cold, expanding, gravitating gas whose particles are now clusters of the original "particles." As the medium continues to expand, the process is repeated. Overdense regions in the distribution of clusters expand more slowly than the medium as a whole and eventually separate out as autonomous self-gravitating *clusters of clusters*. These, in turn, act as particles in the next stage of the process (which I call *gravitational clustering*). In this way, a hierarchy of nested self-gravitating clusters gradually comes into being (Figure 9.5).

This account is somewhat oversimplified. Clusters of a given order (or on a given level of the hierarchy) do not form simultaneously but during some finite time interval, and the intervals during which clusters on successive levels of the hierarchy form probably overlap considerably. It remains, true, however, that clustering occurs on progressively larger scales as the Universe expands, and that smaller self-gravitating systems are formed earlier than larger ones. For example, the oldest stars were formed earlier than the oldest galaxies, which, in turn, were formed before clusters of galaxies. At least, that is what the present theory predicts.

Gravitational clustering is a genuinely historical process. The hierarchy of self-gravitating systems is not implicit in the law of gravitation (as the structures of molecules, atoms, and subatomic particles are implicit in the laws of quantum physics), nor was it prefigured in the structure of the early Universe. Each stage in its evolution was the product of earlier stages and laid the groundwork for later stages. Thus clusters formed at one stage of the clustering process play the role of "particles" at the next stage. Unlike all other forces, gravitation has no built-in scale. The range of the gravitational forces that act in gravitational clustering is determined by the scale of the prevailing density variations, which increases steadily as progressively larger clusters are formed.

If one translates the preceding account of the evolution of a cold gravitating gas into mathematical language, one finds that the scale of clustering increases like the square of the radius of a sphere of constant mass.[2] This result, together with the known rate at which the magnitude of the internal energy increases, determines the form of the *clustering spectrum*—the relation between internal energy and clustering scale. The predicted spectrum has a simple form: the average internal energy of clusters of a given mass is proportional to the cube root of that mass. Now, the internal energy is just the negative of the quantity that we earlier called binding strength and plotted in Figure 9.1. Thus the predicted form of the clustering spectrum agrees with the observed form over a range of more than twelve powers of ten in mass and four powers of ten in energy.

The coefficient of proportionality in the relation between internal energy and mass is determined by the internal energy of the first clusters, whose "particles" are planet-size chunks of solid hydrogen. The theory predicts that the internal energy of these clusters is comparable with, but somewhat smaller than, that of the chunks of solid hydrogen of which they are composed. Figure 9.1 shows that this prediction also agrees well with observation.

The Large-Scale Structure of the Universe

A collection of gravitating particles with negative total energy and zero total spin, left to itself, settles into a sphere. If it has finite spin, it settles into a flattened

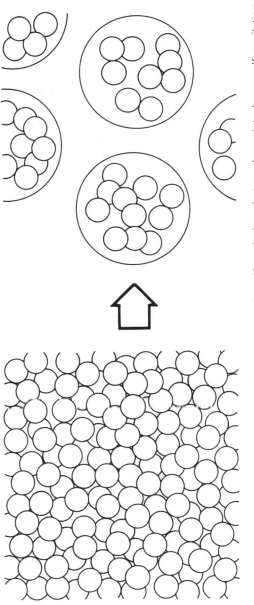

FIGURE 9.5 Gravitational clustering. As the Universe expands, self-gravitating clusters clump into larger self-gravitating clusters. Because gravitation has no intrinsic scale, the process can now repeat itself. In this way, a hierarchy of self-gravitating systems is gradually built from the bottom up.

sphere, or spheroid. Most compact self-gravitating systems—planets, stars, rich star clusters, galaxies, and rich galaxy clusters—are indeed spheres or slightly flattened spheroids. At one time, spiral galaxies were thought to be an exception to this rule, but we now know that most of the mass in a spiral galaxy, like our own Milky Way, doesn't appear in photographs because it gives off very little light. Although we don't know what form this dark matter is in, we can observe its gravitational effects, and from such observations infer that its distribution is roughly spherical.

In our own planetary system, most of the mass is concentrated in a thin disk that contains the Sun, the planets, and their satellites. But the comets and meteors form a nearly spherical system. This suggests that the Solar System began as a sphere and that its present form is the outcome of *frictional dissipation*—the conversion of organized motion into heat and ultimately into radiation—which, in the primeval Solar System, would have selectively attenuated motions other than pure rotational motion about the system's original spin axis.

Although spherical and nearly spherical systems are the rule over virtually the entire astronomical hierarchy, the rule breaks down suddenly and dramatically when we reach the top level—the largest structures in the astronomical Universe. Galaxy clusters, which are themselves spherical, are not clustered in spheres or anything resembling spheres. Instead, they are strung out in filaments, which lie on the surfaces of huge bubblelike regions in which there are no clusters and few or no bright galaxies (Figure 8.8). It is as though figure and ground had been reversed: instead of clusters of galaxy clusters with empty (or underdense) space between them, as we might have expected, we find empty (or underdense) regions in the form of huge bubbles, with galaxy clusters and bright galaxies relegated to their interstices. Why?

The explanation hinges, I believe, on another important finding of modern observational cosmology: that cosmic space has *negative curvature*.

What does it mean to say that space is curved? The surface of a sphere is a curved two-dimensional space. We can *see* that it is curved. Analogously, we can imagine—or at least construct mathematically—a sphere in an imaginary four-dimensional Euclidean space. The surface of such a sphere is a three-dimensional space. There is no *logical* reason why physical space shouldn't have the same geometric properties as the three-dimensional surface of a sphere in four-dimensional Euclidean space. A fourth spatial dimension need not actually exist. It is merely a convenient mathematical device for working out the properties of a curved three-dimensional space. But we can, if we choose, describe those properties without invoking a fourth dimension.

Consider, for example, the properties of triangles. In Euclidean space, the angles of every triangle add up to exactly 180°. In spherical space, they add up to more than 180° (Figure 9.6). Moreover, the "angular excess"—the difference between the angle sum and 180°—increases with the area of the triangle. This is true, for example, of triangles drawn on the surface of an idealized spherical Earth. The straight lines on this surface are arcs of circles whose centers coincide with the center of the sphere. The surfaces of three- and four-dimensional spheres are said to have positive curvature. We can detect and measure the curvature of such a space by measuring the angles and areas of triangles.

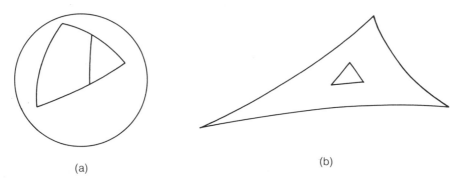

(a)   (b)

FIGURE **9.6** (a) Triangles in a space of positive curvature (the surface of a sphere) have angle sums greater than 180°. The "angular excess" increases with the area of the triangle. (b) In a space of negative curvature, triangles have angle sums less than 180°, and the "angular defect" increases with the area of the triangle.

Although the surface of three- and four-dimensional spheres have finite area and volume, respectively, they are—like Euclidean space—perfectly uniform and isotropic. No property of the surface of a sphere (in any number of dimensions) serves to define a preferred position or a preferred direction at a given position.

In the early nineteenth century, three mathematicians—Janos Bolyai, Carl Friedrich Gauss, and Nikolai Lobachevsky—discovered another logical possibility. They succeeded in constructing uniform and isotropic spaces with negative curvature. Unlike the surfaces of spheres, these negatively curved spaces can't be embedded in Euclidean spaces, and that is probably why it took mathematicians so long to find them. Also unlike the surfaces of spheres (but like Euclidean spaces), negatively curved spaces are infinite. In a negatively curved space, the angle sum of a triangle is less than 180° and diminishes as the vertices of the triangle move farther apart (Figure 9.6). If we let all three vertices recede to infinity, we get a triangle with an angle sum of zero—and finite area!

The Strong Cosmological Principle implies that physical space is uniform and isotropic, but it doesn't tell us whether the curvature of space is positive, negative, or zero. Local measurements don't help much, because all uniform and isotropic spaces mimic Euclidean space in small regions. But cosmological observations, interpreted in the light of Einstein's theory of gravitation, offer a way of *measuring* the curvature of space.

Consider a spherical region of a uniform, expanding medium. It is a consequence of Einstein's theory that if this sphere was isolated in an otherwise empty Newtonian universe, it would expand at exactly the same rate as it does when it is part of the cosmic medium. The behavior of a uniform, expanding, self-gravitating sphere in empty, static Newtonian space is determined by its energy. A sphere of positive or zero total energy expands forever; a sphere of negative energy expands to a maximum radius and then begins to contract. *Einstein's theory predicts a connection between the energy of such a sphere and the curvature of space.* If the energy is positive (so that the Universe expands forever), the curvature of space is negative; the geometry appropriate to cosmic dimensions is the non-Euclidean geometry of Bolyai, Gauss, and Lobachevsky. If the energy is negative (so that the Universe expands to a minimum average density and then begins to contract), the curvature of space is positive; the appropriate geometry is then spherical geometry. If the energy is exactly zero (so that the Universe expands forever, but at an ever slower rate), space is Euclidean.

It will be convenient to have a name for the energy of a spherical region of the expanding cosmic medium, regarded as an isolated Newtonian system. I will call it *curvature energy*. To calculate the curvature energy, we have to know the rate at which space is expanding (this determines the kinetic contribution to the curvature energy) and the average density of mass (this determines the gravitational contribution). Modern estimates of these quantities strongly suggest that *the curvature energy is positive*. We will now see that this conclusion—which, incidentally, contradicts a prediction of the inflationary hypothesis mentioned in Chapter 8—is just what we need to understand the figure–ground reversal that takes place at the largest scales of cosmic structure.

We saw earlier that the internal energy of a newly formed cluster is proportional to the cube root, or one-third power, of its mass (Figure 9.1). It turns out that the curvature energy is proportional to the two-thirds power of the mass. Thus it increases more rapidly with increasing mass than the internal energy. During the early stages of the clustering process, when clusters of relatively small mass are forming, the curvature energy is much smaller than the internal energy. But as progressively more massive clusters form, the relative importance of the curvature energy increases, as illustrated in Figure 9.7. At a certain critical mass, the two energies become equal in magnitude. Would-be clusters of this mass have zero total energy, so they can never evolve into true clusters. What happens to them?

We might at first suppose that they would simply expand, but at a slower rate than the cosmic medium as a whole. But this won't happen, *because the medium is still unstable against the growth of density fluctuations*. The instability condition and the energy condition are independent. The curvature energy doesn't quench the instability; it merely prevents the structures created by it from separating out as clusters. The instability tends to enhance the density contrast between these structures and their surroundings, but because they are not gravitationally bound (or are only weakly bound), they must nevertheless be pulled apart by the expansion of space. The combined effect of the instability and the expansion must be to create irregular filamentary structures. I suggest that these are the filamentary structures illustrated in Figure 8.8.

When the curvature energy is just large enough to keep the largest-scale density fluctuations from condensing, the cosmic medium is in a state analogous to that of a *superheated liquid* at the point where it becomes unstable. Continued expansion beyond this point causes part of the liquid to turn into gas. This occurs through the formation and rapid expansion of gas bubbles. The bubblelike voids that are so conspicuous a feature of the large-scale distribution of galaxies may be analogous structures.

If this hypothesis is correct, the scale of the large-scale structure—the typical diameter of a bubble or the typical dimension of a filamentary supercluster—

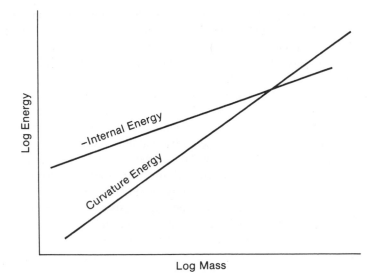

FIGURE 9.7 The curvature energy increases faster with increasing mass than the magnitude of the internal energy. Curvature plays no role in the early stages of the clustering process, but gradually becomes more important. Eventually, it causes the process to terminate.

should be related in a definite way to the curvature energy, because the instability that creates filaments and bubbles sets in when the curvature energy becomes equal to the internal energy of the largest overdense regions. The observational evidence is consistent with the predicted relation.

Thus gravitational clustering in an initially cold, structureless universe seems capable of accounting for all the major features of cosmic structure.

It also offers a novel starting point for theories that seek to explain a relatively minor feature of the astronomical Universe, our own planetary system. If the earliest self-gravitating systems were objects of planetary mass, the primordial Solar System must have been a cluster of clusters of these objects—rather than a structureless gas cloud, as most current theories postulate. Its present hierarchical structure—the fact that the giant planets and their satellites are miniature solar systems—then becomes much easier to understand.

The Origin of Cosmic Blackbody Radiation

But can the hypothesis of a cold initial state be reconciled with the existence of cosmic blackbody radiation? According to the standard model, this uniform sea of photons is the relic of a primordial fireball, generated under physical conditions far beyond the range of current physical laws. But if the Universe began to expand from a cold initial state, this sea of photons must have been produced by processes described by existing physical laws under conditions much less exotic than those attributed by the standard model to the early hot universe.

In 1974 Ray Hively and I suggested that the cosmic radiation background originated as starlight emitted by an early generation of massive stars that formed and burned out when the Universe was about 10 million years old. If these stars had contained 90 percent of the mass of the Universe they would have emitted enough radiation to account for the energy density of the present background. The same stars would have manufactured and ejected heavy elements, which would have condensed into solid grains whose interaction with the starlight could have transformed it into blackbody or nearly blackbody radiation. The consequences of this theory agree well with astronomical observations. As we saw earlier, at least 90 percent of the mass in the present-day Universe is indeed nonluminous. And there is strong evidence that stars now shining were formed from recycled matter—matter that has been "processed" in more massive stars.

In its present stage of development, the hypothesis of a cold beginning seems promising. It explains more and encounters less formidable difficulties than its more popular rival, the standard model. But it is still too early to decide between the hypotheses on purely scientific grounds. New evidence and new theoretical arguments could easily overthrow one—or both—of them.

*   *   *

The theory of gravitational clustering shows how structural order might have evolved in an expanding Universe. We need not assume, as Clausius and Boltzmann did in the nineteenth century and as many modern astronomers and physicists still do, that the Universe started out with a huge store of order that it has been gradually dissipating ever since. If the hypothesis outlined in this chapter is correct, the initial state of the Universe was wholly lacking in order. Structural order began to emerge after the cosmic medium froze and shattered into planet-size fragments of solid hydrogen. The resulting cold gas of mutually gravitating "particles" became unstable; density fluctuations grew and evolved into self-gravitating clusters. These, in turn, became the "particles" in a repetition of the process. In this way, a hierarchy of nested self-gravitating systems was gradually built up.

The expansion of space is responsible for both chemical and structural order, but it produced these two kinds of order in quite different ways. Chemical order was created (in both the cold Universe and the standard model) when the rates of equilibrium-maintaining nuclear reactions fell below the cosmic expansion rate. Structural order resulted, according to the hypothesis sketched in this chapter, from the instability of a gravitating "gas" with negative internal energy, an instability driven and maintained by the cosmic expansion.

We turn now to the most important consequence of chemical and structural order in the Universe: life.

10

Molecules, Genes, and Evolution

Four Views of Life

To a naïve observer, no distinction seems plainer than that between the animate and the inanimate, between objects that initiate motion and change and objects that respond passively to external influences. Understanding the deeper meaning of this obvious distinction has been a central task of Western science and philosophy since their beginnings in ancient Greece. There have been four main approaches.

The first approach is rooted in prescientific and prephilosophical thinking. It is summed up by a saying attributed to Thales, the first Greek philosopher-scientist: "There are gods in everything." This suggests that the animate and the inanimate differ in degree rather than in kind, that even "inanimate" objects have an inner being.

This doctrine is easier to believe than it may seem at first sight. Children and many adults assume that dogs, cats, and other animals experience the world more or less as we do. Some children include trees and flowers in the company of sentient objects. In *Through the Looking-Glass,* a tiger lily explains to Alice that flowers spend most of their time sleeping because their beds are too soft. From there it is a short step to the conviction that rivers and stones also have an inner being. Various forms of this belief have been held by such philosophers as Gottfried Leibniz, Arthur Schopenhauer, Alfred North Whitehead, Pierre Teilhard de Chardin, and even a few modern biologists, notably the embryologist C. H. Waddington and the evolutionary biologist Bernhard Rensch. Most recently, it has been defended on logical grounds by the philosopher Thomas Nagel.

Panpsychism's great attraction is that it offers a unified view of reality. Its great flaw—more evident today than in earlier times—is the complete absence of supporting evidence. While it is true that physicists don't understand everything about the basic structure of matter, their theories are certainly not incomplete in any way that would be helped by the postulate that subatomic particles have an inner life.

The second approach is the mirror image of the first. Its proponents assert, in effect, that the physicists' picture of the world (whatever it happens to be at the moment) is complete. Anything that doesn't fit into it isn't real in the same way that atoms and molecules, say, are real. In particular, the distinction between the animate and the inanimate must be illusory because every object, whether animate or inanimate, is nothing but ———. The words that complete the sentence change as science grows. Democritus might have said "a collection of atoms whose motions are governed by Necessity"; Laplace, "a collection of particles whose motions are governed by Newton's laws." Some contemporary molecular biologists say that life is nothing but chemistry. Most people, however, find it easier to imagine that rivers and stones have some rudimentary form of inner life than to admit that our own inner life is no more real than that of a stone.

The first two approaches argue that the common-sense distinction between the animate and the inanimate, or between mind and matter, is illusory. The third approach posits that it is real and irreducible: matter and energy are described by natural science; they are the stuff of physical reality. Mind (or soul or spirit or consciousness or life-force) is the stuff of a second kind of reality, underlying the phenomenon of life. It can never be fully captured by science.

This was the point of view of Socrates, who vigorously opposed the world-views of philosophers like Thales (who saw gods in all things) and Democritus (who saw them in none). Socrates, his pupil Plato, and Plato's pupil Aristotle all held that soul and matter are distinct and that mechanical explanations are incomplete. Matter, they argued, is passive. It responds to being pushed and pulled but cannot initiate motion. For both Plato and Aristotle, the ability to initiate motion is the defining property of soul. Since motion, left to itself, tends to decay, it must be continually renewed by some nonmaterial agency. As for mechanical explanations of natural phenomena, they may explain *how* something happens but not *why*.

In Plato's dialogue *Phaedo,* Socrates, on the last day of his life, ridicules such explanations. A reductionist philosopher, he says, would explain that

> I sit here because my body is made up of bone and muscles; and the bones are hard and have joints which divide them, and the muscles are elastic, and they cover the bones . . . ; and as the bones are lifted at their joints by the contraction or relaxation of the muscles, I am able to bend my limbs, and that is why I am sitting here in a curved posture—that is what he would say; and he would have a similar explanation of my talking to you, which he would attribute to sound, and air, and hearing, and he would assign ten thousand other causes of the same sort, forgetting to mention the true cause, which is, that the Athenians have thought fit to condemn me, and accordingly I have thought it better and more right to remain here and undergo my sentence.[1]

Mechanical explanations are equally inadequate for astronomical phenomena, says Socrates. He is not content to know how the Sun, the Moon, and the planets are situated and how they move; he wishes to know why they are where they are and why they move as they do. An adequate answer to this question would explain "what was best for each and what was good for all." In a similar vein, Aristotle explains that a stone falls because this act restores it to its natural place: as close as possible to the center of the world.

In modern times, the most eloquent advocate of metaphysical dualism has been the French philosopher Henri Bergson, whose widely read books—especially *Creative Evolution,* published in 1907—influenced such diverse writers as William James, George Bernard Shaw, Marcel Proust, and Jean Piaget. Bergson pictured life as an impulse "traversing" matter, a wave spreading outward from a center, transmitted by genetic material from generation to generation. The vital impulse is creative, expressing itself in the production of increasingly organized forms. It is opposed by the entropic tendency of the matter it traverses. The conflict between these opposing tendencies—rather like that between Love and Strife in the philosophy of Empedocles—is the mainspring of cosmic evolution:

> Life as a whole, animal and vegetable, in its most essential aspect, appears to be an effort to accumulate energy and then to release it in flexible, deformable channels, at the extremities of which it will accomplish infinitely varied tasks. That is what the vital impulse, traversing matter, would like to do in one stroke. It would undoubtedly succeed if its power was unlimited or if it could get some aid from outside. But the impulse is finite, and it was given once and for all. It cannot surmount all the obstacles. The movement that it expresses is sometimes diverted, sometimes split, always opposed, and the evolution of the organized world is just the unrolling of this struggle.[2]

The fourth approach to understanding the phenomenon of life and its relation to physical laws—the approach we will explore in the remainder of this chapter—is that of modern biology. It is scientific but not reductive. Because it recognizes that present scientific theories are incomplete, it doesn't dismiss as unreal phenomena that seem to lie outside their scope. At the same time, it rejects as needlessly pessimistic the claim that there are aspects of objective reality that science can never hope to understand.

We now know that every living organism is made up of relatively simple molecular building blocks whose linkages and interactions are governed by the same physical laws that operate in test tubes, stars, and interstellar gas clouds. There is no vital impetus, no "second kind of reality," that transcends physical reality. Living matter differs from nonliving matter not because it contains an extra, nonphysical ingredient but because it is organized in a special way. But that difference is crucial. The reductionist view that life is nothing but chemistry is inadequate because it fails to recognize that order is no less basic an aspect of the world than energy. What is distinctive, and indeed unique, about living organisms resides in their order. What is biological order? And where does it come from? Our main aim in this chapter and the next will be to try to answer these questions.

Biological order manifests itself on many levels, from the level of molecules and their interactions to that of whole organisms and their interactions. Which level you find most interesting depends largely on your temperament and training. There are naturalists who couldn't care less about what goes on at the molecular level, and molecular biologists who scorn natural history. There are excellent books on biological evolution that don't mention molecules, and books on biochemistry that make little or no reference to plants or animals. The following discussion will focus on the molecular level—not because I find that level more congenial or more interesting than higher levels but because the molecular level is most relevant to an inquiry into the *origins* of biological order.

Viewed at the molecular level, life consists of the coming together and coming apart of molecules, pieces of molecules, and groups of molecules. These are the elementary movements in an unimaginably complex ballet, performed continuously in every living cell. The score of the ballet resides in the cell's genetic material, long threadlike molecules of deoxyribonucleic acid (DNA). Molecular biologists and biochemists have learned to read the score. They have discovered how the information it contains is encoded and how it is translated into the thousands of exquisitely regulated and coordinated couplings and decouplings that make up the ballet. While the most basic parts of the story are very well known, it will be useful for the subsequent discussion to review them briefly.

DNA and All That

The DNA molecule consists of two long chains twisted together in the form of a double helix (Figure 10.1). Each unit of both chains is composed of three building blocks: a sugar (deoxyribose) linked to a phosphate group and to a nitrogen-rich base (Figure 10.2). The sugar and the phosphate group are identical in each unit, but there are four distinct bases, denoted by the letters A (for adenine), G (for guanine), T (for thymine), and C (for cytosine) (Figure 10.3).

The backbone of each chain is a strictly repetitive structure in which only the sugars and the phosphate groups participate, adjacent sugars being linked by phosphate groups, as illustrated in Figure 10.4. The backbones of the two chains run along the outside, like the rails of a twisted ladder. The rungs of the ladder are formed by *pairs* of complementary bases: G and C, and T and A, linked as shown in Figures 10.3 and 10.4. The members of each pair fit snugly together, and the two kinds of rungs (GC and TA) have the same width. The resulting structure (Figures 10.1 and 10.4), worked out in 1953 by James Watson and Francis Crick, is a double helix of uniform width and pitch.

The structure of DNA is adapted to replication. Each strand of the double-stranded molecule serves as a template for the assembly of a complementary strand—that is, a strand whose bases are complementary to the corresponding bases of the template strand, as in the following example:

$$\ldots \text{GTAACTTG} \ldots$$
$$\ldots \text{CATTGAAC} \ldots$$

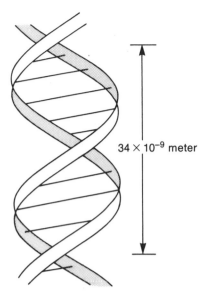

34×10^{-9} meter

FIGURE **10.1** Schematic diagram showing the geometry of a section of a DNA molecule. The molecule resembles a long twisted ladder. There are ten rungs per complete turn. Adjacent rungs are separated by 3.4 nanometers (1 nm = 10^{-9} meter). The units that make up the rails of the ladder are shown in Figure 10.2; the units that make up the rungs are shown in Figure 10.3.

As far as physics and chemistry are concerned, the sequence of the base pairs (AT, GC, TA, CG) in a DNA molecule seems to be completely arbitrary. Up until now, no one has been able to detect any change in the molecule's energy or in any of its other physical or chemical properties resulting from the replacement of one base pair by another. But such a substitution may have important *biological* consequences if it occurs in a cell from which an organism develops. Sickle-cell anemia, for example, is caused by a single substitution of this kind.

DNA encodes a program for the development of an organism. Viewed microscopically, the process of development consists of regulated and coordinated chemical reactions. All these reactions, along with their regulations and coordinations, are mediated by *proteins*. Some proteins, called *enzymes*, act as highly specific catalysts for reactions that wouldn't otherwise occur at an appreciable rate in the normal cellular environment. Each of the thousands of reactions that involves the making or breaking of a strong chemical bond is turned on by a specific enzyme. Other proteins serve a regulatory function by inhibiting or promoting the action of specific enzymes with which they form relatively weak chemical bonds. Still others serve regulatory and other functions by binding weakly to specific regions of DNA itself. The synthesis of each of these development-mediating proteins is directed by a segment of DNA, called a *gene*.

A protein molecule consists of one or more chains, called *polypeptides*. The units that come together to form a polypeptide are *amino acids*, whose chemical

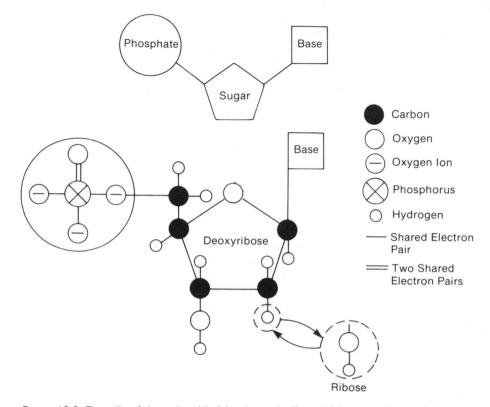

FIGURE 10.2 The rails of the twisted ladder shown in Figure 10.1 are made up of identical units, each consisting of a five-carbon sugar (deoxyribose) linked to a phosphate group. RNA is a single-stranded molecule that closely resembles one of the two strands of DNA. The sugar in RNA is ribose, which has one more oxygen atom than deoxyribose, as illustrated in the lower part of the figure.

structure is shown schematically in Figure 10.5. The variable part of an amino acid's structure is the *side chain*, denoted by R in the figure. Twenty different amino acids (that is, amino acids with twenty distinct side chains) are encoded directly by DNA. The side chains don't participate directly in the linkage between adjacent amino acids in a polypeptide. Thus polypeptides, like strands of DNA, have strictly repetitive backbones (Figure 10.6).

In their natural cellular environment, the enzymes and regulatory proteins encoded by DNA, which typically contain from a hundred to several thousand linked amino acids, fold spontaneously into compact and exceedingly complex shapes (Figure 10.7). The shape that a protein assumes in its natural environment is completely determined by the sequence(s) of side chains in its constituent polypeptide(s). Gentle heating causes a protein to unfold, and its constituent strands (if there are more than one) to separate. On cooling to the normal cellular temperature, the strands spontaneously recombine and refold, and the protein resumes its original shape.

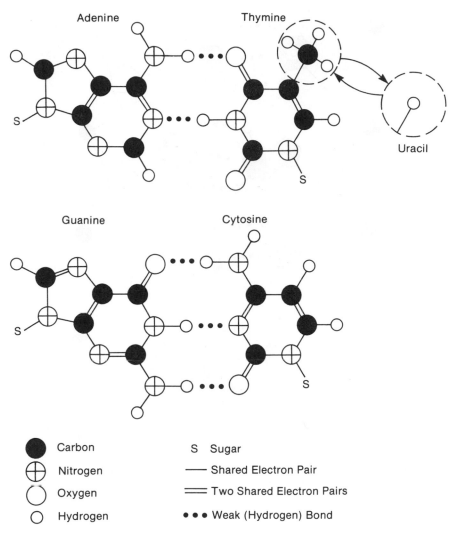

FIGURE 10.3 The rungs of the twisted ladder are pairs of complementary bases held together by weak bonds (rows of three dots). In DNA, there are four bases: adenine, thymine, guanine, and cytosine. In RNA, uracil replaces thymine, from which it differs in the way indicated in the figure.

The shape of a protein is what enables it to fulfill its highly specific catalytic or regulatory function. An enzyme spontaneously forms weak chemical bonds with the molecule or molecules whose reaction it catalyzes. These bonds can form only if the reactant molecule or molecules fit closely into specific binding sites on the enzyme, the way a key fits a lock. After the reaction has taken place, the product molecule or molecules, which don't fit into the binding sites, move off (Figure 10.8). Many enzymes have auxiliary binding sites to which regulatory proteins or products of chemical reactions can spontaneously attach if they have

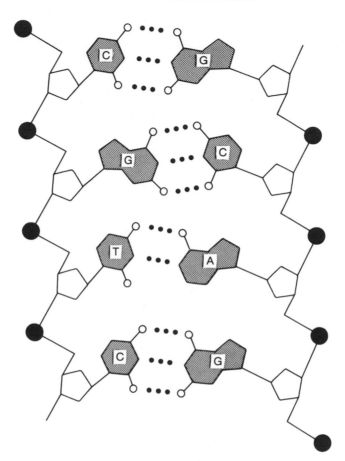

FIGURE 10.4 A fragment of the twisted ladder, showing how the structure pictured in Figures 10.2 and 10.3 fit together. The twist isn't shown.

exactly the right shape, thereby decreasing or enhancing the enzyme's catalytic efficiency. Other regulatory proteins bind spontaneously to specific DNA segments.

We may think of the twenty amino acids, or their side chains, as letters in an alphabet. The sequence of amino acids in a protein then represents a sentence whose meaning is the enzyme's specific catalytic or regulatory function. A protein, like a sentence, is an orderly structure because it has a meaning, and the more precise and specific its meaning, the more order it contains. Just as the meaning of a sentence may depend more strongly on some words in it than on others, the ability of a protein to carry out its catalytic or regulatory function may depend more strongly on some of the ''letters'' in its ''sentence'' than on others. But there is abundant, if indirect, evidence that every ''letter'' contributes to the meaning of the ''sentence.''

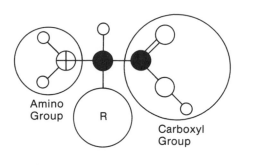
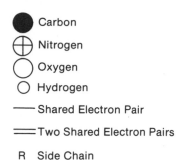

(a)

(b)

FIGURE 10.5 Structure of an amino acid. (a) An amino group (NH_2), a carboxyl group ($COOH$), a hydrogen atom, and a side chain each share a pair of electrons with a central carbon atom. (Side chains are more or less complex chemical groups. The simplest side chain is a single hydrogen atom. Twenty different side chains are encoded by DNA.) (b) The carbon atom is at the center of an equal-sided tetrahedron, whose corners are occupied by the nitrogen atom of the amino group, the carbon atom of the carboxyl group, a hydrogen atom, and an atom belonging to the side chain. (The corners of an equal-sided tetrahedron are noncontiguous corners of a cube. The center of the cube is also the center of the tetrahedron.)

The sequence of amino acids in an enzyme or a regulatory protein is specified by the sequence of bases in the gene that codes for it. Each of the twenty amino acids is specified by one or more sets of three consecutive bases ($=$ *triplets*). Since there are four bases, there are $4 \times 4 \times 4 = 64$ distinct triplets. All but three of them code for specific amino acids; the remaining three code for the analogue of a period: they indicate the end of the protein-sentence. Thus a gene contains exactly the same information as the protein it encodes, written in an alphabet of DNA triplets instead of amino acids.

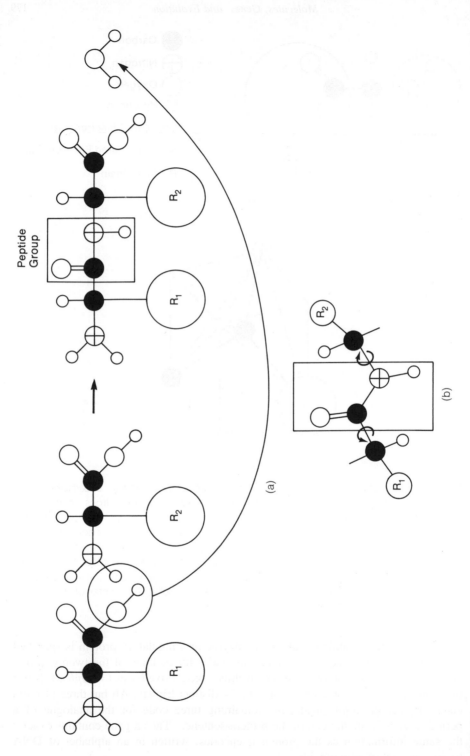

Peptide
Group

(a)

(b)

FIGURE **10.6** (a) Two amino acids combine to form a dipeptide and a water molecule (H–O-H). A tripeptide is formed when a dipeptide combines with a third amino acid, and so on. The four atoms that make up a peptide group (framed by a rectangle in both diagrams) lie in the same plane, which also contains the central carbon atoms of the tetrahedral amino-acid configuration shown in Figure 10.5. (b) Each tetrahedral configuration can rotate freely around the bond that links it to the peptide group, as indicated by the curved arrows. Thus a polypeptide is a very flexible structure. Its actual shape is determined by the sequence of side chains.

FIGURE **10.7** The backbone of the enzyme carboxypeptidase A. Each small circle represents the central carbon atom of the tetrahedral amino-acid configuration. The shaded circle near the center of the protein represents a zinc atom, which plays an important role in the molecule's enzymatic function. (From W. N. Lipscomb, *Proceedings of the Robert A. Welch Foundation Conference on Chemical Research* 15 [1971], fig. 134. Reprinted by permission)

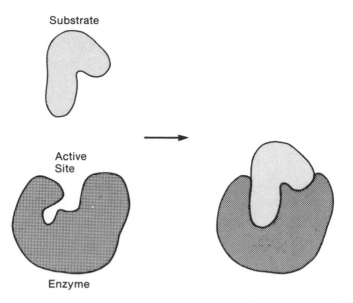

FIGURE 10.8 An enzyme and its substrate (the reactant molecule or molecules) fit together like a lock and a key. The union of the substrate and the enzyme changes their shapes to a greater or lesser degree. The reaction that the enzyme catalyzes destroys the fit, forcing the substrate and the enzyme to move apart.

Genes direct the synthesis of the proteins they encode through a sequence of steps involving three kinds of RNA (ribonucleic acid). RNA is a single-stranded molecule whose structure closely resembles that of one of the two strands of DNA. The only differences are that the sugar unit in RNA is ribose instead of deoxyribose, and the base uracil (U) replaces thymine. Like thymine, uracil forms a base pair with adenine. Thus one strand of DNA can serve as a template for the assembly of a complementary strand of RNA.

This in fact is the first step in protein synthesis: one strand of the DNA encoding a protein (or several proteins) is *transcribed* into a complementary strand of RNA (Figure 10.9) carrying exactly the same information, encoded slightly differently. This RNA molecule, called *messenger RNA* (mRNA), migrates from the nucleus of the cell, where the DNA permanently resides, to the cytoplasm. There it encounters and binds to one of many large and complex molecular structures called *ribosomes,* made up of proteins and *ribosomal RNA* (rRNA). The ribosomal RNA contains a sequence of bases that enables the ribosome to recognize and bind to a complementary leader sequence of messenger RNA immediately preceding the triplet that codes for the first amino acid of the polypeptide.

Also present in the cytoplasm are molecules of *transfer RNA* (tRNA), each linked at one end to an amino acid. The opposite end of the molecule contains a sequence of three bases that are complementary to the sequence of three bases that codes for that amino acid in messenger RNA (Figure 10.10). A specific enzyme joins each transfer RNA molecule to the appropriate amino acid. Loaded tRNAs bind to the complementary segments of the messenger RNA at a specific site on the ribosome. Special enzymes then direct the detachment of the amino acid from its tRNA and its attachment to the growing polypeptide chain that is being synthesized at the ribosome. The ribosome then moves three bases down the messenger, bringing its next triplet into position to receive a loaded tRNA. The *translation* process continues, amino acids being added one by one to the growing polypeptide, until one of three triplets that doesn't code for an amino acid shows up at the binding site. A protein that recognizes and binds to one of these triplets then causes the completed polypeptide to cut loose from the ribosome.

"The Secret of Life"

Thanks to the efforts of thousands of investigators in laboratories around the world, each step of the process just sketched, by which genetic information is translated into proteins, is now understood in great (although still incomplete) detail. Many other basic processes are also understood in considerable detail: how energy is extracted from foods and from sunlight; how this energy is stored and used to make muscle fibers contract, to operate chemical pumps, and to synthesize biomolecules and their building blocks; how antibodies are manufactured. The detailed mechanisms of these and other biological processes, and the molecular structures that mediate them, have proved to be more intricate than their discoverers could have foreseen, but all of them obey the same laws that apply to nonliving matter. Biochemical reactions, like all other chemical reactions, rearrange atoms and electrons while conserving energy and generating entropy. The molecules that living organisms synthesize are built from about sixty common building blocks in accordance with the same rules that govern the structure of nonbiological molecules.

What distinguishes a living cell from a nonliving aggregation of the same molecules is its organization, and the organization of the processes that go on within it and at its boundary. These kinds of organization, we have seen, express information encoded in the cell's DNA. DNA is thus, in a sense, the modern counterpart of Democritus's smooth round atoms whose presence was supposed to distinguish living from nonliving matter, although it plays this role in ways that no scientist, philosopher, or poet ever anticipated. DNA directs the assembly of cells and multicellular organisms. Its replication ensures the continuity of life from one generation of cells and organisms to the next. In all known forms of life, from bacteria to higher plants and animals, DNA molecules are assembled from the same four nucleotides, and the proteins whose assembly they direct contain the same twenty amino acids. All known forms of life use the same genetic code, linking a specific amino acid (or a stop signal) to every sequence of three consecutive nucleotides. They also use essentially the same molecular strategy to translate the information encoded in DNA into proteins. And this strategy, as well as the genetic code itself, is encoded in the DNA of living organisms. (The genetic code is embodied in transfer RNA "adaptors" and in the enzymes that attach

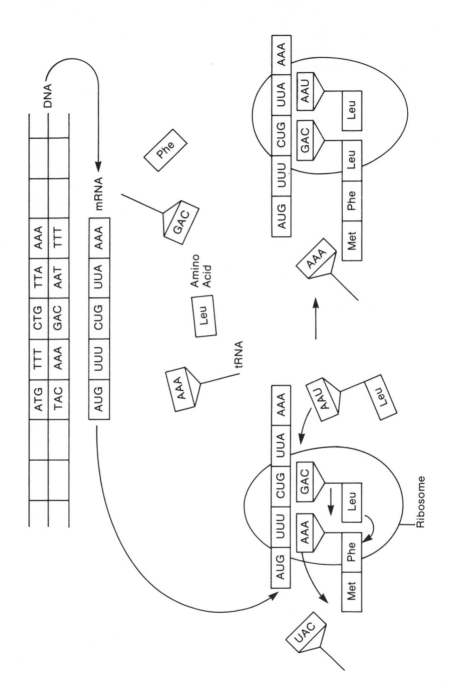

activated amino acids to the appropriate adaptors. The assembly of these adaptors and enzymes is directed by DNA.)

Thus DNA, and the information it encodes, is the "secret of life." But how did the information get there? And how did it come to be encoded in DNA?

Copying Errors and Differential Reproduction

The short answer to these questions is, of course, evolution. The processes by which DNA is replicated and transmitted from one generation of cells and organisms to the next guarantee its evolution. The DNA text is copied letter by letter by "copyists" that are skillful and conscientious (although not infallible) and that don't understand what they are copying. Occasionally they mistake an A for a G or a C for a T. Sometimes they leave out or repeat a whole word. Most of these accidental changes don't "improve" the text, but every now and then one of them does. (In the spirit of Humpty Dumpty, who defined *glory* as "a nice knock-down argument," I will take *improvement* to mean "a genetic change that, on average, increases the reproductive success of its carriers.") Thus an "improved" gene is likely to displace the original gene because it gets replicated more frequently.

Let's consider an example. Suppose that a certain protein acts as a catalyst for two unrelated but chemically similar reactions. Suppose that, through a copying error, the DNA of some organisms in a population has come to contain two identical genes coding for this protein. Now consider the consequences of copying errors that change individual base pairs. A copying error that makes the product of one of the genes a better catalyst for the first reaction but a worse catalyst for the second reaction improves the combined performances of the two genes. Be-

FIGURE 10.9 Transcription and translation of DNA. A process that hinges on the pairing of complementary bases produces a single-stranded copy of the sequence of DNA bases that specifies a protein or part of a protein. This copy, messenger RNA (mRNA), migrates to another part of the cell, where several large molecular complexes called ribosomes attach themselves to it. Each ribosome moves along the mRNA molecule, helping to attach one amino acid after another to a growing polypeptide in accordance with the genetic code. The amino acids are brought to the ribosome by molecules of transfer RNA (tRNA). Each of these small molecules bonds to an amino acid specified by a strategically located sequence of three nucleotide bases complementary to one of the mRNA sequences that specifies that amino acid (Figure 10.10). A "charged" tRNA (carrying the amino acid designated Leu) bonds to the second of two sites on the ribosome. The first site is already occupied by a tRNA carrying the portion of the polypeptide that has already been synthesized at that ribosome (the dipeptide MetPhe). The ribosome now advances three bases, exposing a new codon and causing the charged tRNA to occupy the site previously occupied by the tRNA to which the growing polypeptide is attached. The latter detaches itself from its tRNA and bonds to the amino acid carried by the tRNA that now occupies the first site. A new charged tRNA comes to fill the now-empty second site, and the process repeats. Special start and stop codons initiate and terminate the translation process.

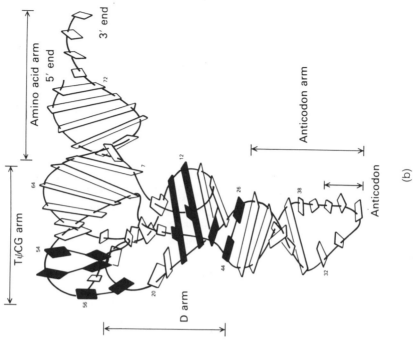

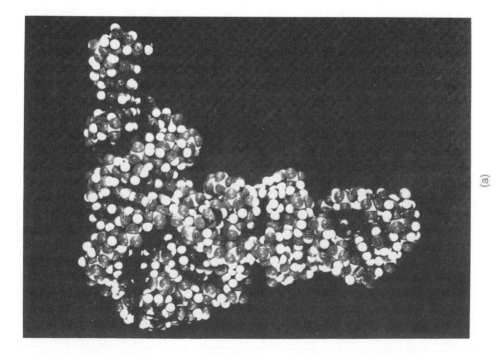

(a)

(b)

cause the second reaction continues to be efficiently catalyzed by the product of the second gene, the combined catalytic efficiency of the two genes increases. Subsequent copying errors that make either gene product more efficient in catalyzing its own reaction and less efficient in catalyzing the other reaction improve the reproductive success of organisms that inherit them. The two gene products therefore evolve into increasingly specific enzymes. This in turn opens the way for the evolution of independent mechanisms for modulating their activity.

This example illustrates how differential reproduction can shape new genes and combinations of genes from the raw material furnished by accidental copying errors. Since both copying errors and differential reproduction are facts, evolution, their logically necessary consequence, must also be a fact. The only remaining question is whether evolution accounts for *all* the order and diversity of the biological world. Did all past and present forms of life evolve from a single ancestral population through reproduction coupled with accidental genetic variation—that is, chance errors in the copying of genetic material (DNA)?

The Origin of Species

The hypothesis that evolution results from differential reproduction and *undirected* genetic variation was framed by Charles Darwin in 1838. Evolution itself wasn't a new idea when Darwin began to think about it. It had been discussed by his own grandfather, Erasmus Darwin, and detailed evolutionary hypotheses—although very different from Darwin's—had been put forward earlier by the French naturalist Jean-Baptiste Lamarck and by the English philosopher Herbert Spencer. So an important part of the intellectual community was well prepared for Darwin's ideas. At the same time, Darwin's hypothesis contradicted widely held and seemingly well-established scientific beliefs about the immutability of species and the age of the Earth. It also struck at the core of a religious belief shared by nearly all the members of Darwin's profession, of his social class, and of the English nation: that God created human beings and set them apart from brute creation; that human life has a divine purpose and meaning as well as a divine origin.

The Origin of Species, published in 1859, meets these challenges head-on. As Darwin remarked, the book is "one long argument." Its thesis is that the available evidence bearing on the origin of biological order and diversity can be explained more coherently and more economically by a "theory of descent with modification through variation and natural selection" than by the hypothesis that biological species were created by God. ("Natural selection" was the name coined by Dar-

FIGURE 10.10 (a) A space-filling model of the tRNA for phenylalanine (Phe) in yeast. (b) A diagram of this molecule, showing the anticodon and the region where the amino acid attaches. Bases and base pairs are represented by rectangles. (From J. L. Sussman and S. H. Kim, "Three Dimensional Structure of a Transfer RNA in Two Crystal Forms," *Science* 192 [28 May 1976]: 853–58. Copyright © 1976 by the American Association for the Advancement of Science. Reprinted by permission)

win for the tendency of accidental genetic "improvements" to spread by differential reproduction.) In support of this thesis, Darwin marshals evidence and arguments from natural history, taxonomy (biological classification), geology, paleontology (the study of fossils), animal and plant breeding, embryology, animal and human behavior, sociology, and ecology. He piles detail on detail; yet, as in a Beethoven symphony, every detail illuminates and enriches the work's main themes.

To say that *The Origin of Species* offers the first unified vision of the biological sciences is to underestimate Darwin's achievement, because before *The Origin* there were no biological sciences in the modern sense. There were collections of data, hypotheses, doctrines, and dogmas, but no solid theoretical foundation. That was what the theory of evolution was intended to supply, and in so doing it transformed biology into a science:

> When we no longer look at an organic being as a savage looks at a ship, as something wholly beyond his comprehension; when we regard every production of nature as one which has had a long history; when we contemplate every complex structure and instinct as the summing up of many contrivances, each useful to the possessor, in the same way as any great mechanical invention is the summing up of the labour, the experience, the reason, and even the blunders of numerous workmen; when we thus view each organic being, how far more interesting—I speak from experience—does the study of natural history become![3]

The Origin of Species taught biologists to see their world with Darwin's eyes. To a modern biologist, "nothing in biology makes sense except in the light of evolution." In this respect, *The Origin* may be compared with Newton's *Principia*, which brought about an equally profound transformation in the way physicists see the physical Universe.

Genetics and Darwin's Theory

One area of biology that Darwin didn't include in his grand synthesis was genetics, which hadn't yet been born. He was therefore in no position to reject out of hand the Lamarckian hypothesis that use or disuse of an organ, continued over many generations, produces heritable modifications of that organ. Darwin conceded that congenital flightlessness in birds like the penguin and congenital blindness in animals like the mole might have been caused by disuse—as indeed, in a certain sense, they are (natural selection doesn't weed out mutations that reduce the effectiveness of an organ that isn't used). But in *The Origin,* he consistently stresses the role of natural selection as the chief architect of evolutionary change and, especially in the first edition, downplays the possible role of use and disuse.

Mendelian genetics, rediscovered in 1900 (Mendel's now famous studies of peas had been published in 1866), filled an important gap in Darwin's theory. Mendel's experiments led him to postulate that certain traits of the pea plant, such as the shape and color of its seeds, are determined by pairs of factors, one from each parent, and that these factors are transmitted unchanged from generation to generation. The discovery that heritable traits are determined and transmitted by discrete entities (genes) with fixed properties stands in much the same relation to modern biology as the discovery of the atomic structure of matter does to modern physics. Genes are, indeed, the atoms of heredity. Before Mendel, most people had assumed that parental traits are blended in the offspring, like paint from two cans. But this would make it impossible for new traits to establish themselves through natural selection: any variation arising in a few members of a large population would in a few generations be diluted into significance, like a drop of red paint added to a vat of white paint. Of course, plant and animal breeders knew perfectly well that selection works; and everybody knew that particular traits—an abbreviated earlobe, a jutting jaw—run in families. Mendelian genetics explains why. Variant genes are not diluted, and if they increase their carriers' reproductive success, they gradually spread throughout a population and eventually displace less fit variants.

The discovery, in the early twentieth century, that genes undergo spontaneous mutations resolved another difficulty. Natural selection acts on the raw material supplied by genetic (= heritable) variation. Darwin's own observations had convinced him that plenty of variation is present in natural populations of plants and animals (although he didn't know where it came from). But differential reproduction tends to deplete genetic variation. Fitter genes become increasingly frequent; less fit genes, increasingly infrequent. Why does genetic variation persist? Mutations provided an answer. While natural selection drains the reservoir of variability, mutations replenish it.

Thus Mendelian genetics, *as we understand it today*, greatly strengthened the theoretical underpinnings of Darwin's theory. The pioneer genetics of the early twentieth century didn't see it that way, though. As late as 1915, the Danish geneticist Wilhelm Johannsen considered it "completely evident that genetics has deprived the Darwinian theory of selection entirely of its foundations."[4] Johannsen and other prominent geneticists, including William Bateson in England and Thomas Hunt Morgan in the United States, embraced a theory of evolution proposed by the Dutch geneticist Hugo de Vries, who had carried out breeding experiments similar to Mendel's, using evening primroses instead of peas. De Vries inferred from these experiments that a new species can be founded by a single mutant, and he immediately extrapolated this inference (later shown by the American geneticist B. M. Davis to be incorrect) from the evening primrose to all genera of plants and animals. New species do not evolve by small steps over thousands of generations, as Darwin had proposed, said de Vries, but in single mutational leaps.

De Vries, Bateson, Johannsen, and Morgan stuck to this hypothesis for many years, in spite of its shaky experimental basis, its inherent implausibility, and the fact that it contradicted many kinds of evidence, including voluminous observations bearing on speciation that had been accumulated by naturalists from Darwin onward. Darwin himself had already explained, in the sixth edition of *The Origin*, why the hypothesis that evolution occurs by "abrupt and great changes of structure" is untenable. A proponent of this hypothesis will, he said, be compelled to believe that many structures beautifully adapted to all the other parts of the same creature and to the surrounding conditions, have been suddenly produced; and of such complex and wonderful co-adaptations, he will not be able to assign a shadow of an explanation. He will be forced to admit that these great and sudden transformations have left no trace of their action in the embryo. To admit all this is, as it seems to me, to enter into the realms of miracle, and to leave those of Science.[5]

Geneticists weren't alone in rejecting Darwin's account of evolution. The evolutionary biologist G. Ledyard Stebbins recalls his introduction to evolutionary theory as a student at Harvard in 1926:

At that time biology students were often asked to read what was considered to be the most authoritative history of biology, by Erik Nordenskiöld of Sweden. He wrote: "To raise the theory of natural selection, as has often been done, to the rank of a natural law, comparable to the law of gravity established by Newton is, of course, completely irrational, as time has already shown; Darwin's theory of the origin of species was long ago abandoned. Other facts established by Darwin are all of second-rate value."[6]

Not until the 1930s and 1940s, a century after Darwin formulated his theory, did it gain general acceptance by biologists. In France, Darwinism remains controversial to this day, according to the French biologist Ernest Boesiger:

France today (1974) is a kind of living fossil in the rejection of modern evolutionary theories: about 95 percent of all biologists and philosophers are more or less opposed to Darwinism.[7]

Anti-Darwinism

Why has Darwinism evoked such fierce and protracted opposition from scientists and philosophers? Initially, of course, many people refused to accept the *fact* of evolution—some for scientific reasons, others for religious reasons. After the publication of *The Origin,* however, support for the creationist view dwindled among scientists. Debate about the reality of evolution as a phenomenon gradually gave way to debate about its causes and mechanisms.

Many people who have accepted evolution as a fact have nevertheless rejected Darwin's account of how it occurs, because that account has seemed inimical to their belief that life and its history have meaning and value. A purely naturalistic account of evolution seems at first sight to exclude meaning and value. (Recall Socrates's complaint against the natural philosophers of his day.) I think this explains the enormous popularity of metaphysical theories like those of Bergson and Teilhard de Chardin that purport to explain the deeper meaning of evolution.

Not all critics of Darwinism have been hostile to naturalistic explanations, however. Like the geneticists of the early twentieth century, many of them have believed as firmly as Darwin that evolution is a natural process but have refused to believe that natural selection alone could have shaped the order and diversity of the biological world. Many, although by no means all, of Darwinism's scientific critics have been people trained in mathematics and the physical sciences. In the following quotation, Werner Heisenberg recalls a conversation that took place in the early 1930s:

> But even if we agree that selection leads to the emergence of particularly fit or viable species, it is very difficult to believe that such complicated organs as, for instance, the human eye were built up quite gradually as the result of purely accidental changes. Many biologists obviously take the view that this is precisely what did happen. . . . Others are more skeptical. I have been told about a conversation between von Neumann, the mathematician, and a biologist. The biologist was a convinced neo-Darwinist; von Neumann was skeptical. He led the biologist to the window of his study and said: "Can you see the beautiful white villa over there on the hill? It arose by pure chance. It took millions of years for the hill to be formed; trees grew, decayed and grew again, and then the wind covered the top of the hill with sand, stones were probably deposited on it by a volcanic process, and accident decreed that they should come to lie on top of one another. And so it went on. I know, of course, that accidental processes through the aeons generally produce quite different results. But on just this one occasion they led to the appearance of this country house, and people moved in and live there at this very moment." Needless to say, the biologist disliked this line of reasoning. But though von Neumann is no biologist, and though I myself can't judge who was right, I take it that even among biologists there is some hesitation as to whether Darwinian selection provides an adequate explanation of the existence of the most complicated organisms.[8]

What We Do Understand About Evolution

As a first step toward understanding the reasons for contemporary dissatisfaction with Darwinism (in its modern form), let's review the aspects of evolution that are now well understood and securely established:

• Plants, animals, and bacteria are descended from a common ancestral population. Before the advent of molecular biology, there were rational grounds for skepticism concerning this claim. Such grounds no longer exist. The fact that every living organism carries the same kind of genetic material (DNA) and uses the same code to translate the information it contains into the proteins that mediate the organism's development is *prima facie* evidence of descent from a common ancestral population.

Many other kinds of molecular evidence point to a common origin for all existing life forms—ATP, for instance. Every complex economy, including the economy of living cells, requires a standardized medium of exchange, or currency. In all cells, without exception, the currency is ATP (adenosine triphosphate); molecules of ATP deliver the high-grade energy needed to drive the chemical reactions that maintain the structure of a cell against entropic decay and enable it to do work on its surroundings. ATP may have come to play its present role even before DNA and the genetic code came on the scene.

In higher cells (cells whose DNA is enclosed by a membrane), ATP is synthesized in complex intracellular organs called *mitochondria*. A key participant in the process is a protein called *cytochrome c*. The sequence of amino-acid residues in this protein is different in different organisms, but the key parts of the sequence—those in which any change would greatly impair the protein's functioning—are identical in organisms as diverse as yeast, wheat, tuna, and humans. The remaining parts of the sequence differ much less between organisms like humans and monkeys, or donkeys and horses, than between distantly related organisms like humans and wheat. The evolutionary interpretation of this finding is that accidental substitutions of one nucleotide for another in the gene for cytochrome *c* have occurred at a roughly constant rate. Thus the number of differences between the nucleotide sequences of two organisms ought to be roughly proportional to the time that has elapsed since they diverged from their common ancestor.

Bacteria lack a nuclear membrane and intracellular organs, including mitochondria. Nevertheless, bacteria that use oxygen to break down fuel molecules and synthesize ATP contain cytochromes whose structure closely resembles that of cytochrome *c* (Figure 10.11). The bacteria themselves are similar in size and structure to mitochondria. Mitochondria, moreover, contain DNA that codes for some of the proteins needed for their own synthesis. They also contain ribosomes, on which proteins are synthesized. These mitochondrial ribosomes are more like bacterial ribosomes than "normal" higher cell ribosomes.

These and other findings strongly suggest that mitochondria evolved from bacteria that took up residence in cells ancestral to modern higher cells. Analogous findings suggest that *chloroplasts,* the intracellular organs in plant cells that extract energy from sunlight and use it to synthesize ATP, evolved from photosynthetic bacteria. Biologist Lynn Alexander Margulis has argued that higher cells have evolved from *symbiotic communities* of bacteria. Whether or not this hypothesis is eventually confirmed, the biochemical evidence already in hand clearly demonstrates that there is an evolutionary link between bacteria and higher cells.

• Biological processes at the molecular level are governed by the same laws that govern molecules and their interactions in lifeless matter. Life is a state of matter that results from the *organization* of molecules and molecular interactions.

• Biological organization is based on sequences of nucleotide bases in DNA. These sequences are not strongly influenced (if they are influenced at all) by physical laws. Physical laws have a lot to do with the language of life, but little or nothing to do with the story that evolution has written in that language.

• DNA is altered in its passage from generation to generation.

• These alterations are not selectively influenced by the organism's life experience. During the development of an organism, information flows from its DNA to the proteins encoded by the DNA, but there is no known or suspected mechanism by which information could flow in the reverse direction. Thus there seems to be no way in which the lessons of experience could be transmitted to or interpreted by the genes that an organism passes on to its progeny.

• Thus biological order has been shaped entirely by differential reproduction.

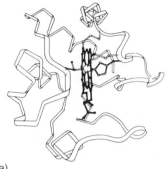

(a)

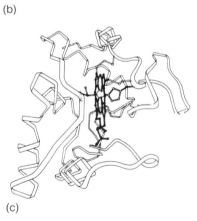

(b)

(c)

FIGURE **10.11** The shape of the molecule cytochrome *c* is much the asme in (a) tuna, (b) a photosynthetic bacterium, and (c) a denitrifying bacterium. (From *Biochemistry*, 3rd ed., by Lubert Stryer. Copyright © 1975, 1981, 1988. Reprinted with the permission of W. H. Freeman and Company)

No other conclusion is compatible with what is known about the molecular processes underlying life and evolution.

Differential Reproduction and Natural Selection

I must now distinguish between two terms that I have used almost interchangeably until now: *differential reproduction* and *natural selection.* Darwin's central insight was that heritable variations that improve an organism's ability to survive and reproduce in a given environment tend to spread within a population, while variations that diminish this ability tend to die out. This process of gradual and systematic change in the genetic composition of a population, resulting from genetic differences that affect their possessors' reproductive success, is called natural selection. *But not all long-term changes in the genetic composition of a population reflect genetic differences that affect their possessors' reproductive success.* Some long-term changes are the result of historical accidents. For example, as we will discuss in Chapter 11, genetic changes that don't significantly affect their possessors' biological fitness nevertheless have a finite chance of becoming fixed in a population. Again, a cataclysm may decimate a population, leaving behind a genetically unrepresentative remnant. So, as evolutionary biologist Richard Lewontin has emphasized, we aren't justified in assuming that every biological trait is adaptive—that it has been shaped by natural selection:

> Indeed, the fundamental issue in evolutionary biology which is unsolved is how much of what we see is a consequence of common ancestry . . . and how much . . . is a consequence of direct natural selection.[9]

To what extent is biological order shaped by natural selection and to what extent by random and accidental processes? The answer is simple: biological order is shaped entirely by natural selection—because that's how we chose to *define* it. This may seem an unduly legalistic, if not downright underhanded, way of dealing with what seems to be an important substantive question, but it accords with scientific practice. Terms are defined by the theories in which they figure. One might argue, for example, that the Newtonian definition of force doesn't adequately represent our intuitive conception of force, and perhaps it doesn't. Nevertheless, such an argument would have no bearing on the validity of the scientific definition. Analogously, the question whether biological order, defined as order generated by genetic variation and natural selection, adequately captures intuitive notions of biological order, although interesting in itself, has no bearing on the validity or scientific utility of the present definition.

The Root of the Difficulty

Knowing that natural selection has shaped biological order is not the same thing, however, as understanding *how.* The task of science is not only to discover the causes of natural phenomena but also to explain how they produce their effects. A demonstration that natural selection must be the sole begetter of biological order, however cogent, can never, by itself, allay skeptical doubts like those expressed by von Neumann; and before von Neumann, by Bergson; and before Bergson, by Darwin's earliest critics. Nor should it, for science thrives on skeptical doubts.

The following crude analogy illustrates the skeptics' main argument. Evolution by natural selection may be compared to a word game that I will call Darwin. You start with a word, say *ape,* which you may modify by adding, deleting, or changing a single letter; but each change must produce a real word, as in the sequence *ape, tape, tap, tan, man.* Skeptics argue that complex adaptations are more like meaningful sentences than like single words, and that it isn't possible to produce a sequence of meaningful sentences (= functional variants) by adding, deleting, and changing single letters.

The difficulty becomes even more severe when we consider examples of *convergent evolution,* in which similar organs have evolved along independent routes. The octopus has a camera eye whose structure resembles that of the vertebrate eye. Both kinds of eye have a cornea, a lens, and a retina on which an inverted image is formed. Yet mollusks (the phylum to which octopuses belong) and vertebrates diverged over 500 million years ago. Their last common ancestor undoubtedly had a much simpler eye. Referring to such examples, Bergson asks:

> How can we suppose that accidental causes, occurring in an accidental order, have on several occasions had the same outcome, the causes being infinitely numerous. . . .

It is as though two players of Darwin, both starting with "The quick brown fox jumps over the lazy dog" and working independently, had produced *Hamlet* and a Spanish translation of *Hamlet.*

Bergson also argued that Darwin's theory doesn't explain why evolution has produced increasingly complex adaptations and organisms:

> [A] glance at the fossil record shows us that life need not have evolved at all, or evolved only within narrow limits, if it had chosen to ensconce itself in its primitive forms.

The biologist Peter Medawar, a staunch supporter of Darwinism, makes the same point. Modern evolutionary theory, he asserts, gives

> no convincing account of evolutionary progress—the otherwise inexplicable tendency of organisms to adopt ever more complicated solutions to the problem of remaining alive.[10]

Medawar attributes this failure—correctly, in my opinion—not to any defect in the theory's basic premises but to its lack of a fully worked out theory of variation, that is, of candidature for evolution, of the forms in which genetic variants are proffered for selection.[11]

In the same vein, the evolutionary biologist G. Ledyard Stebbins has written:

> The theory that organisms evolve through the interaction of mutation, genetic recombination, and natural selection is now accepted by most biologists and by nearly all of those who have made intensive studies of evolution in progress. In its most elementary form, however, this theory gives us no basis for understanding how organisms have evolved greater complexity through geologic ages of time.[12]

But the growth of complexity is surely the most interesting and important feature of evolution. It is what nonspecialists *mean* by "evolution." If the current theory doesn't explain the "tendency of organisms to adopt ever more complicated solutions to the problem of remaining alive," what *does* it explain?

Darwinism's critics say that current evolutionary theory explains *microevolution* but not *macroevolution*. Microevolution encompasses the kinds of evolutionary change presided over by plant and animal breeders. More generally, microevolutionary changes result from changes in the relative frequencies of genes in a population's gene pool, as when a population of flies develops resistance to DDT. Before the population was exposed to DDT, one or more flies already carried a gene that conferred resistance to it. The introduction of DDT into the population's environment depleted the ranks of noncarriers of the gene and enabled carriers to multiply. It didn't create a new gene.

Macroevolution encompasses more spectacular kinds of evolutionary change, such as the emergence of higher cells, of multicellular organisms, of nervous systems, and of human intelligence. Macroevolutionary changes result from major changes in the developmental program encoded in DNA. Most serious students of evolution are convinced that the basic processes underlying the well-understood, but relatively modest, kinds of evolutionary changes involved in microevolution are exactly the same as those involved in macroevolution. I think they are right. But I agree with Medawar and Stebbins that current accounts of evolution don't provide an adequate basis for understanding macroevolutionary change. In Chapter 11, we will take a closer look at the difficulties involved in understanding the growth of biological complexity, and try to resolve them.

11

Evolution and the Growth of Order

Bergson regarded biological evolution as the embodiment of a nonmaterial life-force striving continually to overcome matter's entropic tendency by creating ever more complex forms of order. This is a poetic image rather than a scientific explanation. But it is a compelling image. A perennial source of dissatisfaction with Darwinism has been its apparent failure to supply a scientifically based picture of evolution that is equally compelling. What does Darwinism have to offer in place of the life-force? And why does evolution by random genetic variation and natural selection generate ever more complex "solutions to the problem of remaining alive"?

The physicist Paul A. M. Dirac once remarked that it isn't the business of physics to supply us with pictures. Quantum mechanics, for example, enables us to predict the observable properties of atoms and subatomic particles but not to visualize them. Can we extend Dirac's remark to evolutionary theory and say that its business is to make predictions, not to supply us with pictures? I think that that would be asking both too much and too little of evolutionary theory. Too much because, as Bergson insisted and as I will argue in this chapter, evolution is a genuinely creative process, giving rise to novel, and hence unpredictable, forms of order. Too little because an adequate theoretical account of evolution should do more than describe the elementary processes underlying evolutionary change. It should also explain why these processes have given rise to such varied and complex forms of life.

The Driving Force of Evolution

For Bergson, evolution and life were inseparable, two sides of a coin. In *Chance and Necessity* the molecular biologist Jacques Monod argued against this view:

> But where Bergson saw the most glaring proof that the "principle of life" is evolution itself, modern biology recognizes, instead, that all the properties of living beings rest on *a fundamental mechanism of molecular invariance.* For modern theory *evolution is not a property of living beings,* since it stems from the very *imperfections* of the conservative mechanism which indeed constitutes their unique privilege. And so one may say that the same source of fortuitous perturbations, of "noise," which in a nonliving (i.e., nonreplicative) system would lead little by little to the disintegration of all structure, is the progenitor of evolution in the biosphere and accounts for its unrestricted liberty of creation, thanks to the replicative structure of DNA: that registry of chance, that tone-deaf conservatory where the noise is preserved along with the music.[1]

Monod considered the central property of life to be the *precise* replication of genetic material. Every species, he says, has "an essential project": "the transmission from generation to generation of the invariance content characteristic of [that] species." (The "invariance content" of mouse DNA or of human DNA, according to Monod, is the part of the mouse genome or the human genome that makes a mouse a mouse or a person a person.) Evolution, in Monod's view, results from *imperfections* of the replicative mechanism that serves each species' essential project. It is a by-product of the universal entropic tendency toward disorder. Monod assigns the precise replication of genetic material to the realm of necessity. Errors in replication belong to the realm of chance. Evolution results from an interaction between the two realms. Necessity, in the form of invariant replication, snaps up and amplifies the fortunate accidents of chance.

I would like to argue for a different view: that the central property of life is not reproductive *invariance* but reproductive *instability,* and that reproductive instability is also the "driving force" of evolution. Reproductive instability is thus the scientific counterpart of Bergson's life-force.

I believe this view is close to Darwin's, although he didn't use the phrase "reproductive instability." Darwin said that the idea of natural selection came to him while he was reading Thomas Robert Malthus's *Essay on the Principle of Population,* published in 1798. Malthus, in turn, credits Benjamin Franklin with that essay's germinal idea:

> It is observed by Dr. Franklin that there is no bound to the prolific nature of plants or animals but what is made by their crowding and interfering with each other's means of subsistence. Were the face of the earth, he says, vacant of other plants, it might gradually be sowed and overspread with one kind only, as for instance with fennel; and were it empty of other habitants, it might in a few ages be replenished from one nation only, as for instance with Englishmen.[2]

Malthus argued that the "constant tendency of all animated life to increase beyond the nourishment prepared for it [is] the one great cause [that has] hitherto impeded the progress of mankind toward happiness.'' Darwin saw in the same tendency the mainspring of evolution, leading inevitably to a "struggle for existence" in which the fittest survive. (Although the phrase "survival of the fittest" was coined by the philosopher Herbert Spencer, Darwin adopted it in later years as a synonym for "natural selection.") Independently of Darwin but some twenty years later, the naturalist Alfred Russel Wallace arrived at exactly the same idea by much the same route.

Modern organisms owe their "prolific nature" to organization encoded in their DNA. But reproductive instability is a far more primitive property of living matter than the possession of DNA. DNA and RNA must have evolved from molecules that could be replicated without the specialized enzymatic help that modern nucleic acids require. The chance formation of the first such molecules changed the world forever. Their descendants invaded the Earth's seas, rivers, and lakes, its solid surface, and its atmosphere, organizing inert matter and mobilizing energy for the production of more and yet more molecules possessing this novel recursive property. Life is an infection of inert matter. Just as a virus transforms its living host into a factory for producing more virus, so life transforms its inert host into a factory for producing more life.

Genetic material is inherently unstable against modifications that increase its reproductive rate. This is one way of stating the principle of natural selection. As long as genetic material is susceptible to changes that increase its reproductive instability, its expansion can never be permanently checked. It can never achieve a truly stable equilibrium. The infection can never be permanently contained.

This inherent instability of genetic material is, I suggest, the "driving force" of evolution. It plays the role of Bergson's life-force. Unlike the life-force, however, it is not a finite impulse, "given once and for all." Each conquest of inert matter paves the way for fresh conquests, making available new sources of high-grade energy, new environments, and new ways of exploiting them for the reproduction of genetic material. Like the life-force, reproductive instability is inherently dynamic and inherently directionless; it has no purpose or project. But whereas Bergson's life-force is an aspect of reality that transcends intelligence and can be apprehended only from within by an act of pure intuition, reproductive instability is a perfectly intelligible property of molecules that are organized in a particular way.

Reproductive Instability and Evolution

How does reproductive instability lead to evolution? Consider a self-replicating molecule in an environment that is rich in building blocks and high-grade energy. In these circumstances, the molecule multiplies rapidly. In forty generations, it has 1,000 billion descendants. As Malthus remarked, such an exponentially growing population must eventually outstrip its food supply. However, if the food supply is constantly renewed, a *steady state* may result in which replicas are formed at the same rate as they are destroyed, so that the number of self-replicating molecules remains constant.

The first reproductively unstable molecules were almost certainly not perfect replicators. A molecule needn't replicate itself perfectly to spawn a growing population, however. A molecule that catalyzes the formation of imperfect replicas that are themselves imperfect replicators will have an exponentially increasing number of descendants as long as the population has access to enough building blocks and nutrients to support exponential growth. If nutrients are supplied at a constant rate, the population will eventually stop growing and settle into a steady state. The steady-state population will contain several variants of the original molecule, each a viable mutant of the original molecule or of one of its descendants.

Under the conditions we have been considering, a population of imperfect replicators evolves toward a steady state, which, once achieved, persists as long as the environment doesn't change. Thus reproductive instability doesn't necessarily lead to evolution in the broadest sense. Under the assumed conditions, it leads to evolutionary stagnation.

The argument assumes, however, that the environment is uniform in space and time. This assumption can't be true. Even if *external* factors of the environment change slowly, the supplies of building blocks and high-grade energy must dwindle as the population grows. Organisms that thrived in an environment of plenty may languish in an environment of scarcity, thereby creating evolutionary opportunities for mutants that would have been out-reproduced in the initial environment.

But the assumption of uniformity has an even more serious defect. A reproductively unstable population spreads out in all directions until it encounters conditions in which it can't thrive. It follows that *the environmental conditions that prevail at the edges of a population's range are marginal also in a biological sense;* that is, they are marginally capable of sustaining the given population. The margins may thus be hospitable to mutants that wouldn't have done well under the initial conditions of growth. If a small peripheral group becomes isolated from its parent population, it may found a new species. The fundamental importance of this aspect of the evolutionary process was first explicitly recognized by Ernst Mayr in 1954. It is part of the *founder principle,* to which we will return later.

Reproductive Invariance as an Evolutionary Strategy

Reproductive invariance, far from being opposed to evolution, as Monod maintained, is a key evolutionary strategy. As part of a larger strategy, it enhances the reproductive instability of genetic material. The reason is that the length of a self-replicating molecule is limited by the accuracy of the replicative process. For example, consider a self-replicating molecule (DNA, RNA, or a precursor of modern RNA) consisting of N units (nucleotides in a molecule of nucleic acid). Let f stand for the frequency with which a single unit is miscopied. For example, if $f = 0.01$, then any unit will be miscopied 1 percent of the time on average. The fraction of error-free copies *of the whole molecule,* which has length N, is then $(1-f)^N$. Figure 11.1 shows how this fraction decreases with N when f is a small number— that is, how the fraction of error-free copies of a molecule decreases with the size

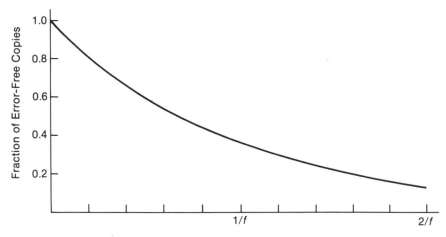

FIGURE 11.1 The fraction of error-free copies of a self-replicating molecule decreases with increasing length of the molecule. It is significanntly less than unity when the number of units that have to be replicated is comparable with the reciprocal of the error frequency *f*.

of the molecule. As long as the number of units is small compared with the reciprocal of the error frequency *f*, nearly all the copies will be error free. But when the number of units is comparable with the reciprocal of the error frequency, half or more of the copies have at least one error. If a single error has lethal consequences—if it results in a molecule that can't replicate itself—then the longest text that can be passed on from generation to generation is about $1/f$ units long. For example, if the error frequency is 1 percent, the longest self-replicating molecule contains about 100 units.[3]

Because evolutionary innovations that enable organisms to exploit their environments more effectively require expanded capacity to store and process genetic information, and because this capacity is limited by the fidelity with which genetic material can be replicated, natural selection promotes genetic changes that increase copying fidelity. The complex apparatus that all present forms of life use to store, translate, and transmit genetic information has resulted from genetic changes of this kind.

The physical chemist Manfred Eigen and his colleagues have carried out extensive experimental and theoretical studies of copying fidelity and its role in the origin of life. Eigen argues persuasively that the first self-replicating molecules were molecules of RNA similar to modern transfer RNA (the adaptor molecules that mediate the translation of each DNA triplet into a specific amino acid).[4] In the absence of specific enzymes, RNA nucleotides pair correctly (G with C, A with U)[5] about 99 percent of the time. The error frequency is thus about 0.01, so the longest reproductively viable RNA molecules should contain fewer than 100 nucleotides, which is consistent with experimental findings.

The genetic material of some viruses is RNA. The enzymes that mediate the replication of viral RNA reduce the copying error by a factor of 10 to 100, allowing RNA molecules of 1,000 to 10,000 nucleotides to be reproductively viable. The information needed to construct these enzymes is contained in the viral RNA itself and couldn't be encoded by substantially smaller segments of RNA. In short, RNA molecules 1,000 or so nucleotides long can be accurately replicated only with the help of enzymes whose assembly is directed by RNA molecules of this length. It follows that the RNA molecules and the enzymes that help to replicate them must have evolved together. Neither could have come into being without the other. How this joint evolution came about is still an unsolved problem.

The genetic material of the least complex of modern organisms, bacteria, contains about 100 times as much information as could be encoded in RNA. (Viruses are not autonomous organisms; they use the chemical machinery of the cells they invade to replicate themselves.) The evolutionary invention of DNA and its associated proofreading and repair mechanisms made possible the exceedingly high fidelity of replication on which all existing forms of life depend. How DNA evolved from RNA is also an as-yet-unanswered question.

Genetic Variation

"Reproductive invariance," or high-fidelity copying, isn't a perfect evolutionary strategy. A strategy whose sole ingredient was copying fidelity would, to the extent that it was successful, stifle evolution. Monod argued that evolution occurs precisely because a strategy of reproductive invariance *can't* be completely successful. Contrary to this view, I will argue that suppression of genetic variation is but one aspect of a more comprehensive evolutionary strategy. The complementary aspect of this strategy has to do with the *management* of genetic variation. I will argue that genetic variation, although blind to its consequences for the developing organism, isn't a purely random process but is regulated by a genetic program that has been shaped by natural selection. We will see, for example, that the mechanisms that suppress and repair copying errors don't operate equally in all parts of an organism's DNA. In some parts, copying errors are highly suppressed; in others, they are mildly suppressed; and in still others, they may be actively promoted. I will argue that such differences are genetically controlled and that the genetic systems that control these differences have been shaped by natural selections.

Let's begin by reviewing the *resources* for genetic variation that modern organisms deploy. These may be divided into two classes: genetic processes that speed up the process of natural selection by creating fresh combinations of genes, and processes that create new genes.

One of the benefits of sex is that it exposes fresh combinations of genes to natural selection. Consider a population of organisms that reproduce without sex. Each organism passes on an exact copy of its *genome* (genetic material) to its daughters, which in turn pass on exact copies to *their* daughters. If the environment is uniform in space and constant in time, small differences between the reproductive rates or *fitnesses* of different genomes give rise to gradually increasing differences between the numbers of their carriers. The fittest genomes (those that reproduce fastest) crowd out those that reproduce more slowly. Eventually all members of the population carry copies of the fittest genome.

This genome, however, will almost certainly not consist of the fittest genes. It will contain a preponderance of fitter-than-average genes, but, as in bridge or poker, the best possible hand rarely gets dealt.

Sexual reproduction exposes to selection segments of DNA smaller than an entire genome. The DNA in the nucleus of an egg or a sperm cell is divided into several segments (four in fruit flies, twenty-three in humans, fifty in goldfish), called *chromosomes*. Each chromosome contains many genes. The fertilized egg from which a new organism develops, as well as all the cells that arise from it in the course of development *apart from the organism's egg or sperm cells,* contains a maternal and a paternal version of each chromosome—hence twice as many chromosomes as egg and sperm cells. Egg and sperm cells contain *either* a maternal *or* a paternal version of each chromosome, and any given chromosome in such a cell is equally likely to be of maternal or paternal provenance. New combinations of chromosomes therefore appear in every generation. As a result, chromosomes are *individually* exposed to selection, and in a variety of genetic contexts— that is, in company with many different combinations of other chromosomes. Over many generations, chromosomes that increase their carriers' reproductive success displace less successful variants in the chromosome pool. Natural selection increases not only the average fitness of the genomes in the population, but also the average fitness of individual chromosomes.

An analogy may help to make this clear. Twelve tennis players, six men and six women, get together to play mixed doubles. For the sake of variety, they decide to form new two-person teams every day by drawing names out òf a hat. Every day the six newly constituted teams are randomly matched and play a fixed number of games. A record is kept of each player's wins and losses. On a given day a strong player may team up with a much weaker player and lose more games than he or she wins, but after a month or so the total number of games that a player has won will accurately reflect his or her relative skill at doubles (including the ability to adjust his or her game to different partners). The tennis players represent chromosomes; the teams represent genomes; winning represents replication. After many generations the fittest chromosomes will be most heavily represented in the chromosome pool.

Genetic recombination, which occurs during the formation of egg and sperm cells, exposes *parts* of chromosomes, or even individual genes, to natural selection. The cell that gives rise to an egg or a sperm cell contains a maternal and a paternal version of each chromosome. At a certain stage in the process of forming the egg or sperm cell, corresponding maternal and paternal chromosomes come into close contact. Contiguous strands of DNA, one from each parent, may then *cross over* and reconnect to form two new strands, each with its full complement of genes. A second cut must then occur to separate the two double-stranded DNA molecules, and the cut ends must be suitably reconnected. All such cuts and reconnections are mediated by appropriate enzymes. Possible outcomes are illus-

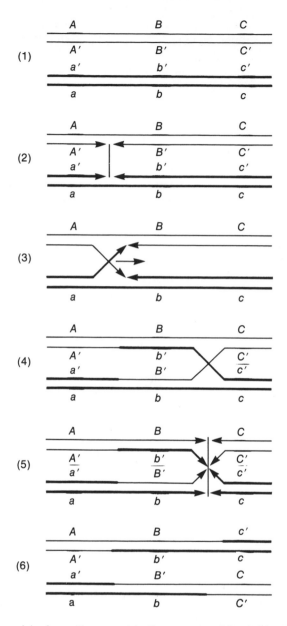

FIGURE 11.2 A model of genetic recombination proposed by Robin Holliday. (1) Two chromosomes, one from each parent, are aligned so that homologous genes lie opposite one another. The pair of heavy lines represents the maternal chromosome; the pair of light lines, the paternal chromosome. *AA'*, *BB'*, *CC'* are distinct regions of the paternal chromosome; *aa'*, *bb'*, *cc'* are the corresponding regions of the maternal chromosome. (2) One strand of each chromosome is cut. The cuts occur in corresponding positions. (3) The cut ends cross over and rejoin. (4) The crossover point migrates to the right. (Migration of the crossover point requries the breaking of bonds joining complementary strands of DNA and the formation of new bonds.) This is the crucial step in the Holliday

trated in Figure 11.2. In each outcome, pieces of DNA are exchanged between maternal and paternal chromosomes.

The processes we have considered thus far expose fresh combinations of genes to natural selection. They don't, however, create new genes, or even permanent new combinations of genes. They exchange sections of text between existing drafts, but they don't create new text. The actual composition of the drafts is accomplished by genetic mutations and natural selection. Mutations furnish the raw material; natural selection shapes it into meaningful text. The mutational resources of the least complex living organisms, bacteria, are comparable to the text-editing capability of a modern word processor. Mutations can change, delete, or insert a single character of genetic text (a single pair of complementary nucleotides). In addition, larger units of text—pieces of genes, whole genes, groups of genes—can be deleted, duplicated, and copied from one part of the genome to another, even from one chromosome to another.

Gene duplication has played an especially prominent role in evolution. It isn't yet known how or when the initial duplication occurs. Some, although not all, geneticists hypothesize that it results from an asymmetrical kind of genetic recombination, *unequal crossing over,* illustrated in Figure 11.3a. One strand of DNA ends up with two copies of a gene or set of genes, side by side, and the other strand ends up with no copies. Unequal crossing over is represented symbolically by the "reaction"

$$ABC + A'B'C' \rightarrow ABB'C + A'C'$$

where *ABC* and *A'B'C'* represent corresponding sequences of genes on the recombining strands of DNA, and *ABB'C* and *A'C'* represent gene sequences in the recombinant strands: one recombinant strand has an extra *B;* the other has no *B* at all. After the initial duplication, *oblique pairing* could produce a strand of DNA with three *B*s, a third could produce a strand with four or five *B*s, and so on, as illustrated in Figure 11.3b.

Molecular biologist Roy Britten discovered that a large (and variable) fraction of the DNA in the cells of plants, animals, and fungi resides in repeated sequences. Repeated sequences account for 30 percent of human DNA and 80 percent of salamander DNA. By contrast, the DNA of bacterial cells contains few repeated sequences. Some repetitions have an economic function: they match supply to demand. For example, higher cells contain hundreds of consecutive copies of the genes coding for ribosomal RNA, enabling them to meet the huge demand for this essential ingredient of ribosomes (the workbenches on which proteins are

---

model, because the *B* regions of both double helices are now mismatched. DNA repair processes correct the mismatches. In so doing, they may create three *BB'* pairs and one *bb'* pair, or three *bb'* pairs and one *BB'* pair. These "anomalous" ratios near crossover points are actually observed. Earlier models omitted this step and therefore couldn't explain these ratios. (5) All four strands are cut at corresponding points, indicated by the vertical line. (6) The strands rejoin. Cutting and splicing are chemical reactions catalyzed by enzymes.

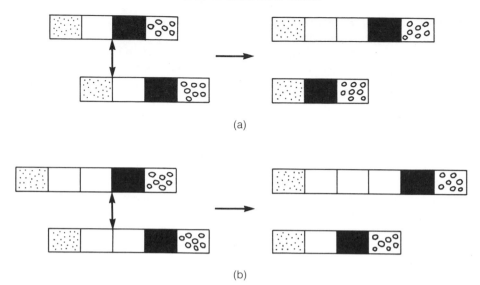

FIGURE 11.3 Possible origin of duplicated genes. (a) Unequal crossing over between two contiguous chromosomes. The chromosomes exchange the segments of DNA to the right of the double-headed arrow. Because they haven't lined up properly, one chromosome ends up with an extra copy of one segment, and the other with no copy of that segment. There is no direct evidence that this process occurs. (b) Oblique pairing, in which there is *local* alignment of the chromosomes: the segments to the left of the double-headed arrow are identical. This kind of unequal crossing over is more plausible than the first kind, but it obviously cannot give rise to the first gene duplication.

synthesized) during the early stages of development when cells, each containing tens of thousands of ribosomes, are dividing rapidly. Even bacterial DNA contains a few extra copies of the genes coding for ribosomal RNA. For similar economic reasons, higher cells contain multiple copies of the genes coding for the protein cores around which their DNA is wound, like stripes around a barber pole.

Evolution Through Gene Duplication

Gene duplication also has a creative function. Consider bacterial ferredoxin, an iron-containing protein that plays a key role in nitrogen fixation (the conversion of atmospheric nitrogen into ammonia, which plants can use to synthesize nucleic acids). The two halves of its polypeptide chain have similar sequences of amino-acid residues, which suggests that the DNA segments coding for them evolved from copies of the same gene. In plants, a similar molecule whose polypeptide chain is twice as long as that of bacterial ferredoxin plays a chemically similar role in photosynthesis. The gene coding for this protein evidently arose from two successive duplications and joinings.

Extra copies of genes have often evolved into genes coding for new proteins.

According to the usual description of this process, the original gene continues to code for the same protein, leaving the duplicate free to accumulate mutations that give the protein it codes for a new structure and function. But it may also happen that duplication enables *both* the original gene and its copy to evolve in such a way that their products assume more specific functions. In the first scenario, the duplicate gene *diverges* from the original; in the second, the two copies become *differentiated*. The second scenario is more general; it includes the first as a special case.

The evolution of hemoglobin offers an excellent example of gene duplication in its creative role. The cells of animals get their energy by burning fuel molecules such as glucose in the presence of oxygen. Fuel molecules are relatively loosely bound; it takes relatively little energy to disrupt them. The chemical reactions that constitute burning rearrange the atoms of the fuel molecules and link some of them to oxygen atoms to produce more tightly bound molecules—in particular, water and carbon dioxide. Some of the energy released in the process is used to synthesize ATP, the universal currency of energy in living organisms; the rest is wasted. In very tiny organisms and in insects, oxygen diffuses to individual cells directly from the air or water in which the organism lives. The evolution of larger oxygen-consuming organisms went hand in hand with the evolution of a circulatory system that could carry oxygen to cells far from those able to take it up directly from the environment. In vertebrates (fish, amphibians, reptiles, birds, and mammals), oxygen is carried from gill or lung tissue by molecules of hemoglobin in red blood cells. (Hemoglobin is what gives these cells their color.) Muscle tissue, which requires large supplies of oxygen on short notice, has a supplementary oxygen carrier, myoglobin, which, like hemoglobin, readily binds and releases oxygen molecules.

Myoglobin evolved first, probably from a similar oxygen-carrying molecule in the primitive ancestors of fish. It consists of a polypeptide chain folded into a very compact structure (Figure 11.4). Nestled in a crevice of the folded polypeptide is a *heme*, a flat disk shaped molecule with an iron atom at its center. (Hemes with exactly the same structure are found in several other molecules, including cytochrome *c,* a much earlier evolutionary invention than myoglobin.) The iron atom in heme has two possible states, in one of which it can bind an oxygen molecule, in the other not. An isolated heme reverts spontaneously to its nonbinding state. The folded polypeptide in which the heme nestles keeps the iron atom in its oxygen-binding state by inhibiting the chemical reaction that brings about this transition. (It also discourages, although it doesn't entirely prevent, carbon monoxide molecules from binding to the iron atom. On their own, hemes bind more readily and more strongly to carbon monoxide than to oxygen. Since reactions within cells produce some carbon monoxide, *free* hemes couldn't function as oxygen carriers even if they could be kept in their oxygen-binding state.)

The hemoglobin of lampreys and hagfish, descendants of the most primitive jawless fishes, closely resembles their own myoglobin. Like myoglobin, it consists of a single polypeptide chain containing a single heme. The hemoglobin of all other vertebrates—jawed fishes, amphibians, reptiles, birds, and mammals—consists of four polypeptide chains, each holding a heme identical with the heme

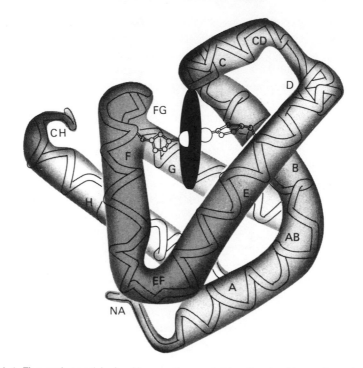

FIGURE 11.4 The polypeptide backbone of myoglobin. The backbone has eight helical sections, labeled A through H. The heme group is represented by a disk. It is centered on an iron atom that is held in place by residues of the amino acid histidine bonded to helices E and F. The nearly complete circle opposite the iron atom represents an oxygen atom. (From *Molecular Cell Biology,* by James E. Darnell, Harvey Lodish, and David Baltimore. Copyright © 1986 Scientific American Books, Inc. Adapted by permisison from Dickerson, R. E., and Geis, I. *The Structure and Action of Proteins.* Benjamin/Cummings, Menlo Park, Calif. Reprinted with permission of Scientific American Books and Irving Geis)

in myoglobin, weakly bonded together to form a symmetrical structure (Figure 11.5). Two of the chains are of one kind (designated alpha); the other two are of a different kind (beta). In mammals, there is a further refinement. Mammals have two main kinds of hemoglobin. Fetal red blood cells manufacture hemoglobin with a pair of gamma chains in place of the beta chains, which appear after birth. Fetal hemoglobin takes up oxygen more readily than postnatal hemoglobin. This promotes the transfer of oxygen from mother to fetus across the placental barrier.

A hemoglobin molecule is not simply a package of four myoglobin-like molecules. Because of interactions among its components, the whole is biologically more effective than the sum of its parts. The binding of one oxygen molecule to a hemoglobin molecule makes it easier for the second and third oxygen molecules to bind, and the binding of the fourth molecule still easier. The proportion of oxygen-carrying hemoglobin molecules therefore increases steeply with the concentration of oxygen in their immediate environment. This enables hemoglobin to deliver more oxygen to oxygen-poor tissue than its four myoglobin-like compo-

FIGURE 11.5 A model of hemoglobin at low resolution. The alpha chains are shown in light gray, the beta chains in darker gray, and the hemes in black. (Drawing by James Engleson, in M. F. Perutz, "The Hemoglobin Molecule," *Scientific American*, November 1964, p. 65. Reprinted with the permission of W. H. Freeman and Company)

nents could do if they were working separately. Again, hemoglobin has the remarkable property of releasing oxygen molecules more readily in an environment where the concentrations of carbon dioxide and hydrogen ions are high (as in working muscle tissue) and taking up oxygen molecules in an oxygen-rich environment (as in the tiny branching capillaries of the lungs). Myoglobin has no such property.

Using a technique known as X-ray crystallography, which played an important part in the discovery of DNA's double helical structure, scientists have been able to determine the three-dimensional structure of several kinds of myoglobin and hemoglobin. Myoglobin and the individual chains that make up hemoglobin turn out to have remarkably similar three-dimensional structures (Figure 11.6). The various chains are also similar in length: myoglobin contains 153 amino-acid residues; the alpha chain of hemoglobin contains 141; the beta and gamma chains, 146. The sequences of amino acids in the beta and gamma chains are quite similar, but the amino-acid sequences of human myoglobin and the alpha chain of human hemoglobin are *dissimilar* to each other and to the amino-acid sequence of the beta chain, despite the similarity of their three-dimensional structures. The amino-acid sequences of the 3 chains agree in only 24 of 141 positions. This strongly suggests that (1) the three-dimensional structure of a polypeptide chain—its biological meaning—is determined mainly by a small subset of its amino-acid

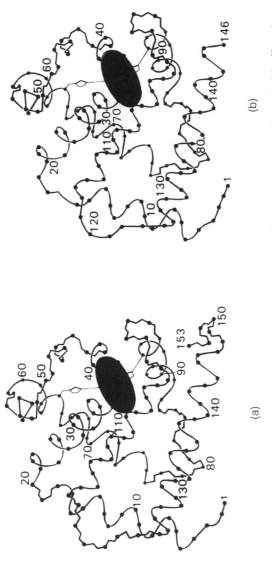

(a)

(b)

Figure 11.6 The polypeptide backbones of (a) myoglobin and (b) the beta chain of hemoglobin. (Drawing by Alex Semenoick, in M. F. Perutz, "The Hemoglobin Molecule," *Scientific American*, November 1964, p. 59. Reprinted with the permission of W. H. Freeman and Company)

residues, and (2) this subset is much more resistant to evolutionary change than the rest of the molecule.

These facts and inferences suggest the following evolutionary picture. In the ancestors of vertebrates, myoglobin played both the roles now played by myoglobin and hemoglobin. Or rather, the two roles were not yet differentiated. Duplication of the myoglobin gene made it possible for the two copies of the gene to evolve independently by substitutions, deletions, and insertions of single pairs of complementary nucleotides, and for their products to take on more specialized functions: hemoglobin to deliver oxygen to cells in every part of the body, myoglobin to dispense oxygen to muscle cells. One copy of the gene continued to code for myoglobin; the other copy eventually came to code for the alpha chain of hemoglobin. Later that copy was duplicated, and the second copy evolved into a gene coding for the beta chain. Finally a duplicate of the beta chain evolved into the gamma chain. (Two more, relatively rare, chains, designated delta and epsilon, have also evolved from duplicates of the beta chain.)

Another example of evolution by gene duplication and subsequent differentiation is provided by a set of digestive enzymes manufactured in the pancreas. The building blocks of proteins serve as fuel molecules. The function of digestive enzymes is to break proteins in food into their amino-acid building blocks by catalyzing reactions that sever the chemical bonds between adjacent amino acids in a polypeptide chain. Among these enzymes are *trypsin, chymotrypsin,* and *elastase,* all synthesized in the pancreas. Each of these enzymes bonds with only certain kinds of amino-acid residues and cleaves a polypeptide chain only at sites adjacent to these residues. The three enzymes cleave polypeptides at different kinds of sites, but acting together (in concert with other digestive enzymes), they cut up proteins into their constituent amino acids. Although trypsin, chymotrypsin, and elastase have different, nonoverlapping specificities, their amino-acid sequences and three-dimensional structures are very similar. Their differing specificities result from small structural differences between their binding sites (the parts of the enzyme that bind to the polypeptide during the reaction that cleaves the chain). These small structural differences result, in turn, from a small number of critical differences between the amino-acid sequences of the enzymes.

We may infer that the genes coding for the three enzymes evolved from a common precursor by gene duplication and subsequent mutations. The enzyme encoded by the precursor was presumably less specialized—and less efficient— than any of its descendants. As the three versions of the primordial gene diverged, natural selection promoted increased *but coordinated* specialization of their enzymatic activity. By narrowing the scope of its task, each enzyme was able to perform that task more efficiently; and the three enzymes were able to perform their joint task more efficiently. The three genes constitute a *battery,* in that the activities of the enzymes they code for are both complementary and coordinated. These enzymes are in fact activated by the same molecular mechanism and turned on by the same signal.

The tasks of enzymes descended from a common ancestor may also diverge. A key process in the clotting of blood is mediated by an enzyme called *thrombin.* Thrombin's specific task (to cleave polypeptide chains at a particular site) is very similar to that of the digestive enzyme trypsin, and thrombin itself closely resembles trypsin both in its three-dimensional structure and in its amino-acid sequence. The genes coding for the two enzymes have evidently diverged from copies of a common precursor. But the activities of the two enzymes are not coordinated because the biological processes they mediate—digestion and blood clotting—are not closely related.

Evolution as Hierarchic Construction

These examples show that gene duplication followed by mutations can have a variety of outcomes. Gene duplication can give rise to a single gene coding for a polypeptide chain twice as long as the original one, as in the example of ferredoxin; to genes coding for individual chains in a protein consisting of several chains, as in the evolution of hemoglobin; to genes coding for enzymes that fulfill complementary, specialized functions, such as the digestive enzymes chymotrypsin, elastase, and trypsin; and to genes coding for enzymes that fulfill biologically unrelated functions, such as trypsin and thrombin. In every case, new structures and functions arise through combined and simultaneous processes of *differentiation* and *integration*. Duplication of a gene G coding for a protein P that performs a function F yields two copies of G, whose products P' and P'' *jointly* perform the function F, an improved version of it, or two separate functions.

Evolutionary changes in a given structure or function often require, or open the way for, coordinated changes in related structures and functions. Consider, for example, the evolution of the three small bones of the mammalian middle ear, which transmit sound waves from the eardrum to the oval window. The middle ear of reptiles has only one bone. The two additional bones in the middle ear of mammals evolved from two bones that in reptiles form a hinge between the upper and lower jaws. The fossil record shows that these two bones became redundant when reptiles developed a new jaw articulation. Their subsequent integration into the mammalian three-bone structure made possible, and was driven by, improvements in the hearing of early mammals. These evolutionary changes in turn stimulated a variety of structural and behavioral changes that allowed mammals to exploit the increased flow of auditory information about their environment.

This example illustrates another important evolutionary generalization, emphasized by Ernst Mayr: the evolution of a new structure is usually initiated by a change in function of an existing structure whose original function is duplicated by another structure. To cite another example, both the mammalian lung and the swim bladder in fish evolved from a simple baglike lung in a primitive fish that was also equipped with gills.

These examples illustrate a hypothesis put forward in the 1940s by the evolutionary biologist Ivan Ivanovich Schmalhausen. Studies of comparative anatomy and comparative embryology led Schmalhausen to recognize that biological organization is a *functional hierarchy* and to hypothesize that biological evolution is a process of *hierarchic construction* in which new functional units result from the

TABLE 11.1   Levels of Biological Organization

Levels	Examples
Molecular building blocks	Amino acids Nucleotides Sugars Fatty acids
Macromolecules	Proteins DNA, RNA Starch Lipids
Macromolecular complexes	Ribosomes Enzyme complexes Membranes
Bacterial cells, cellular organs (organelles)	Cell nuclei Mitochondria Chloroplasts
Higher cells	
Tissues	Xylem, phloem Blood Smooth muscle Bone
Organs	Stem Taproot Kidney Heart
Organ systems	Root system Circulatory system Digestive system Nervous system
Higher organisms	
Communities	

differentiation and integration of existing functional units. Differentiation and integration are closely linked.

> Progressive differentiation is always accompanied by integration. New differentiations are integrated by the system of internal [regulatory] factors of individual development.[6]

The outcome of hierarchic construction is hierarchical organization. The major levels of biological organization are indicated in Table 11.1. The structural, functional, and evolutionary aspects of this hierarchy are all closely related.

Structurally, the hierarchy is based on aggregation. *Systems on a given level are aggregates of systems on the preceding level.* Thus macromolecules are aggregates of molecular building blocks; macromolecular complexes are aggregates of macromolecules; and so on.

This structural hierarchy is also a *temporal* hierarchy: it evolved level by level, each level supplying structures and functions needed for the construction of the next higher level. At least, the available evidence is consistent with this hypothesis. We don't know for sure that the first self-replicating objects were molecules of RNA, and it is pure conjecture that self-replicating complexes of macromolecules preceded the first cells. We do know, however, that bacterial cells preceded higher cells, and most biologists agree that some of the parts of higher cells—the nucleus, mitochondria (where energy stored in fuel molecules is converted into energy stored in ATP), and chloroplasts (where sunlight is converted into chemical energy)—are descendants of bacterial cells. Subsequent stages in the evolution of the hierarchy are very well documented. Multicellular organisms evolved from colonies of identical cells. Tissues gradually differentiated and were integrated into organs, which, in turn, became progressively more highly differentiated and more highly integrated into organ systems. In plants, the differentiation of organs hasn't gone as far as it has in animals. Some biologists argue that higher plants have only two organs, the root and the shoot. Others regard the shoot as an organ system, containing stem, leaves, and reproductive equipment as separate organs.

The units of the biological hierarchy are primarily units of function and only secondarily units of structure. For example, in eukaryotes (organisms with nucleated cells) proteins aren't encoded by single, continuous segments of DNA but by two or more segments separated by "meaningless" intervening sequences of DNA bases. Also encoded in the DNA are signals that cause these intervening sequences to be excised when the DNA is transcribed into messenger RNA. Such split genes are functional but not structural units.

Higher levels of organization also furnish examples of functional unity without structural unity. The immune system is no less a functional unit than the nervous system, although less integrated structurally. And a community of bees or of human beings remains a community when its members are physically separated.

Hierarchical organization is encoded in DNA. Consider the hemoglobin molecule. As we saw earlier, it contains four folded polypeptide chains, two of one kind (designated alpha) and two of a slightly different kind (beta). Both kinds of chains have evolved from the single-stranded molecule myoglobin, whose folded shape closely resembles that of the alpha and beta chains. The three-dimensional shape of a polypeptide chain is determined by its sequence of amino-acid residues, which in turn is determined by the linear sequence of DNA triplets in the gene coding for that polypeptide. How does the sequence of amino-acid residues determine the shape of the folded polypeptide?

One important factor is the positions in the sequence of water-repelling amino-acid residues. Like oil droplets, and for the same physical reason, these residues tend to cluster together in a tight ball, squeezing out water molecules and leaving the other residues, which mix readily with water, on the outside. The amino-acid residues also interact with one another. The formation of relatively weak chemical bonds between distant residues helps to shape the folded polypeptide the way a dressmaker shapes a piece of fabric by gathering and pinning it. These and other factors that determine the shape of a folded polypeptide chain in its natural cellular environment are all specified by the sequence of amino-acid residues.

Similarly, the way in which the four folded strands of hemoglobin fit together to form the complete molecule is determined by interactions between amino-acid residues in different chains. If a solution of hemoglobin is gently heated, the four strands of each molecule come apart and unfold. When the solution cools to its normal temperature, the molecules spontaneously reassemble themselves. Thus both the conformation of the molecule and the linkages between its four components are encoded in the genes that specify the amino-acid sequences of the two kinds of chains.

We saw earlier that the alpha chain, the beta chain, the gamma chain of fetal hemoglobin, and their common ancestor, the myoglobin molecule, all fold into very similar shapes. Yet their amino-acid sequences appear at first sight to be quite dissimilar. Careful comparisons show that *the shape of the folded polypeptide is sensitive to certain features of its amino-acid sequence but insensitive to others, and the features of the sequence that matter most for the shape of the folded polypeptide are indeed very similar in all the chains.* For example, the water-repelling amino-acid residues, which tend to cohere in a tight central ball, occur in nearly the same places in all the chains. Similarly, the way in which two alpha and two beta chains fit together to form a hemoglobin molecule depends sensitively on specific features of the amino-acid sequences of the two kinds of chains, *and these are not the same features that are important for the shapes of the individual folded chains.* Otherwise, the folded hemoglobin chains couldn't have the same shape as myoglobin.

The structural features that myoglobin and the individual hemoglobin chains have in common—their shape and the structure of the heme group—are critical for myoglobin's oxygen-carrying function. During the evolution of the alpha, beta, and gamma chains of hemoglobin, these structural features and the corresponding gene text were preserved, allowing the mutant chains to continue performing the basic function of their myoglobin-like precursor. Eventually, some of the mutant chains acquired the rudiments of a new function. By forming weakly bound complexes of four chains, they were able to function slightly more efficiently than the four chains acting independently. Natural selection gradually improved this rudimentary cooperative function.

Recognizing that biological organisms and their genetic programs are hierarchically organized is a first step toward resolving a classic difficulty in the theory of evolution. Imagine a jigsaw puzzle with 10,000 pieces. Could we improve the picture by randomly varying the shapes of individual pieces? Clearly, this would be a hopeless project, because changing the shape of a single piece would require coordinated changes in the remaining 9,999. Analogously, random genetic variation acting on a complex genome would hardly ever produce suitable candidates for selection. Critics of Darwinism have been making this point for well over a century.

The hierarchical organization of genetic programs makes it possible to change a few pieces of the genetic text, or even to add new pieces, without disrupting its tightly integrated structure. The evolution of higher cells from bacteria left intact basic molecular machinery for storing, replicating, transcribing, and translating genetic information, and much else besides. The subsequent evolution of many-celled plants and animals left intact the basic design of their cells. And we our-selves have inherited nearly the whole of our genetic wealth from our prehuman ancestors.

The evolution of the giant panda illustrates how a small number of genetic changes can bring about a major evolutionary change. For many years, taxono-mists debated whether the giant panda is more closely related to bears or to rac-coons. In 1964, D. D. Davis published an exhaustive study of this animal and its evolutionary provenance.[7] He concluded that, in the words of Steven Stanley, "the giant panda is essentially an aberrant bear specialized for feeding upon coarse vegetable matter (primarily bamboo)"[8] and that the divergence of giant pandas from their ursine ancestors required only a handful of genetic changes—perhaps five or six. These changes were focused on high hierarchical levels in the animal's developmental program, such as relative rates of development. Organs connected with chewing hypertrophied at the expense of the postcranial anatomy; the animal acquired a large head and a weak back.

As we have seen, hierarchical organization is an outcome of hierarchic con-struction, which in turn is an evolutionary strategy that must itself have been shaped by natural selection. We have already discussed several components of this strategy: duplication and rearrangement of whole words and phrases of genetic text (segments of DNA); and insertions, deletions, and substitutions of single characters (nucleotides). Do these components add up to a strategy of hierarchic construction?

Imagine trying to construct meaningful text with the help of a device that randomly duplicates words and phrases and randomly alters single characters. Suppose we begin with a short but meaningful sentence. We make a copy of this sentence and allow our device to perform a single random operation on it. If a meaningful piece of text results, we copy it and feed the copy into our device. If not, we try again with another copy of the original sentence, and keep trying until we are satisfied with the result. Proceeding in this way, we are bound to produce, eventually, a lengthy piece of meaningful text. But as the length and quality of the piece increase, the probability that a random change will improve the text decreases. Hence the average number of trials needed to produce a slightly longer or better piece increases as the text grows. Indeed, the waiting time between successive additions to the text increases exponentially with the length of the text. Analogously, if genetic variation were a strictly random process, the pace of evo-lutionary change would diminish rapidly with increasing genetic complexity, and evolutionary stagnation would eventually result. What does the evidence show?

Evolutionary Rhythms and Tempos

Relevant evidence comes from three main sources: the fossil record, genetics, and studies of biochemical evolution. The fossil record contains the most direct evi-dence about the history of life. Geologists have developed increasingly reliable methods for dating the geologic strata and individual rocks in which the fossils of plants, animals, and bacterial colonies are found, and paleontologists have become increasingly adept at reconstructing from fragmentary petrified remains not only the anatomy but also the behavior of ancient animals. Geneticists study the biological processes and molecular mechanisms underlying heredity and genetic variation. Their studies have taught us how genetic material is transformed in the course of evolution, and have yielded estimates of mutation rates in bacteria, fungi, plants, and animals. Finally, comparisons of the amino-acid sequences of a given protein in different species yield information about evolution that, in a curious and interesting way, complements the information provided by the fossil record.

The evidence shows clearly and unmistakably that *genetic variation is not a random process*. As evolution has given rise to increasingly complex organisms, it has speeded up rather than (as the hypothesis of random variation predicts) slowed down. The earliest steps—the evolution of higher cells from bacteria and of multicellular organisms from unicellular organisms—were also much the slowest. Among vertebrates, more recent and more complex groups have evolved faster than older and less complex groups. Compare frogs with placental mammals:

> Although there are thousands of frog species living today, they are so uniform phenotypically [that is, anatomically and behaviorally] that zoologists put them all in a single order (Anura), whereas placental mammals are divided into at least 16 orders. The anatomical diversity represented by bats, whales, cats, and people is unparalleled among frogs, but frogs are a much older group than placental mammals. The present-day frogs are not easily distinguishable morphologically from those living 90 million years ago. In contrast, during this same period, mammals have become extremely different in morphology from their progenitors.[9]

Darwin believed that major evolutionary changes resulted from the accumulation of many small, individually advantageous changes, and must therefore have occurred slowly and gradually. In fact, the fossil record shows that major evolutionary changes usually occur rapidly during relatively brief periods. Human evolution is a case in point. Our earliest tool-making ancestors appeared only 2 million years ago. Their brain cases had about half the capacity of ours, and they hadn't yet learned the uses of fire. During the next 1.5 million years, species that built fires and had brain cases nearly as large as our own evolved. About 500,000 years ago, Neanderthals appeared. There is some controversy about how closely they resembled modern humans. According to paleoanthropologist David Pilbeam, modern fossil evidence suggests that they were

> less cultural, less symbol-nimble, more biology-bound, than us. Everything suggests a system in which communication was less effective, and the amount of stored and transmitted information less.[10]

According to Pilbeam, recognizably modern (human) behavioral patterns of many kinds—cave painting and sculpture, sophisticated tools and elaborately planned cooperative hunting, increased population size and broadened geographical and ecological range—all emerged during a period that lasted only about 100,000 years

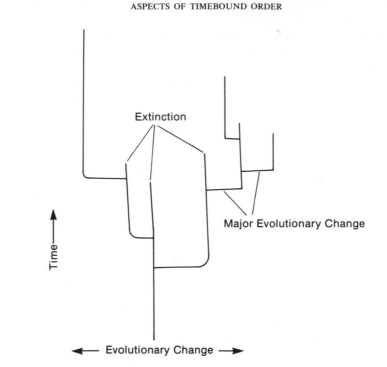

FIGURE 11.7 Schematic representation of the pattern of macroevolutionary change, as viewed by most students of the fossil record. Major changes usually take place during relatively short time intervals and are followed by long intervals of very slow change. The vast majority of evolutionary lines become extinct.

and that ended as recently as 35,000 years ago. Pilbeam and others believe that language was the major cause of this "last big jump in human evolution."

The temporal pattern of evolution, as it is now conceived by paleontologists, is illustrated schematically in Figure 11.7. In this diagram, time runs vertically, and evolutionary change is represented by horizontal displacements. New lineages branch off almost horizontally and then rise almost vertically. Thus the diagram represents the differentiation of evolutionary lineages (speciation) as taking place during a short fraction of the available time and being followed by much longer periods of slow change.

Some of the evidence on which this picture rests was available in Darwin's day. If new species diverged slowly and gradually from their parent species, as Darwin believed, we would expect to find fossils representing intermediate stages in the evolution of a species, from its ancestral to its "final" form.

> Why then [Darwin asks in *The Origin*] is not every geological formation and every stratum full of such intermediate links? Geology assuredly does not reveal any such finely-graduated organic chain; and this, perhaps, is the most obvious and serious objection which can be urged against the theory. The explanation lies, as I believe, in the extreme imperfection of the geological record.[11]

In this conclusion Darwin was mistaken. The fossil record is certainly imperfect, although less so than in Darwin's day. It does, however, provide a basis for secure statistical inferences about rates of evolution. The statistical approach to the interpretation of the fossil record was pioneered by George Gaylord Simpson, whose *Tempo and Mode in Evolution*, published in 1944, showed that evolution is not, as Darwin had imagined, a process of gradual improvement occurring usually and in most places at a steady, uniform rate. On the contrary, evolution usually alternates between two extreme tempos, *adagio* and *presto*.

There is plenty of direct evidence for the slow tempo. Well-established species usually exhibit little or no evolutionary change apart from, in some instances, a slow increase in overall size. Among modern animals that have undergone little or no discernible evolutionary change for long periods are notostracan crustaceans (the record holders, at 305 million years), horseshoe crabs (230 million years), sturgeons and garfishes (80 million years), snapping turtles (57 million years), and alligators and New World porcupines (35 million years).[12] Darwin referred to such animals as living fossils. He regarded their failure to evolve as exceptional, but Niles Eldredge, Steven Jay Gould, Steven Stanley, and other contemporary paleontologists have argued persuasively that sluggish evolution is the rule rather than the exception for well-established species. As illustrated in Figure 11.7, revolutionary change seems to be concentrated in relatively short bursts that accompany the branching off of new species.

Why do intermediate forms appear so rarely in the fossil record? The rapidity with which speciation occurs is part of the reason, but it can't be the whole reason. The most widely accepted (and, in my opinion, correct) explanation was put forward in 1954 by Ernst Mayr. Mayr argued that new species are most likely to evolve from small peripheral groups of organisms that become isolated from their parent population. For example, a pair of birds may colonize an island previously unvisited by members of their species, or a small group of fish may be cut off from its parent population when the body of water it lives in becomes a separate lake. More generally, as we discussed earlier, every population tends to expand and fill the region in which it can survive and multiply. The edges of this region will, by definition, be marginally inhabitable, and the organisms that live there will be only marginally adapted to their environment. They will therefore experience much stronger selection pressures for evolutionary change than organisms living in environments to which they are well adapted. At the same time, groups living at the periphery of their species' range are most likely to become geographically isolated. Such groups are also likely to have unrepresentative gene pools, just because they are small. For these reasons, Mayr has argued, evolutionary novelty is most likely to appear in small peripherally isolated groups. According to this picture, new species form by budding off established species. It is because the growth of the bud is both rapid and, at the outset, highly localized that fossil evidence for the process is so hard to come by.

In *Populations, Species, and Evolution*, Ernst Mayr summarizes his views on macroevolution in the following terms:

> The evolutionary significance of species is now quite clear. Although the evolutionist may speak of broad phenomena, such as trends, adaptations, specializa-

tions, and regressions, they are really not separable from the progression of the entities that display these trends, the species. The species are the real units of evolution, as the temporary incarnation of harmonious, well-integrated gene complexes. And speciation, the production of new gene complexes capable of ecological shifts, is the method by which evolution advances. Without speciation there would be no diversification of the organic world, no adaptive radiation, and very little evolutionary progress. The species then is the keystone of evolution.[13]

Macroevolution, then, proceeds by speciation. And speciation, as we have seen, is most likely to occur in small, isolated, peripheral colonies. Often it is spearheaded by the evolutionary invention of a single new structure, function, or behavior. A frequently cited example is the feather, whose evolution from the reptilian scale enabled a colony of small dinosaurs, the progenitors of modern birds, to establish the first gliding and flying club. The new structure or function or behavior may appear at any level of the biological hierarchy, from molecules to communities. It may provide access to a more abundant food supply, afford increased protection from predators, or enable its possessors to have more surviving progeny.

This story has been fleshed out in great detail and with a multitude of examples by Mayr and other evolutionary biologists. As an account of macroevolution at the level of visible structures and functions, I find it entirely convincing. To armchair skeptics' questions of the form "How could X possibly have evolved from Y?" students of evolution have now been able to supply answers of the form "Here's how it happened, and here is the evidence to support this story." The evidence hasn't been easy to come by and important details remain to be filled in, but the account of macroevolution just sketched fits a very wide array of diverse observations and has stood up well under theoretical analysis. It has the ring of truth.

But solutions to important scientific problems often raise new problems at deeper levels. The invention of feathers entrained a host of related genetic modifications, including, for example, those responsible for the distinctive features of avian bones and the avian skeleton. The harmoniously integrated genome of birds' saurian ancestors had to be profoundly reorganized, and it had to maintain its integrity throughout the process. How did the Darwinian process of blind genetic variation and natural selection accomplish this feat? Classical population genetics—the mathematical study of evolutionary processes at the level of genes and chromosomes—has been notoriously unsuccessful in its attempts to explain macroevolutionary processes of this kind. Later in this chapter, I will analyze the difficulty and propose a solution to it, but first we should look at molecular evidence concerning rates of evolution.

Evolutionary Rates: Molecular Evidence

The development of methods for rapidly determining the amino-acid sequences of proteins made it possible to compare the sequences of a given molecule like hemoglobin or cytochrome *c* in different species. Such comparisons showed that the amino-acid sequences of a given protein in closely related species, such as dogs and foxes or humans and chimpanzees, differ much less than they do in distantly related species, such as yeast and humans. Biochemists immediately set out to discover the temporal pattern of protein evolution. Does it mimic the temporal pattern of organismic evolution? Do proteins evolve in spurts, and are they evolving faster today than they did 1 billion years ago?

Apparently not. For example, the gene that codes for hemoglobin has accumulated mutations (substitutions of one nucleotide for another) at about the same rate in frogs as in mammals, although mammals have evolved far more rapidly in body structure and function. There are many other similar examples. Human evolution, as we noted, has been unusually rapid; but human proteins have evolved no faster than the corresponding proteins of slowly evolving species. And the cytochromes of living fossils such as the cyanobacteria, or blue-green algae (whose rate of organismic evolution is 1 billion times slower than that of higher animals), seem to have accumulated sequence changes at much the same rate as those of rapidly evolving species. Each of the many proteins for which comparative studies have been made accumulates sequence changes at a rate that fluctuates over short periods of time (a few million years) but is remarkable constant over very long periods.

Protein-sequence evolution has another striking and important property. Although each sequence evolves at a fixed average rate, different sequences evolve at different average rates. For example, the rate at which myoglobin accumulates mutations is two and a half times as great as the rate for cytochrome *c*, but only one-sixth the rate for some of the polypeptide chains of antibody molecules. In some antibody chains, the amino-acid sequence changes by 1 percent in less than 1 million years. By contrast, the amino-acid sequences of proteins in the core around which the DNA in higher cells is wound take several hundred million years to change by the same fraction.

What do these and our earlier conclusions about evolutionary rates tell us about genetic variation? We have seen that some evolutionary changes in a protein's amino-acid sequence significantly affect an organism's ability to survive and reproduce, while others affect its biological fitness only slightly, if at all. Evolutionary biologists disagree about the relative importance of these two classes of mutations. One school postulates that most of the mutations that have been fixed in evolution are neutral; that is, they don't significantly influence the organism's ability to survive and reproduce. The opposing school disputes this assumption. Using a modern version of an argument that goes back to Darwin, they point out that apparently trivial changes in the structure of a protein can affect biological fitness in subtle ways. For example, the molecular biologist Max Perutz has shown, in the words of Richard Lewontin, that

> many different hemoglobins in different species of vertebrates have amino acid changes in certain positions in the molecules to tune them precisely for different environmental circumstances. There is an intricate interactive relationship between the oxygen pressure, the carbon dioxide concentration in the blood, and

the ability of the hemoglobin molecule to pick up and release oxygen. This inter-active relationship is very sensitive to the particular amino acid composition of hemoglobin and can be changed by amino acid substitutions. Thus, deep sea mammals have a hemoglobin that is different in precisely the right way from the hemoglobin of high altitude mammals so that oxygenation is efficient at the two very different pressures that these different organisms experience.[14]

At first sight, it may seem strange that a neutral mutation could ever become fixed in the course of evolution. Why should a mutation appear in all, or nearly all, members of a population if it doesn't contribute to its carriers' reproductive success?

The answer was given by the population geneticist Sewall Wright in the 1930s, long before the phenomenon of protein evolution was discovered. Imagine that a single gene for blue eyes suddenly appears in a population of brown-eyed people, and assume that possession of that gene in no way affects a person's reproductive success. Will it spread? Or will it die out? The chances are that it will die out, but there is a slim chance that all members of the population will eventually carry the mutant gene—that the mutant gene will become *fixed*. Wright showed by a mathematical argument that the average rate at which a neutral mutant becomes fixed doesn't depend on the size of the population, but only on the mutation rate. In fact, the average rate at which a neutral mutant becomes fixed is equal to the mutation rate. For example, if the chance that an egg or a sperm cell undergoes a particular mutation is 1 in 100,000, that mutation is likely to become fixed after 100,000 generations.

So much for neutral mutations. Let's now consider mutations that significantly affect an organism's prospects of contributing to its population's gene pool. Dar-win and Wallace made natural selection the shaping factor of evolution. De Vries and other geneticists of the early twentieth century, rebelling against Darwinism, belittled natural selection and argued that evolution is really driven by mutations. Eventually, as we have seen, genetics and Darwinism came together and produced what Julian Huxley called the Modern Synthesis, which restored natural selection to its former preeminent place as the shaper of evolutionary adaptations and as-signed to mutations the passive and subsidiary role of the supplier of raw mate-rials. Since the 1940s, mainstream evolutionary biologists have strongly resisted suggestions that mutations might, after all, play an active role in evolution.

It is true that the gene pools of natural populations of sexually reproducing organisms always contain plenty of variability on which natural selection can act to fine-tune structures and functions. But the deeper problem is one of quality rather than quantity. Consider the bat. The oldest bat skeleton is about 50 million years old. George Gaylord Simpson describes it in this way:

This skeleton does have some features more primitive than those of later bats, pointing back to ancestry in nonflying, ecologically shrewlike earlier mammals. Nevertheless, it was already fully batlike in essentials shared with all later and recent bats. Its anatomical adaptations to flying were complete and were *sui ge-neris* for bats, radically unlike those of either flying reptiles (the extinct ptero-

saurs) or flying birds. The subsequent evolution of bats involved great prolifera-
tion of species, genera, and families.[15]

If Mayr's account of speciation is basically correct, as I believe it is, bats evolved
from a small founder population of shrewlike mammals—a population with a correspondingly small gene pool. Did that gene pool already contain the raw material
from which natural selection fashioned bats' distinctive anatomical adaptations to
flight? Could it have contained, in addition, enough variety to produce 1,000
distinct species? Presumably not. But how could purely random mutations have
given rise to just the right kinds of genetic variation during the period when bats
were evolving and diversifying?

Human evolution raises analogous questions (although our genus has only one
surviving species). Were the genetic raw materials needed to construct our capacity for speech and reflection already present, waiting to be assembled by natural
selection, in the ancestral population from which apes and humans diverged 4 to
8 million years ago? Obviously not. But it seems as hard to believe that just the
right kinds of genetic variation could have been produced in such a short time by
purely random processes.

Critics of contemporary population genetics argue that it accounts splendidly
for *microevolution*—gradual shifts in the relative frequencies of genes in a population's gene pool—but fails utterly to account for *macroevolution*—the emergence of true evolutionary novelties. I believe they are right. Certainly, speciation
isn't driven by "mutation pressure," as the early-twentieth-century geneticists
maintained. But neither, I believe, does mutation play a purely passive role, as
mainstream population geneticists currently maintain. Instead, I suggest, *speciation is the outcome of a special kind of interaction between mutation and natural
selection.* Of course, this interaction must be consistent with the basic tenet of
current evolutionary theory—that genetic variation is blind to its consequences for
the organism—and we will see that it is.

Alpha and Beta Genes

Before explaining my hypothesis, I have to define some terms. By *gene,* I will
mean any segment of DNA that has an identified function. The genes that figure
in classical genetics are *structural genes*. They encode whole proteins, individual
polypeptide chains in a protein, and, in some cases, "functional domains," or
parts of polypeptide chains. The transcription of structural genes into RNA is
regulated by nearby segments of DNA that perform various functions. These segments of DNA have descriptive names, such as operator, promoter, initiator, en-
hancer, attenuator; collectively, they are known as *regulatory genes.*

Structural and regulatory genes together specify an organism's developmental
program. I will call these genes, which are expressed directly in development,
alpha genes. There are also genes that aren't part of the developmental program,
and I will call them *beta genes*. Beta genes direct the "management" of genetic
material. Some of them encode the elaborate chemical machinery involved in the

replication, proofreading, and repair of DNA. Others regulate mutation rates and rates of genetic recombination.

My thesis is that *natural selection has shaped systems of beta genes that make evolution a more efficient process by focusing and modulating genetic variation, suppressing variations that are likely to diminish fitness, and permitting or promoting variations that are likely to increase fitness.* I will argue that the system of beta genes encodes a *strategy for evolution,* one aspect of which is hierarchic construction (new functional units arise in the course of evolution through differentiation and integration of existing functional units). We will see that this hypothesis accounts for the most conspicuous temporal properties of the evolutionary process. It explains why major evolutionary changes occur rapidly and why increasing genetic complexity has not resulted in evolutionary stagnation. It also explains why the amino-acid sequences of proteins evolve at different but roughly uniform rates, and why these rates are related to rates of visible change in structure and function. Before discussing the implications of the hypothesis, however, we should examine its scientific credentials. Is it consistent with the basic tenets of evolutionary theory? What evidence is there that regulatory mechanisms of the kind envisaged by the hypothesis actually exist?

Let's begin with the second question. As we discussed earlier, observed rates of protein-sequence evolution span a wide range. We also saw that the rates of protein-sequence evolution seem to be unconnected with the rates of conspicuous evolutionary changes such as those involved in speciation. For example, mammals have evolved far more rapidly than frogs, but the "protein clocks" of mammals don't go faster than those of frogs. We saw that these findings are consistent with the (still controversial) assumption that protein-sequence evolution is dominated by mutations that don't significantly affect biological fitness. We also saw that the three-dimensional shapes and chemical functions of some proteins have not evolved noticeably since these proteins first appeared. The large and systematic differences between rates of genetic change revealed by these and other findings suggest that mutation rates are themselves under genetic control and that they have been strongly influenced by natural selection.

How is genetic control of genetic variation exercised? In 1977, when I formulated the beta-gene hypothesis,[16] only a few hints were available. One of these came from discoveries that had been made in the 1950s by the geneticist Barbara McClintock, working with corn. McClintock found that certain mutations in corn are caused by a gene, which she named *Dissociator,* that inactivates genes on either side of it—but only if another gene, *Activator,* is present somewhere else, anywhere else, in the genome. If *Activator* isn't present, *Dissociator* has no effect. McClintock discovered that both genes have another remarkable property, never previously observed in a gene: mobility. Ordinary genes have fixed positions. Geneticists have devised techniques for mapping genomes, and such maps don't change. *Activator* and *Dissociator,* however, do change their positions after a few generations.

Another mobile gene in the corn genome, *Dotted,* regulates the mutability of a particular structural gene. The rate at which the gene mutates is proportional to the number of copies of *Dotted* present in the genome.

Geneticists were as shocked by McClintock's discovery of mobile genes as astronomers would have been by an announcement that stars hop from constellation to constellation. Most geneticists dismissed McClintock's "jumping genes" as aberrations peculiar to corn. During the following two decades, this assessment underwent a dramatic revision. New techniques in molecular biology led to the discovery of many new varieties of mobile genes, first in bacteria, then in yeast and in flies. *Dissociator* and *Activator* proved to be not isolated curiosities but the tip of a submerged continent. A recent textbook introduces a discussion of mobile genes in fruit flies with these words:

> It is now estimated that most spontaneous mutations and chromosomal rearrangements in *Drosophila* are caused by transposable elements [mobile genes]. As much as 10 percent of the *Drosophila* chromosome may be composed of families of dispersed, repetitive DNA sequences, which move as discrete elements![17]

Mobile genes play a central role in the "immune system" of bacteria. The introduction of antibiotics in the 1940s created a new environmental challenge for disease-causing bacteria. In responding to the challenge, they revealed unsuspected resources for rapid evolutionary change. In 1955, during an outbreak of bacterial dysentery in Japan, one bacterial strain proved to be resistant to four common antibiotics.

The genes that confer resistance to these antibiotics are carried on a small auxiliary chromosome whose structure is illustrated in Figure 11.8. (A bacterium's normal complement of genetic material is contained in a single large loop of DNA, but one or several auxiliary chromosomes in the form of small loops called *plasmids* may also be present.) The resistance-conferring plasmid has a modular structure, consisting of a transfer module and several resistance-conferring modules, each of which imparts immunity to a single antibiotic. Each module is a mobile gene, or *transposon,* consisting of a segment of DNA flanked by two identical *insertion sequences,* so called because they can be inserted at any point in a genome (Figure 11.9). A transposon may be copied into another genome or into another place in the same genome. Resistance-conferring plasmids are assembled, module by module, from their mobile components.

The transfer module directs a complex sequence of processes that cause one strand of the plasmid to be transferred to another bacterium, where a complementary strand is synthesized (Figure 11.10). Thus the second bacterium acquires a copy of the plasmid. Meanwhile, as illustrated in Figure 11.10, the strand left behind also directs the synthesis of a complementary strand. Remarkably, the second bacterium may belong to a different species from the first. Resistance to antibiotics is thus an infectious property. The genes that confer this property are indeed similar in other important ways to certain viruses that infect bacteria.

The "immune system" of bacteria has been shaped by their strategy for genetic variation, which includes the ability to incorporate specific genes—in this case, genes that confer resistance to specific drugs—into larger, mobile modules (transposons) that can in turn incorporate themselves into segments of DNA that can cause copies of themselves to be inserted into the genomes of other bacteria.

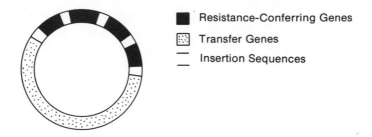

- ■ Resistance-Conferring Genes
- ▦ Transfer Genes
- — Insertion Sequences

FIGURE 11.8 A bacterium's immune system. Genes conferring resistance to specific drugs are combined in a small auxiliary chromosome (plasmid) with genes that enable this genetic information to be transferred to another bacterium.

Bacteria use the same strategy to exchange other kinds of genes as well. As we have seen, genetic recombination tends to enhance the reproductive instability of genetic material by exposing variant genes to selection in fresh combinations, thereby making it possible for the fittest genes eventually to come together. Hence natural selection would have favored the evolution of mechanisms that promote genetic recombination.

At first sight, it seems miraculous that the resistance-conferring plasmids of bacteria that have managed to survive in an environment poisoned by antibiotics should contain just those genes that enable their carriers to block or metabolize just these antibiotics. We know that a new antibiotic may be able to kill nearly all

FIGURE 11.9 A segment of DNA before and after the insertion of a transposon. The *target sequence* (black box) is a half-dozen or so bases long. It is duplicated during the process illustrated by this diagram. When the process is completed, the transposon is flanked by copies of the target sequence. The transposon's functional genes are themselves flanked by identical copies of an *insertion sequence,* several hundred bases long. The ends of the insertion sequence (arrows) approximate inverted copies of each other. The base sequences in the regions marked by oppositely directed arrows are approximate mirror images: abc . . . cba.

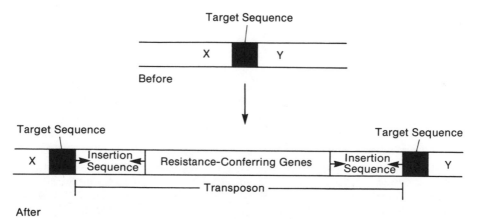

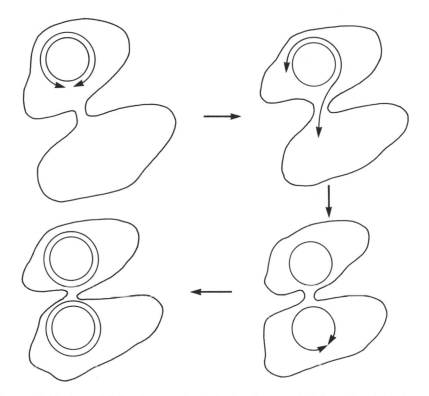

FIGURE 11.10 A model for the way bacteria transfer genetic information. Male bacteria contain appendages called *sex pili;* female bacteria contain receptors for these appendages. While the male and female cells are joined by a pilus, one strand of a plasmid (or of the main bacterial chromosome) in the male cell gets cut and unwinds. This strand makes its way into the female cell through the pilus. The strand left behind acts as a template for the formation of a replica of the departed strand, which in turn acts as a template for the formation of a complementary strand in the female cell. At the end of the process, each cell has a complete double-stranded copy of the plasmid or chromosome.

the bacteria exposed to it for the first time. We also know that a few bacteria in a culture may carry a gene that confers resistance to an antibiotic to which the culture hasn't been previously exposed. In light of these facts, it seems safe to assume that before the population of dysentery-causing bacteria was exposed to antibiotics, it already contained a few members that were resistant to *individual* antibiotics. Once we make this assumption, it is easy to understand how plasmids containing several resistance-conferring genes arose and spread. *But how did genes conferring resistance to four specific antibiotics happen to be present in the bacterial population to begin with? And how did these particular genes come to be mobilized?*

I suggest that the gene pool of a normal bacterial population may contain a vast library of different, readily mobilizable genes, a fraction of which confer resistance to the products of modern pharmacology. Each member of the population carries only a few of these genes, embedded in readily detachable (or already detached) transposons. Appropriate signals from the environment trigger the process by which a bacterium transfers its resistance-conferring plasmid to other bacteria.

This suggestion makes the "immune system" of a bacterial *population* analogous, in a crude way, to the immune system of an individual vertebrate. The vertebrate immune system contains genes that code for an immense variety of antibodies. Invasion of the organism by foreign cells triggers the rapid multiplication of a relatively small number of antibodies whose members recognize and bind to the invading cells. In both vertebrates and bacteria (if my suggestion is correct) the beta system both promotes and focuses genetic variation, and in both cases natural selection strongly favors the evolution of mechanisms that serve these functions.

In other circumstances natural selection favors the suppression of genetic variation. Consider a gene that has been "optimized" by natural selection, so that every possible mutation either diminishes the gene's fitness (the average reproductive success of its carriers) or leaves it unaltered. Let's focus attention on a short segment of DNA; it could be a single base pair. Suppose there is a beta gene that affects the mutation rate of this segment and that it exists in two versions, one of which promotes mutations and the other of which suppresses them. We have to consider two cases.

Case 1.   The DNA segment contains base pairs that can't be altered without reducing the gene's fitness. Organisms that carry the mutation-suppressing version of the beta gene then out-reproduce organisms that carry the mutation-promoting version. Eventually, the mutation-suppressing version of the gene must spread through the entire population. The beta-gene hypothesis therefore predicts that critical segments of "optimized" genes have relatively low mutation rates. And they do.

Case 2.   The gene's fitness is insensitive to mutations of the DNA segment. Then organisms that carry the mutation-promoting version of the beta gene have the same reproductive success in the long run as organisms that carry the mutation-suppressing version. Averaged over a large population, the mutation rates of base pairs in such segments should be greater than those of base pairs in segments that play a critical functional role and should remain roughly constant over a long period of time. These predictions are supported by the evidence on protein-sequence evolution that we discussed earlier.

Macroevolution Revisited

The preceding considerations support the hypothesis that natural selection has shaped genetic systems that regulate genetic variation by *focusing* and *modulating* it. Aspects of the developmental genetic program in which changes would almost certainly reduce fitness are protected from variation; functionally neutral variations are not suppressed; and variability whose potential benefits are great, as in the generation of a large and diverse library of antibodies or resistance-conferring genes, is actively promoted. This hypothesis enables us to understand two phenomena that are central to macroevolution: hierarchic construction and the enormous variability of evolutionary rates.

Hierarchic Construction

We argued earlier that the hierarchical character of biological organization is the result of an evolutionary strategy in which new functional units arise in the course of evolution through differentiation and integration of existing functional units. We also saw that random duplications and rearrangements of genetic material, together with insertions, deletions, and substitutions of single base pairs, do not add up to a strategy of hierarchic construction. There is a missing ingredient. If evolutionary stagnation is to be avoided, genetic variation must be focused on segments of DNA in which mutation and recombination will not disrupt existing functions. Regulated genetic variability supplies this ingredient, making possible an evolutionary strategy that is simultaneously conservative and innovative. The system of beta genes suppresses variability that threatens to disrupt existing functions and promotes variability that promises to accelerate evolutionary changes that increase fitness. Hierarchic construction is the necessary outcome of this evolutionary strategy, as the following analogy may help to make clear.

The "evolution" of a complex piece of machinery—a car, say—relies on a process of hierarchic construction in which regulated variability is crucial. Consider the wheel. Modern car wheels retain some features of the one-piece carriage wheels from which they "evolved." For instance, they are round; designers have never experimented with elliptical or octagonal wheels. Again, modern wheels, like their progenitors, have three functionally distinct components: rim, spokes, and hub. The introduction of solid-rubber tires (an example of evolutionary differentiation) brought about a modest improvement in "fitness" and didn't necessitate other changes, but it opened a new evolutionary pathway. "Microevolutionary" changes (improvements consistent with the basic design) exploited the new possibilities and paved the way for further "macroevolutionary" changes, such as the inner tube. At every stage experimentation was focused on aspects of the design susceptible to improvement. And at every stage changes had to be coordinated. Thus improvements in tire design sometimes required changes in rim design. "External" requirements—speed, load, durability—provided the highest level of hierarchic regulation.

The hierarchic character of automotive evolution is a direct and automatic consequence of the focusing and coordination of evolutionary change in ways that subserve "fitness" (the functioning of the car as a whole, as judged by the population of owners and drivers). Analogously, the hierarchic character of organic evolution is, I suggest, a direct and automatic consequence of the focusing and coordination of evolutionary change in ways that subserve fitness (reproductive success). And the genetic mechanisms that focus and regulate genetic variation themselves evolved through blind genetic variation and natural selection.

The conclusion that evolution proceeds by hierarchic construction explains the "tendency of organisms to adopt ever more complicated solutions to the problem of remaining alive." It is important to bear in mind, however, that this tendency (whose ultimate source is the reproductive instability of genetic material) is only rarely able to express itself. The fossil record suggests that most lineages evolve very little once they have become established. Usually they fail to meet the challenge of a rapidly changing environment, going extinct when hard times come. Major evolutionary innovations are conspicuous but exceedingly rare occurrences.

Evolutionary Rates

The hypothesis that genetic variation is focused and modulated enables us to understand the enormous variations in evolutionary rates inferred from the fossil record. Each stage in a process of hierarchic construction requires genetic changes at only a few genetic loci (as Davis showed in his study of the evolution of the giant panda), and the beta-gene system is able to focus genetic variation on these loci. Thus a small population can evolve rapidly once an evolutionary opportunity suited to its genetic makeup presents itself.

Major evolutionary innovations such as the acquisition of flight by bats open up a wide range of opportunities, whose exploitation opens up still more opportunities for the expansion of genetic material. Thus we may expect major innovations to be accompanied by branching and splitting of lineages in many directions.

Human evolution is a conspicuous exception to this rule. Whereas the acquisition of flight by the shrewlike ancestors of bats resulted in the formation of dozens of new species and genera, the acquisition of the capacity for language and culture resulted in only a single new species. The reason may be that this capacity, by its very nature, makes genetic specialization unnecessary. In humans, cultural diversification has taken over the evolutionary role normally played by speciation, accomplishing many of the same "ends" more rapidly, more efficiently, and without the price of irreversible genetic change that speciation exacts.

What is the relation between genetics and culture? To what extent are patterns of human behavior biologically constrained? To what extent are they culturally determined? What is left over for individual freedom? During the past two decades, these questions—all of which are aspects of a broader question: What is the nature of human nature?—have been debated by biologists, anthropologists, and philosophers. In trying to answer them, we will be guided by the picture of biological evolution sketched in this chapter and the previous one.

12

Language, Thought, and Perception

At this moment the King, who had been for some time busily writing in his
note-book, called out "Silence!" and read out from his book "Rule Forty-
two. *All persons more than a mile high to leave the court.*"
Everybody looked at Alice.
"I'm not a mile high," said Alice.
"You are," said the King.
"Nearly two miles high," added the Queen.
"Well, I shan't go, at any rate," said Alice: "besides, that's not a regular
rule: you invented it just now."
"It's the oldest rule in the book," said the King.
"Then it ought to be Number One," said Alice.

LEWIS CARROLL, *Alice's Adventures in Wonderland*

A scientific picture of the world would be incomplete without an account of our
own place in it. Modesty may compel us to admit that the human race represents
a tiny detail of the big picture. We are one species among millions on an undistinguished planet circling an undistinguished star that travels along an undistinguished orbit in an undistinguished galaxy. But we must acknowledge that our
place in the picture, however small, is not insignificant, because it is we who
make and remake the picture. And the strongest and most persistent reason for
doing so—for trying to understand what the Universe is really like—has always
been the desire to understand how we fit into it.

Almost 2,400 years ago, Socrates defended his belief in the self as a free agent
against the scientific reductionism of his day. In modern times Socrates's view
of human autonomy has been attacked by adherents of two opposing scientific
schools, behaviorism and biological preformationism.

Behaviorism

Founded by J. B. Watson in the early twentieth century, behaviorism found its most eloquent spokesman in Watson's disciple, the experimental psychologist B. F. Skinner. Skinner argued that the notion of an autonomous human being is an anachronism, a holdover from prescientific ways of talking about human behavior. In Skinner's view, attempts to relate human actions to internal states ("John ate the pork pie because he was hungry") are analogous to Aristotle's attempts to relate the motions of physical objects to their inner nature ("A falling body seeks to regain its natural place"). Behavior is determined by two factors: genetic endowment, which we can't do anything about; and interactions between the organism and its environment, which we can control to some extent. The proper task of a science of behavior, according to Skinner, is to discover rules connecting specific environmental factors with specific behaviors. This task is hindered rather than helped by references to hypothetical internal factors such as motives, character, personality, goals, emotions, desires, and will. A behaviorist would say that Socrates declined to flee Athens not because such an act would have betrayed his principles but because he had in the past been "positively reinforced" (rewarded) for law-abiding behavior.

Animal trainers have long exploited the plasticity of animal behavior and its capacity to be shaped by appropriate rewards and, to a lesser degree, punishments. Everyone knows that our own behavior is also shaped in part by our experiences, especially early in life. Skinner went further, arguing that we should first find out which techniques are most effective in shaping human behavior and then use these techniques to further desirable social goals. The latter suggestion outrages people who fail to perceive its utter lack of originality. Human societies have been trying to shape the behavior of their members for thousands of years. That is the main purpose of law, religion, and education. Skinner is merely suggesting that there may be more effective (and more humane) ways of accomplishing ends we have always pursued.

As a *theory* of human behavior, however, behaviorism is on shakier ground. Behaviorists claim that by not talking about what goes on in people's heads they have avoided introducing untestable hypotheses about the causes of behavior. In fact they haven't dispensed with hypotheses, only with stating them. Behaviorism's central hypothesis is that *learned* behavior (as distinct from instinctive, or genetically programmed, behavior) is shaped mainly by experience. More specifically, an animal's genetic endowment defines and limits the kinds of behavior it can learn, but within these boundaries the relation between the learning experience and what the animal learns is relatively direct and unproblematic. That is behaviorism's underlying premise. How well does it hold up?

Pretty well in most of the animal world, from single-celled organisms (which can learn to swim toward new food) to our cousins, the chimpanzee and the gorilla (which can even be taught rudimentary forms of verbal behavior). Many aspects of human behavior can also be shaped by appropriate "schedules of reinforcement." The premise doesn't hold, however, for behaviors that have a significant *creative* component. A conspicuous example is human language. One might imagine—and Skinner claimed—that children's use of language is shaped by their verbal experience. However, as Noam Chomsky pointed out in a famous review of Skinner's *Verbal Behavior* (1957), the rules that govern the utterances of young children can't plausibly be interpreted as inductive generalizations from the samples of language the children have been exposed to. There is more order in the outcome than in the experiences that contributed to it.

The same is true of other forms of creative behavior. The most obvious examples are works of art. An original piece of music contains, by definition, a kind of order that didn't exist before the piece came into being. An artistic performance of a piece of music contributes additional order. Imagine a piano equipped with computer controls that would enable it to reproduce precisely the notes, note values, tempo indications, and dynamic markings in a piano score. Such a device would reproduce everything in the score, but the results wouldn't be musically interesting. The musical interest of a performance is a form of order that a gifted performer adds to the order inherent in the score. The disparity between the order inherent in "experiential inputs" and the order exhibited by "behavioral outputs" is, I believe, the common factor and the defining property of creative behaviors. And because behaviorism in effect denies that such disparities exist, creative behavior lies forever outside its scope.

Genetics and Behavior

Behaviorism is closely allied to a philosophical stance known as empiricism. Just as behaviorists contend that learned behavior is completely shaped by previous experience, so empiricists contend that all knowledge derives from experience. Among philosophers opposed to empiricism, some—most notably Plato—have maintained that reason rather than experience is the source of true knowledge. Others, notably Kant, have maintained that both experience and the way we think about the world owe their *structure* to internal factors. We experience external objects as extended and deployed in space; we experience events as ordered in time. But space and time aren't aspects of an objective reality, according to Kant; they are "forms of sensibility," aspects of the way we interact with an otherwise unknowable external reality. Similarly, Kant argued, the relation between a cause and its effect doesn't belong to objective reality. Rather, it belongs to the way every human being understands the world; it is a "category of understanding."

Kant's philosophy of knowledge has appealed strongly to biologists, especially students of animal behavior. The information that an animal acquires about the external world is limited and structured by its perceptual apparatus (sense organs and nervous system). The way it acts on that information is governed, at least in part, by a genetic program. Both the perceptual apparatus and the behavioral program were shaped by natural selection. The frog's eye, for example, functions as a "bug detector," registering only small dark shapes in motion; and registration of a small dark shape in motion causes the frog to stick out its tongue and, with a little luck, capture a bug.

Analogously, the human eye and the human brain were shaped by natural selection not to inform us about what the world is really like (whatever that may mean) but to help us in the practical business of survival and reproduction. Because our ways of making a living are more diverse than the frog's, the information furnished by our visual system is less specialized, but it is still tailored to the uses to which it is normally put. For example, our eye and brain pay special attention to edges and boundaries. By exaggerating the discontinuities that occur there, they make it easier for us to separate objects from their background. Our perceptual system even manufactures contours where none are objectively present, as shown in Figure 12.1. Again, our eye and brain don't trouble us with reports about the sizes and colors of retinal images. If an object isn't too close or too far away, we perceive something like its "true" size; and if the lighting conditions aren't extreme or unnatural, we perceive something like its "true" color. By separating the effects of size from those of distance, and the effects of color from those of lighting, our visual system enables us to recognize objects at different distances and at different times of day.

We never perceive things "as they really are." Perception is largely a matter of what the nineteenth-century physicist and physiologist Hermann von Helmholtz called "unconscious inference." Of course, unconscious inferences aren't always correct. This fact is exploited in stage illusions. A magician holds two unbroken metal hoops, one in each hand. He brings them together, and they are suddenly linked. Then he pulls them apart. Our eye and brain have deceived us, and there is nothing we can do about it, even if we know how the trick is done. We can no more help seeing what we know is an impossible sequence of events than the frog can help flicking its tongue at a small, dark, moving shape in its visual field.

The unconscious inferences that underlie perception are based largely on a genetic program. So, too, are many aspects of behavior. Studies of animals in their natural environments have led ethologists to conclude that many aspects of biologically important behaviors are genetically programmed. Konrad Lorenz's studies of *imprinting* in fowl exemplify this generalization. Within a few days of hatching, chicks, ducklings, and goslings learn to approach and follow their mothers and to avoid other objects. Lorenz found that when the mother wasn't present, hatchlings would imprint on some other object—on Lorenz himself, for example, or on a big orange ball. The infant fowl's genetic program is like a form letter with only the name of the addressee left blank, to be filled in by experience.

Experience plays an analogous, although more complex, role in shaping the songs of some birds. The song of the male white-crowned sparrow, for example, has a number of regional dialects. Birds hatched in a given region learn to sing the dialect of that region. Hand-raised male sparrows reared away from other sparrows begin to sing at the normal age, but they sing a stripped-down "basic" sparrow song devoid of regional idiosyncrasies. Thus the male sparrow has the genetic capacity to learn to sing any of a large (or infinite) number of sparrow dialects, but to learn any given dialect he must be exposed to it during a critical learning period (from about two weeks to two months).

In this respect, human speech seems analogous to bird song. We are not programmed to learn to speak a particular language, but we are, in a sense, pro-

FIGURE 12.1 We see a white triangle, slightly darker than the background, covering the three black circles and the outlined triangle.

grammed to learn to speak. W. John Smith, in *The Behavior of Communicating*, carries the analogy further.

> The process of song learning by young birds parallels that of language learning by young humans in several respects. . . . Both song- and language-learning involve selection from myriad available sounds of a set that is copied, and young birds and young humans are genetically predisposed to recognize the set that is appropriate for each. . . . Humans and young male birds are further predisposed to practice reproducing the sounds through a stage of "babbling" or "subsong" until their auditory feedback confirms their competence in copying the model; even though female birds of many species do not sing, they must be familiar with the sound of appropriate songs.[1]

Stereotyped routines and predispositions are not the only aspect of behavior that is genetically programmed. What chiefly distinguishes human beings from other primates is our greater capacity for innovative behavior. That capacity is just as firmly rooted in our genes as the termite's "knowledge" of architecture or the honeybee's waggle dance by which she informs her sisters of the direction and distance of food.

The capacity to learn is adaptive because it enables an animal to cope more flexibly with a wider range of environmental challenges than do "hard-wired" behaviors. But it also has a biological cost. To the extent that a behavior can be learned, it can't be genetically programmed. And while it is being learned it is less useful than it would be if it were hard-wired. Whether the advantage of flexibility outweighs the disadvantage of not having instant access to finely adapted but stereotyped behaviors will depend on the variety and predictability of environmental challenges an animal must face. Conversely, an animal's capacity to adapt to new environmental challenges is limited by its capacity for learning.

Human behaviors learned through voluntary acts are not merely extensions of genetically programmed behaviors. This is brought out clearly in the following passage by the neuropsychologist Alexander Romanovich Luria, which provides an interesting contrast with the passage by Smith quoted earlier.

> It may appear that the language of a small child begins with babbling during infancy and that the development of language simply involves the extension of these initial sounds. Many generations of psycholinguists believed that. However, this is not the case. In effect, babbling is the expression of a state and not the designation of objects. Many sounds found in babbling are not later repeated in the child's speech. The first words uttered by a child are often either less distinct or quite different in their phonological structure from the babbling of an infant. We would argue that in order to learn the sounds in a linguistic system, the child must inhibit the sounds in babbling. This argument applies to many aspects of the ontogenesis [development] of children's voluntary movements. . . . Only a few days after birth, a child can have such a strong grasping reflex that it is possible for an adult to lift the child by offering him/her two fingers. However, it has been demonstrated that nothing emerges out of this grasping reflex. It cannot in any way be taken as the prototype of future voluntary movements. Just the reverse is true. It is necessary to inhibit the grasping reflex before voluntary movement appears.

Luria points out that the incompatibility between innate and voluntary behaviors has a neurological basis:

> The grasping reflex is a subcortical act, whereas voluntary movement is a cortical act. The latter has a quite different origin and occurs only when the grasping reflex is inhibited.[2]

Learned behaviors form a continuum. The simplest kinds of learning are analogous to the hypertrophy of a muscle through exercise. Or learning may simply direct development into one of several genetically predetermined channels. Further along the continuum, learning may elaborate a genetic program, as an actor uses gestures and vocal inflections to interpret a script or as a violinist ornaments a melody in Baroque music. The actor and the violinist add a good deal of information to the script or score, but they don't alter its basic character. The behaviors of higher primates, especially the great apes, fall into this category. So, too, do many human behaviors. Given the overall genetic similarity between humans and chimps—greater, some biologists now believe, than that between chimps and gorillas—how could it be otherwise? Not that the behavior of chimps isn't in many respects remarkably innovative, but it is innovative within well-defined biological limits.

Biological Preformationism

Biological preformationists maintain, in effect, that the continuum of learned behaviors ends here, with the elaboration of genetic programs. Beneath the superficial diversity of human languages, customs, beliefs, and institutions, they say, lies a rigid, unchanging deep structure that was already in place 40,000 years ago, before our ancestors had learned to grow food or keep animals. Anthropologist Robin Fox has invented a thought experiment to illustrate this thesis. He imagines a new Adam and Eve, cut off at birth from all contact with human culture.

> I do not doubt that they *could* speak and that, theoretically, given time, they or their offspring would invent and develop a language despite their never having been taught one. Furthermore, this language, although totally different from any known to us, would be analyzable by linguists on the same basis as other languages and translatable into all known languages. But I would push this further. If our new Adam and Eve would survive and breed—still in total isolation from any cultural influences—then eventually they would produce a society which would have laws about property, rules about incest and marriage, customs of taboo and beliefs about the supernatural and practices relating to it, a system of social status and methods of indicating it, initiation ceremonies for young men, courtship practices including the adornment of females, systems of symbolic body adornments generally, certain activities and associations set aside for men from which women were excluded, gambling of some kind, a tool- and weapon-making industry, myths and legends, dancing, adultery, and various doses of homicide, suicide, homosexuality, schizophrenia, psychosis and neuroses, and various practitioners to take advantage of or cure these, depending on how they are viewed.[3]

I find this list of cultural universals singularly unimpressive. Some of its items are trivial: "rules [of no specified character] about incest and marriage," for example. Others are far from universal. For instance, Fox suggests that the descendants of the new Adam and Eve would have laws about property. But the !Kung San of the Kalihari Desert have little or no personal property and no laws about it. Nor do they wage war or have a tool- and weapon-making industry. Yet anthropologists believe that San culture has evolved relatively little during the past 40,000 years.[4] The correct anthropological generalization would seem to be that organized aggression, weapon-making industries, and laws about property are cultural innovations that *lack* a specific biological basis. Again, San women gather food, while the men hunt. This is the usual pattern in hunter-gatherer societies. But it isn't universal. In the Philippines, Agta women conduct organized hunts.[5]

Biological preformationists retort that universality is too strong a criterion for genetically programmed behavior, because innate predispositions to particular kinds of behavior can always be overriden. But it seems strained to argue that the San have overriden an innate predisposition to have laws about property. And the male propensity for hunting and fighting, although unquestionably based on biology, is hardly a specifically human attribute.

Another reason for doubting the preformationist thesis is that human culture is not only exceedingly diverse but also exceedingly *labile*—subject to rapid change. For example, in industrialized societies, rules, customs, practices, and attitudes relating to gender and sex have changed considerably during the past half-century and are still changing very rapidly. Traditional distinctions between the roles of men and women are crumbling, and "activities and associations set aside for men from which women [are] excluded" are rapidly becoming obsolete. This isn't to deny that there are important biologically based differences between the behaviors and perceptions of human males and females. Such differences are important and well established.[6] What seems questionable is the thesis that we are genetically predisposed to develop specific customs, practices, and attitudes relating to gender and sex.

Belief in the supernatural is another highly labile aspect of human culture. What does seem to be part of the human condition is curiosity about the world and how we fit into it. Responding to that need, human societies have created a variety of myths and systems of belief, in many of which the supernatural (or what *we* call the "supernatural"; the line between the natural and the supernatural is less well defined in other cultures) plays a prominent role. By making life a thoroughly natural phenomenon, the Darwinian revolution diminished the role of the supernatural in the belief systems of biologically literate people. Darwin himself, brought up to hold traditional Christian beliefs, became a naturalist in the literal sense of the word. Do people who learn in childhood to see the world from the standpoint of contemporary biology and physics have to struggle with an innate propensity to believe in the supernatural? I am not aware of any evidence that would support this proposition.

Constructivism

Behaviorism and preformationism offer conflicting answers to the question: what is the origin of the order implicit in human behavioral patterns? Behaviorists assert that behavioral order is shaped mainly by external factors (positive and negative reinforcement) in the course of individual development; preformationists, that it is shaped mainly by evolution. Behaviorism ignores the inner springs of human behavior; preformationism ignores its creative aspects. There is a third possibility, *constructivism*. Constructivists, like preformationists, assume that patterns of behavior are mediated by biological structures. But whereas preformationists assume that these structures are formed entirely or almost entirely in the course of evolution, constructivists maintain that some of them—those that underlie creative behaviors—are built up largely by developmental processes. Constructivists assert that development, as well as evolution, contributes significantly to the growth of biological order and, indeed, that most of the order implicit in creative human behavior and its cultural products is produced by developmental processes. Thus constructivists view the developmental processes that shape creative behaviors as extensions of biological evolution.

Earlier we saw that the biological order generated by evolution is embodied in the structure of DNA. We will see later that the order generated by creative developmental processes is embodied in *patterns of connections between nerve cells in the brain*. These patterns are shaped in part by evolution; that is, their development is specified in part by genetic programs. But much is left unspecified by genetic programs in the case of creative behaviors. Instead, the genes supply strategies for continuing the process left unfinished by evolution. That, at least, is the constructivist thesis.

Constructivists also argue that creative developmental processes employ the same order-generating strategy as evolution: selection from a repertoire furnished by random variation. How this strategy is implemented by processes of neural development is still a speculative question. In the rest of this chapter we consider evidence from psychology and psycholinguistics relevant to the constructivist thesis that a wide range of human behaviors, from perception to language acquisition, are genuinely creative processes.

Perceptual Schemata

Constructivist psychology has its roots in the ideas of two great nineteenth-century scientists: Hermann von Helmholtz, whose treatise on sound and hearing is still required reading for every serious student of the subject; and Henri Poincaré, the outstanding mathematician of his generation as well as one of the founders of relativity theory. Helmholtz argued that perception doesn't consist simply of recording information contained in a visual or an auditory or a tactile stimulus. Rather, it consists of *formulating and testing a perceptual hypothesis*. Thus perception is fundamentally an active process. The eye *collects* visual information in much the same way as a camera. The retina and the brain *analyze* this information in various ways. For example, they extract information about the orientation and motion of boundaries between constrasting regions of the visual field. But that is just the beginning of the perceptual process, according to Helmholtz and modern constructivist psychologists. The brain uses the products of visual analysis to test and modify "hypotheses" about the "meaning" of the visual information.

Poincaré asked himself the question: Why do we see the external world as made up of objects deployed in three-dimensional Euclidean space? Kant had posed the same question, but Poincaré wasn't satisfied with Kant's answer: that Euclidean space is a schema imposed on an inherently unknowable external world by our perceptual apparatus. Poincaré argued that Euclidean space is a *construct* that enables us to extract useful information from visual stimuli. By "positing" that the external world consists of more or less rigid bodies moving about in a three-dimensional Euclidean space, our perceptual apparatus enables us to distinguish reliably between changes in the appearances of objects that result from changes of their sizes or shapes and changes that result from simple displacements—changes in perspective. Euclidean geometry isn't the only hypothesis that achieves this result, but it is the simplest.

Contemporary constructivist psychologists—Jerome Bruner, Ulric Neisser, and Julian Hochberg among them—have developed these ideas and sought to put them on a firm experimental basis. Constructivist accounts of perception argue that perception is a cyclic process mediated by internal structures called *schemata* (or *schemas*). Neisser defines a schema as

> that portion of the entire perceptual cycle which is internal to the perceiver, modifiable by experience, and somehow specific to what is being perceived. The schema accepts information as it becomes available at sensory surfaces and is changed by that information; it directs movements and exploratory activities that make more information available, by which it is further modified. In my view, the cognitive structures crucial for vision are the anticipatory schemata that prepare the perceiver to accept certain kinds of information rather than others and thus control the activity of looking. Because we can see only what we know how to look for, it is these schemata (together with the information actually available) that determines what will be perceived. Perception is indeed a constructive process, but what is constructed is not a mental image appearing in consciousness where it is admired by an inner man.[7]

Figure 12.2 illustrates the role of the schema in the perceptual cycle. Schemata regulate the flow of information inward from sensory surfaces and are themselves modified in the light of the information that reaches them.

That we can see only what we know how to look for is attested by a wealth of everyday experience as well as by many formal studies. A birdwatcher on a nature walk, a proofreader reading a galley, a cellular biologist peering at a cell through a microscope, a neurologist examining an electroencephalogram, an astronomer scanning the spectrogram of a star—all these specialists see details invisible to the untrained eye. A representational artist is trained to reproduce what he or she sees; yet different artists don't see the same things. This is the central theme of Ernst Gombrich's fascinating study of representational art, *Art and Illusion*.

Even size and color constancy—our ability to perceive the true sizes of objects at different distances and their true colors under different qualities of light—are not wholly innate. Studies have shown that size constancy develops gradually. Nor is its development inevitable. Neisser remarks that "pygmies who live in dense tropical forests, where distant objects are rarely visible, are said to make ludicrous mistakes when they first see a herd of animals far away."[8] Color constancy is likewise a schema susceptible to developmental modification. A landscape painter cannot convey the quality of the light illuminating a scene unless she is able to see the actual color of light reflected from objects in the scene. To do that, she must reconstruct her schema for color perception.

For hearing, even more than for seeing, common experiences attest to the link between expectation and perception. We have difficulty hearing unfamiliar names for the first time; the sounds somehow don't register. Similarly, a language we have never heard spoken seems on first hearing a poorly organized succession of

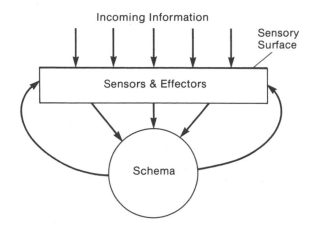

FIGURE 12.2 A perceptual schema actively selects and regulates information incident on sensory surfaces. At the same time, the schema itself is continually modified by the information whose transmission it mediates.

poorly defined sounds. As we listen longer, phonemes gradually emerge as clear and distinct entities. Again, we may be unable to make out the words of an opera sung in a foreign language, even if we are able to read that language. Equipped with a libretto, we hear the words quite clearly.

The construction of perceptual schemata in the course of development doesn't, of course, begin with a blank sheet. A newborn baby follows objects with its eyes. Very young babies show special interest (for example, by suspending rhythmic sucking) in faces and even drawings of faces. It seems safe to assume that all the schemata that a child will elaborate in the course of development already exist in more or less rudimentary form at birth. On this point, constructivists and preformationists agree. What they disagree about is the extent to which the outcomes of the developmental process are prefigured in the genes. Preformationists contend that development merely brings to light order already encoded in the genes. Constructivists maintain that development actually creates most of the order present in perceptual schemata.

Cognitive Schemata

Just as perception is mediated by perceptual schemata, so cognition and cognitive development are mediated by *cognitive schemata*. Jean Piaget, the most influential modern student of developmental psychology, depicted cognitive development as a process of *hierarchic construction* in which—just as in our earlier account of biological evolution—new functional units arise through simultaneous differentiation and integration of existing functional units. Let's consider, from this point of view, Piaget's most famous experiment. Six-year-olds who are shown water being poured from a wide beaker into a narrow beaker insist that there is more to drink in the narrow beaker because the level of the water is higher. Between the ages of seven and eight, most of the children that Piaget observed came to understand that the quantity of water remains the same.

Piaget himself attributed this change to a dawning realization that the pouring of water from one beaker to another is a reversible operation. But it can also be described as an example of hierarchic construction. The child learns to *differentiate* the notion "quantity of liquid" from (1) the notion "height of the liquid" and (2) the appearance of a particular beaker. At the same time, the new notion serves to *integrate* in a single schema (1) the two visible determinants of the quantity of liquid, its height and its cross-sectional area, and (2) the set of liquid-filled beakers with interchangeable contents. The new schema is more complex than its predecessor, and also more adequate.

The child's construction of time, the subject of another famous study by Piaget, admits an analogous description. As long as a child focuses attention on individual examples of uniform motion (for example, toy trains moving along a track), the distance traversed by the moving object serves to measure both the object's speed and the duration of its movement. The child says that the train that has gone farther has also gone faster, and taken longer to do so. Only when the child begins to construct an integrated description of two motions going on at the same time does he or she begin to differentiate the notions of distance traversed, speed, and duration of the motion. This is the way Aristotle described the construction of time and speed in the *Physics,* and it is also the way *Swiss children* construct these notions *under the conditions studied by Piaget.*

Piaget himself would have omitted the italicized phrases. He believed that although the stages of cognitive development are not genetically determined, they *are* universal. They are universal because, as Piaget put it, they represent successive stages in the acquisition of *a "logic inherent in the coordination of actions."* Let me refer to this logic, whose explicit characterization need not concern us, as L. L is a fragment of formal deductive logic, the logic underlying mathematics and theoretical physics. According to Piaget, the child begins by constructing schemata that embody L at a practical level. The conservation schema that we discussed earlier is one example. It emodies reversibility, one of the operations in L. Another example is seriation, the schema underlying the ability to order a collection of sticks by size. Having mastered the practical embodiments of L, the child proceeds to reconstruct L at the formal or abstract level of internal representation. Thus, for Piaget, postsensorimotor cognitive development consists in the elaboration of logico-mathematical structures.

This is an attractive hypothesis, recalling Poincaré's thesis that Euclidean space is a construct that embodies the logic inherent in perceived displacements of rigid bodies. Nevertheless, I think the idea that postsensorimotor development follows a predetermined path is mistaken. Cognitive development, we have seen, is a process of hierarchic construction. In a given culture—itself the product of an evolutionary process that features hierarchic construction—the outcomes of cognitive development are schemata adequate to the needs and values of that culture. I believe that deductive logic is not, as Piaget described it, the "mirror of thought" or the "axiomatics of reason." Instead, it is a cultural invention—an invention whose ramifications, which include mathematics, the natural sciences, and science-based technology, have come to dominate Western culture.

In the West, it isn't only science and technology that cause us to identify rationality with deductive logic. The humanities, too, are rooted in classical Greek thought, which bears the indelible imprint of Greek mathematics and natural philosophy. We are taught very early to recognize, to construct, and—above all—to value logical arguments. But being able to recognize or construct a logical argument isn't at all the same thing as thinking logically. Was Piaget correct in his asertion that fully developed human thought is logical?

Psychological studies begun in the 1960s convincingly demonstrated that most highly educated people don't in fact reason logically. They can, of course, learn techniques for solving logical problems, but their everyday processes of reasoning don't conform to the rules of deductive logic. Among the best known studies that reached this conclusion are those of Peter Wason and Philip Johnson-Laird.[9] They asked undergraduates to solve problems like the following one:

> Some cards are lying on a table. Every card has a letter (*A* or *B*) on one side and a number (1 or 2) on the other side. Some of the cards are face up; some are face down. Those that are face up show either the letter *A* or the letter *B;* those that are face down show either the number 1 or the number 2. Which cards do you have to turn over to test the hypothesis that if a card has an *A* on one side, it has a 1 on the other side?

Typically, only one out of ten students answered such problems correctly. Yet four out of five correctly solved problems like the following.

> Some envelopes are lying on a table. Some are face up; some are face down. Those that are face up show either a ten-cent stamp or a twenty-cent stamp; those that are face down are either sealed or unsealed. Which envelopes do you have to turn over to check that sealed envelopes carry a twenty-cent stamp?

Most of the people who were given the second problem easily reached the conclusion that they would have to turn over sealed envelopes and envelopes carrying a ten-cent stamp, while most of the people who were given the first problem (which is *formally* identical with the second problem) failed to reach the corresponding conclusion. Presumably the second problem is easier to solve because it is more concrete, easier to make a mental model of. (Johnson-Laird has argued that making mental models is the central strategy that people use to solve logical problems.) Whatever the precise explanation may be, it seems clear that if our thought processes mirrored the laws of deductive logic, the two problems would be equally easy (or equally difficult) because they have exactly the same logical structure.

Even more striking are the findings of A. R. Luria in a study carried out in a remote province of the Soviet Union in 1930 and 1931 but not published until much later. The subjects of the study included nonliterate but intelligent adults, "fully capable of managing their agricultural economy, sometimes requiring them to solve extremely complex practical problems connected with the use of irrigation canals.'' Luria found that his nonliterate subjects performed very differently in tests of reasoning (and also in perceptual tests) from schoolchildren and from adults who had just learned to read. Literate subjects easily solved elementary syllogisms. The nonliterate subjects seemed unable to reason abstractly.

> In order to test this hypothesis, we conducted the following experiment. Subjects were given two groups of syllogisms. Syllogisms in one group were drawn from their practical experience, syllogisms in the other were purely abstract. They were taken from a field in which the subjects did not have any practical knowledge. The following is an example of the first type of syllogism: ''Cotton grows in all places that are warm and humid. In the place called N it is humid and warm. Does cotton grow in N or not?'' An example of the second type of syllogism is: ''In the Far North, where there is snow all year round, all bears are white. The place N is situated in the Far North. Are the bears in N white bears or not?'' The [results of the two] cases were quite different. . . . When confronted with the first type of syllogism, our subjects usually answered. ''Yes, of course, cotton probably would grow there. I know that cotton grows only in places where it is warm and humid.'' When confronted with the second type of syllogism, they usually refused to answer, saying that they did not possess knowledge on the subject. Thus they might answer, ''I've never been there and don't know. I don't want to tell a lie, I won't say anything. Ask someone who's been there. He'll tell you.'' [10]

Luria attributed the development of abstract-reasoning ability by his literate subjects to their participation in ''theoretical activity.'' Perhaps it was a specific form of theoretical activity—learning to read.

Speech

Of all specifically human behaviors, speech is surely the most fundamental, for it is the chief instrument by which learned skills and the rules that make complex social structures possible are transmitted from generation to generation. Without speech, our ability to invent and improvise would have considerably less adaptive value. How and to what extent is human speech constrained by biological factors? The behaviorist account of how children learn to speak fails because children's linguistic experience doesn't contain enough information to account for their linguistic performance. Where does the additional information come from?

Noam Chomsky and others have hypothesized that it resides in genetically determined structures in the brain. At first sight, this may seem an absurd notion. An infant growing up in a family whose members speak Chinese also learns to speak Chinese. Its twin, reared in a family that speaks English, learns to speak English. How can a genetically programmed linguistic competence include Chinese, English, and thousands of other languages, including an unlimited number of languages not yet invented?

Chomsky's answer is that natural languages aren't as different, fundamentally, as they appear to be. They all conform to the rules of a "universal grammar" encoded in our DNA, just as the architectural regularities implicit in the elaborate structures that termites build are encoded in *their* DNA. Thus a child comes into the world with a tacit knowledge of language's basic structure, to which he or she gradually gains access in the course of development. Linguistic experience merely fills in the gaps.

Knowing a language means being able to understand not only sentences that one has heard or read before but also sentences, like the present sentence, that one is encountering for the first time. What underlies this kind of understanding? Chomsky's answer is that someone who knows a language knows a set of rules that specifies the structure of every possible sentence. This set of rules constitutes a *generative grammar,* as opposed to a mere descriptive grammar (the kind of grammar that schoolchildren are exposed to in grade school). A descriptive grammar may be compared to a textbook that codifies the practices of, say, nineteenth-century composers of Western music with regard to harmony, counterpoint, and orchestration. By contrast, a generative grammar would be like a computer program powerful enough to generate every musical phrase in that idiom. Chomsky suggests that the generative grammars of different languages have a significant common core, a tacit knowledge of which is encoded in human DNA.

> A theory of linguistic structure that aims for explanatory adequacy incorporates an account of linguistic universals, and it attributes tacit knowledge of these universals to the child. It proposes, then, that the child approaches the data with the presumption that they are drawn from a language of a certain antecedently well-defined type, his problem being to determine which of the (humanly) possible languages is that of the community in which he is placed. Language learning would be impossible unless this were the case.[11]

The notion of *tacit knowledge* plays a central role in Chomsky's argument. Someone who knows how to ride a bicycle may be said to have a tacit knowledge of the dynamic principles that would figure in a physicist's description of bicycle-riding. The mastery of *any* complex skill implies a tacit knowledge of the rules underlying that skill. But what does this statement mean? I think it is best interpreted as a *definition* of "tacit knowledge." Defined in this way, tacit knowledge is very different from explicit knowledge. Few bicyclists are aware of or could explain the rules they follow, just as few speakers of English are aware of or could explain most of the rules that govern their speech. Conversely, full and explicit knowledge of the relevant rules—even if it could be had—would be of little practical use to a beginning bicyclist or speaker of English. The two kinds of knowledge are fundamentally different. We don't acquire complex skills by learning their underlying rules, and having a complex skill doesn't mean that we know those rules, except in a metaphorical sense of "know."[12]

Chomsky writes that "the child approaches the data with the presumption that they are drawn from a language of a certain antecedently well-defined type, his problem being to determine which of the (humanly) possible languages is that of the community in which he is placed." Presumably this doesn't mean that every

two-year-old confronts his mother's utterances in the spirit of investigative phil-
ology, like a precocious Michael Ventris confronting Linear B. The child's "presumption" must be tacit, along with the analytic skills he or she would need to
make effective use of it. But genetically programmed skills don't require a presumption. The greylag gosling doesn't have to presume that Konrad Lorenz is its
mother; it is programmed to attach itself to the most likely candidate that auditions
for that role during a certain critical period of its development. Analogously, if
Chomsky is right, a two-year-old doesn't "presume" anything: he or she is programmed "to determine which of the (humanly) possible languages is that of the
community in which he is placed."

But how? Are we to suppose that a vast number of generative grammars are
genetically encoded or arise during early development, in the manner of antibody
genes? Just as a clone of specific antibodies proliferates when the organism is
challenged by a specific antigen, so, perhaps, competence in a specific language
might burgeon when the two-year old is challenged by utterances in that language.
This is a fantastic idea, and I don't wish to impute it to Chomsky, but I don't
know of any concrete—that is, nonmetaphorical—alternative to it. After thirty
years of active investigation and debate, linguists are still quarreling fiercely about
the *form*—never mind the content—of generative grammars. And although most
linguists acknowledge that the effort to construct generative grammars has had a
strong positive impact on the study of language, the very existence of generative
grammars—and thus of a universal grammar whose principles would constrain
these hypothetical generative grammars—is still warmly debated.

Natural Language and Mathematics

Although no one has yet come close to constructing a generative grammar for any
natural language, there is one area in which generative grammars are commonplace: mathematics. For any given branch of mathematics—Euclidean geometry
or arithmetic, for example—there is a set of grammatical rules that specify the
structure of every "well-formed" statement. Most of these rules are rules of logic
common to all branches of mathematics. These constitute the universal grammar
of mathematics. The remaining rules serve to define notions peculiar to a particular branch of mathematics, such as line, point, and intersection in geometry or
number in arithmetic. In the 1950s Chomsky set out to construct for English (and
eventually other languages) the kind of generative grammar that mathematicians
had at the turn of the century constructed for mathematical languages.

In my opinion, the analogy between natural languages and mathematical languages that inspired the notions of generative and universal grammars is superficial and misleading. Perhaps the most obvious difference between mathematical
and natural languages concerns the meanings of individual words or terms. The
words of a natural language are often fuzzy around the edges. Mathematical terms,
by contrast, have perfectly sharp edges, like ideal geometric figures. Again, natural languages always succeed in breaking out of the straitjackets in which pre-
scriptive grammarians try to confine them, while mathematical discourse, however

complex, always conforms perfectly to the rules of an explicit and easily learned grammar.

Some people interpret these differences between mathematical and natural languages as shortcomings of the latter, even as *remediable* shortcomings. From the time of Aristotle onward, the task of purging natural languages of their ambiguities and inconsistencies has beckoned to philosophers who hankered after the precision and regularity of mathematical language. It is, I believe, an impossible task, because the two kinds of language are fundamentally dissimilar.

As we discussed in Chapter 1, the objects and relations of pure mathematics are defined fully and precisely either by explicit definitions, which allow us to replace the defined term by terms previously defined, or by axioms, which implicitly define the terms they mention. Mathematical objects have no obligatory connection with their counterparts in experience. Pythagoras's theorem is an exact truth about triangles that inhabit a Platonic realm we apprehend with our mind's eye. Interpreted as a statement about physical triangles—for example, triangles whose sides are light-rays in empty space—it is approximately true, at least for small triangles. As applied to triangles whose sides are measured in billions of light-years, however, the theorem may be badly in error.

By contrast, the meanings of words in natural languages are rooted in experience. A dictionary defines words in terms of other words whose meanings are assumed already known to the reader through experience. It couldn't do otherwise. Dictionary definitions, moreover, are usually incomplete. Take the verb *incarnadine;* according to the *American Heritage Dictionary*, it means "to make the color of blood or flesh." An excellent definition, but not quite complete, since a writer who uses this word probably intends to remind the reader of its most famous occurrence:

> Will all great Neptune's ocean wash this blood
> Clean from my hand? No, this my hand will rather
> The multitudinous seas incarnadine,
> Making the green one red.
> *Macbeth*, II.ii.59–62

This context is an important part of the word's meaning. Nor is this an isolated example. We infer (or construct) the meanings of most words from the situations in which we first encounter them, and fresh associations introduce fresh shades and nuances of meaning. It follows that many words in a natural language have significantly different meanings for different speakers, and even for the same speaker on different occasions. And this imprecision (or richness) of natural languages is ineradicable. No natural language can hope to attain, or should aspire to, the precision of mathematical discourse.

This conclusion will not, I think, surprise or disturb many readers. But if most of us readily accept the proposition that the meanings of individual words and phrases have fuzzy edges and are susceptible to growth and change, many of us are less ready to assign the same degree of imprecision and lability to syntax. There are, we tend to believe, right ways and wrong ways of stringing words together in sentences, as well as right ways and wrong ways of using individual words. Some sentences are well formed; others are not. But what determines whether a sentence is well formed?

I suggest that the most basic criterion for a well-formed sentence is that it should sound right to a native speaker of the language. Sounding right isn't the same thing as obeying syntactic rules. Sentences like "It is I" and "Do not wait for me; I shall catch up" conform to the rules of English syntax but no longer sound right to most native speakers of American English. Not that rules don't affect ordinary speech. Efforts to speak and write grammatically are responsible for such dissonant productions as "between you and I" (now so common in some linguistic communities that it *does* sound right to these speakers) and "Give it to whomever shows up."

Although saying that a well-formed sentence sounds right doesn't tell us anything about the structural properties of well-formed sentences, it does help to put us on the right track. It suggests that we shouldn't ask how a child acquires the rules of some hypothetical generative grammar. Instead, we should ask how he or she learns to recognize and use the linguistic patterns that sound right in the particular linguistic community the child happens to grow up in. The two questions are different. The first is like asking how a child comes to know the dynamic principles that underlie the ability to ride a bicycle; the second is like asking how he or she acquires the relevant set of muscular reflexes.

Learning a First Language

How *do* children learn to speak? While linguists have been studying the language of adults (and its historical development), psycholinguists have been observing and trying to make sense of the *development* of linguistic performance and linguistic competence. Chomsky and other linguists of the preformationist persuasion dismiss such studies as largely irrelevant to an understanding of adult performance and competence. They argue that the development of linguistic ability, like the development of a physical organ, is little more than the gradual unfolding of a genetically encoded blueprint. Knowing how the heart or the kidney develops doesn't, they say, help us to understand how it works. I think this argument begs the question. Observation has established that the development of physical organs is insensitive, within broad limits, to external influence. But the development of language is highly sensitive to external influence. One of the main reasons for studying the way children learn to speak is to try to discover just what regularities underlie the developmental process in different children learning to speak diverse languages.

I suggest that language acquisition, like other kinds of cognitive development, is a process of hierarchic construction in which the underlying schemata undergo simultaneous differentiation and integration. In *A First Language,* Roger Brown, a leading student of psycholinguistics, characterizes early linguistic development as follows.

This development is always of the same two kinds. An increase in the number of relations expressed by: 1. concatenating, serially, more relations and omitting redundant terms; 2. unfolding of one term in a relation so that the term becomes itself a relation.

The first kind of development corresponds precisely to what I have called integration, the second kind to differentiation accompanied by integration. The passage continues:

In these data as a whole . . . there is evidence for what I have, not yet very seriously, called a law of cumulative complexity of language development. It is important to realize that as utterances get longer, . . . some sort of increase in complexity is bound to occur, but there is no a priori reason why the increase should take just the forms it does and, in particular, that these forms should be the same for all children studied, whatever the language in question.[13]

Brown's "law of cumulative complexity" is, I believe, identical with what I have called the principle of hierarchic construction.

The preceding passage refers to the first of five stages into which Brown divides linguistic development. The subsequent development can also be described as a process of hierarchic construction. In Stage I, the meanings expressed, for example, by *to Mommy, for Mommy, Mommy's,* are not yet fully differentiated. In Stage II, this kind of differentiation-plus-integration takes place for the basic meanings of Stage I. In English, such "modulations of meaning," as Brown aptly calls them, are accomplished through noun and verb inflections, prepositions, articles, and copulas. "All these," Brown writes, "like an intricate sort of ivy, begin to grow up between and upon the major construction blocks, the nouns and verbs, to which Stage I is largely limited." In Stage III, the child completes the differentiation of sentence modalities (declarative, interrogatives of various kinds, negative, imperative) begun in Stage I. In Stage IV, sentences are constructed in which other sentences function as units ("I hope it doesn't hurt"), and in Stage V two or more sentences are combined into a single sentence, sometimes with the deletion of redundant components ("I bought some gum and a book"). All these kinds of development exemplify the cumulative growth of complexity through hierarchic construction.

One of the main differences between the constructivist and preformationist views of linguistic development concerns its starting point. Chomsky and other preformationists assert that linguistic development is the maturation of a specifically linguistic ability programmed by the genes. Piaget, by contrast, maintained that linguistic structures arise from previously constructed nonlinguistic structures. During the first year and a half to two years of life, a child constructs schemata that enable him or her to structure experience in terms of objects, actions, and simple patterns of action. As in all aspects of cognitive development, each stage builds on earlier stages and paves the way for later ones. In particular, the final achievements of sensorimotor intelligence (as Piaget called this collection of intertwined schemata) pave the way for—and are in fact closely followed by—the beginnings of speech.

Piaget's view has been strongly supported by the observations of psycholinguists. Here, again, is Roger Brown in *A First Language:*

> In sum, I think that the first sentences express the construction of reality which is the terminal achievement of sensori-motor intelligence. What has been acquired on the plane of motor intelligence (the permanence of form and substance of immediate objects) and the structure of immediate space and time does not need to be formed all over again on the plane of representation. Representation starts with just those meanings that are most available to it, propositions about action schemas involving agents and objects, assertions of nonexistence, recurrence, location, and so on. But representation carries intelligence beyond the sensorimotor. Representation is a new level of operation which quickly moves to meanings that go beyond immediate space and practical action.[14]

This passage epitomizes the constructivist account of linguistic development. Reality is not simply "given" to us; we construct it in the first months of life. Before we can begin to associate objects and actions with sounds, we must learn to perceive objects and actions in the world around us. That is, we must construct an external reality populated by stable objects in motion and interaction. The innovation that makes language possible is representation—not just the *association* of sounds (or symbols) with objects and actions, but the construction of a second world of surrogate objects and actions.

Brown emphasized that sensorimotor structures provide the basis for *meanings* rather than for syntactic categories or relations. He and other psycholinguists have had much more success in characterizing the structure of a child's speech from a semantic point of view than from a syntactic one. For example, Brown found that many of a child's first utterances fall into three semantic categories: nomination *(see, there),* recurrence *(more),* and nonexistence *(all gone).* He also found that most of the earliest (Stage I) utterances of the children he studied expressed a small number of semantic relationships, such as agent and action, action and object, entity and location, possessor and possession. Most, but not all. One of the most striking conclusions to be drawn from modern research in children's speech is the negative finding that no generative grammar seems able to encompass even a beginning speaker's range of verbal expression. From its inception speech has a creative, open-ended character that sets it apart from less characteristically human behaviors. And this is as true of form as of content. Fledgling speakers not only say things that haven't been said before, but say them in new ways.

The hypothesis of an innate universal grammar was devised to explain two main observations: the existence of linguistic features common to many—perhaps all—natural languages, and children's ability to master with ease syntactic rules so complex that professional linguists haven't yet been able to give a complete account of them for a single natural language—or even for baby talk. How adequately does the constructivist account deal with these observations?

We have seen that the first stage in language learning consists of *representing* schemata constructed during the period of sensorimotor development, which occupies the first eighteen to twenty-four months of a child's life. These sensori-motor schemata seem to have a universal character. In every culture children learn to perceive and act on the external world in terms of objects, actions, spatial relations, and temporal relations. So, indeed, do young chimpanzees and gorillas, judging from their behavior (which, of course, is also how we assess the sensori-motor intelligence of children). Evidently we share with other primates the genetic capacity and motivation to construct these highly adaptive schemata during infancy and early childhood. So it isn't surprising, in the constructivist account, that the most fundamental semantic categories and relations—those that represent universal sensorimotor schemata—are the same in all human languages. Chimps and gorillas, whose sensorimotor intelligence seems very similar to our own, have also been found capable of understanding and using visual signs and symbols that represent objects, actions, and (perhaps) simple relations like agent–action.

Natural languages use a wide variety of syntactic devices to express complex meanings. What, if anything, do these devices have in common? The common factor, I suggest, is that syntactic complexity always has a hierarchic character. This, of course, is what the constructivist account leads us to expect: hierarchic construction inevitably produces hierarchically organized structures.

Every natural language, however complex its syntax, is learned with ease by children who grow up surrounded by people who speak that language. Why? Chomsky and his followers say it is because children have an innate tacit knowledge of meta-rules that constrain the syntax of every natural language. According to the constructivist view, however, universality is not to be sought in the *outcomes* of language acquisition (that is, in natural languages as they are spoken by adults), but in the *process* by which language is acquired. As Roger Brown remarked, a child's utterances become not only progressively more complex, but also more complex in a particular way: through hierarchic construction. The complexity of every natural language is of just the kind that can be mastered through a step-by-step hierarchic construction that, as Brown demonstrates in *A First Language,* enables children to proceed in easy stages from one word sentences to sentences that exhibit the full complexity of adult speech.

How did natural languages come to have this particular kind of complexity? It seems reasonable to suppose that they evolved by a process of hierarchic construction qualitatively similar to the process that children use to learn their first language. Indeed, until the invention of writing (a relatively recent event in the history of language) the only syntactic changes that could have been reliably transmitted from generation to generation were those that children could learn easily—changes adapted to children's hierarchical mode of constructing and processing language.

* * *

This chapter has discussed language, thought, and perception at a descriptive level. I have used some theoretical notions—schemata, hierarchic construction—to organize the phenomena, but I haven't attempted to relate these ideas to biological structures and processes. We are now ready to address that task.

13

What Is Consciousness?

The Mind–Body Problem

Knowing and perceiving are clearly biological as well as psychological processes. It is less clear, however, how psychological and biological descriptions are related. Do psychological descriptions refer to identifiable biological structures and processes? For example, does ''reading this sentence'' refer to some complicated collection of physical events taking place in the reader's eye and brain? Or does reading a sentence involve something more than physical processes, something that no biological account could ever include?

Such questions pose two kinds of difficulties. The first kind is purely scientific. The nervous system is by far the most complex of all our organ systems. Although its anatomy and physiology are now well understood at the lowest levels of organization, neurobiologists have only just begun to build a detailed account of structure and function at the levels of organization at which perceptual and cognitive processes take place. The second difficulty is philosophical. It is usually referred to as the mind–body problem, and its origins go back to the dawn of Western philosophy (Socrates affirmed and Democritus denied that the mind was distinct form the body).

The philosopher Thomas Nagel has argued that the mind–body problem, the problem of free will, and several other problems in contemporary philosophy have a certain common ground, which he characterizes as follows:

> The problem is one of opposition between subjective and objective points of view. There is a tendency to seek an objective account of everything before admitting its reality. But often what appears to a more subjective point of view

cannot be accounted for in this way. So either the objective conception of the world is incomplete, or the subjective involves illusions that should be rejected.[1]

The "tendency to seek an objective account of everything before admitting its reality" is a natural outcome of individual development. As we pass from infancy to childhood and from childhood to adulthood, we learn to view the world and ourselves more and more objectively. We come to respect objective judgment in others, and we try to cultivate it in ourselves. Even if we never manage to see ourselves as others see us, most of us wish we could.

At another level, the objectifying impulse manifests itself in a growing ability to examine alien societies and systems of belief from the inside—and our own from the outside. Nowhere has the flight from subjectivity progressed farther or reaped greater rewards than in natural science. Galileo, at the dawn of modern science, revived the program launched by Democritus two millennia earlier of replacing descriptions involving "secondary" qualities like red, smooth, and sweet with descriptions involving "primary" qualities like mass and shape. The spectacular success of that program, especially in shedding light on questions that only a few decades ago seemed hopelessly beyond the reach of scientific inquiry—the beginning of time, the infinite reaches of space, the ultimate constituents of matter, the secret of life, the nature of human thought and emotion—has convinced many people that the whole of reality lies within the province of science.

And yet there is something at the core of personal experience that doesn't seem to belong to the external, objective world of natural science—something we find hard to put into words, but even harder to dismiss as illusory. How have modern philosophers dealt with this conflict?

The contrast between the objectivity of scientific descriptions and the subjectivity of immediate experience is the central theme of Bergson's philosophy. Bergson saw in the irreducible subjectivity of experience a manifestation of the life-force, the driving and organizing force of biological evolution. As we saw in Chapter 10, no such force need in fact be postulated to explain the phenomenon of life or the creative character of biological evolution. Nevertheless, in this century many scientists, including neurobiologists Charles Sherrington, John Eccles, and Roger Sperry, have followed Bergson in positing a nonphysical reality that intervenes in certain mental processes. Such hypotheses draw support from our intuition that sensations, perceptions, and thoughts don't belong to the physical world. Yet, as Thomas Nagel has pointed out, they don't achieve their main goal. One can't bridge the gap between the objective and the subjective by postulating two objective worlds, a world of things and a world of thoughts.

> The broader issue between personal and impersonal, or subjective and objective, arises also for a dualist theory of mind. The question of how one can include in the objective world a mental substance having subjective properties is as acute as the question of how a physical substance can have subjective properties.[2]

The philosopher Karl Popper has gone even further than dualists like Descartes and Bergson. He postulates *three* distinct worlds: a physical world of things, a mental world of thoughts and sensations, and a Platonic realm of objects and relations that we "see" with the eye of the mind—numbers, geometric figures, mathematical proofs, and so on. To me, these distinctions make good *logical* sense. They assist clear thinking. For example, when I talk about a cube, it is important for me and my listener to know whether I am referring to a physical object like an ice cube, a mental image, or an ideal geometric object. But does Popper's trisection of reality make *metaphysical* sense? I think not. Where, for example, would he put a defective or an incomplete proof of a mathematical proposition? It certainly doesn't belong in World One (physical objects) or World Two (mental states). If it goes anywhere, it must go into World Three. More generally, if Popper's classification is to be exhaustive, World Three must contain *constructs* of all kinds. But a world of constructs clearly can't be considered to exist independently of human minds.

Many contemporary analytic philosophers have gone to the opposite extreme. Proceeding from the premise that science (perhaps supplemented by philosophy) must be capable of constructing a complete description of reality, they assert that every mental event is *identical with* a physical event in the brain. This assertion is incompatible with the assertion that conscious mental states contain an irreducibly subjective element, because however different the science of tomorrow may be from the science of today, we can be sure that it will offer an objective description of reality. Objectivity is to science what rhythm is to music and color is to painting. How do philosophers who identify mental events with physical events in the brain deal with the subjective element of experience?

Some of them simply ignore it. Others assert that subjectivity is an illusion that will evaporate when we have a better understanding of its biological basis, just as the notion of demonic possession evaporated when people began to understood the nature of mental illness.[3]

The third, and perhaps most convincing, defense of the reductionist thesis consists in attacking the counterthesis that subjectivity is an irreducible part of reality. How can we be certain that a complete objective account of what goes on in a particular person's nervous system—an account that might differ in ways we can't imagine from the fragmentary accounts we now have—wouldn't tell us what it was like to be that person? If we knew enough about a bat's nervous system (to take a favorite example of Nagel's), isn't it at least conceivable that we would know what it was like "for a *bat* to be a bat"?

This argument draws support from the history of science. People have often said, "Such and such a question can never be answered, or even addressed, by science." Usually they have been proved wrong. Both Galileo and Kant were convinced that questions about the finitude of the Universe were in principle unanswerable by science. Maxwell believed that the permanence and stability of atoms could never be explained by science. Well into the twentieth century, many biologists (and more physicists) were convinced that scientific description could never fully capture the difference between life and nonlife. With such examples before them, many philosophers hesitate to assert that the subjective aspects of experience are *irreducibly* subjective. Why assume that science is not only incomplete, as it obviously is, but also incompletable—that the subjective core of experience will always remain outside its domain?

I think there is only one reason to make this assumption, but it is a powerful reason. Science seeks to describe and explain regularities underlying natural phenomena. Equally important, science seeks to uncover new phenomena. These two activities are mutually reinforcing. Efforts to understand the permanence and stability of atoms led to the discovery of the quantum world of crystals, molecules, atoms, and subatomic particles. Efforts to understand life led to the discovery of the world of molecular biology. And efforts to understand the mind are opening up yet another world full of previously unimagined phenomena. But subjectivity, I suggest, *isn't a phenomenon in the sense that scientists understand the term.* I don't mean that such things as sensations, perceptions, thoughts, desires, emotions, dreams, and hallucinations aren't susceptible to scientific exploration and explanation. On the contrary, they are all grist for the neurobiologist's mill. What I wish to suggest—what I take to be Nagel's point, if I understand it correctly— is that objective knowledge, by its very nature, must exclude the subjective aspect of experience.

That doesn't mean we can't acquire subjective knowledge. Obviously, we know our own thoughts and feelings; no knowledge is more direct or secure. Often we know what other people are thinking or feeling, too. A fine poem gives us access to—although it doesn't, of course, *describe*—the poet's (subjective) point of view. Conceivably, a sufficiently detailed knowledge of another person's brain states might enable a skillful interpreter of such information to "get inside" that person's mind the way a good play script enables a skillful actor to get inside a character's mind. But in all such examples, adopting another person's point of view—getting inside that person's skull—means experiencing for oneself something akin to what the other person is experiencing. If a friend says to me, "I know what you must be feeling," I understand him to mean not that he can describe my emotional state but that, by an act of sympathy, he is able in some measure to share it. This ancient and familiar distinction between discursive knowledge and what Bergson called intuition—immediate, nondiscursive knowledge—is what reductionist philosophers of mind deny. Or rather, they assert that however deep the chasm between these two kinds of knowledge may seem in the light of present-day knowledge, it must be bridgeable by science.

This seems to me the same as asserting that music must be capable of representing colors and shapes. It would be nice to have such a bridge between music and the visual arts, but I can't bring myself to believe that it exists. Similarly, I can't bring myself to deny that the subjective aspect of experience—having a point of view, in Nagel's phrase—is a deep property of reality.

Affirming the irreducibility of subjective experience doesn't in any way restrict the scope of scientific inquiry. The following assertions are mutually consistent, although they may not seem to be at first sight:

    1. All events in the external world are governed by physical laws.
    2. There are irreducibly subjective aspects of experience that don't belong to the external world.

The *apparent* contradiction is illustrated by the following example. I see the traffic light change; I decide to cross the street; I step off the curb. The first two events seem to belong to my inner world; the third, to the external world. And if mental events can cause physical events, physical events can't be governed entirely by physical laws. But let's take a closer look.

Although my seeing the traffic light change and my deciding to cross the street are mental events, they have an objective as well as a subjective aspect. Viewed objectively, they take place in the external world—specifically, in that part of the external world known as my brain. And if the first of the above assertions is valid, there is no theoretical reason (although there may be practical reasons) why we shouldn't someday be able to give a complete scientific account of the objective aspect of mental events.

How, then, are the objective and subjective aspects of mental events related? Philosophers who affirm that events in the external world are governed by physical laws but deny that mental states are identical with physical states usually say that mental events, in their subjective aspect, don't influence physical events; yet they themselves are *caused* by physical events. Specifically, (subjective) states of mind are *caused* by (objective) brain states. I am troubled by the use of the word *caused* in this context. Although *causation* isn't a precise scientific concept, it always refers to a relation between objects, events, or phenomena in the external world—for example, the relation between the Moon and ocean tides or between a deficiency of vitamin A and night blindness. Since the irreducibly subjective aspect of mental events doesn't belong to the external world, I think it would be better not to talk about causation in this context and to say instead that mental states have an objective aspect and a subjective aspect.[4] Viewed from the outside, they are brain states; from the inside, states of mind. We know that certain spatio-temporal patterns of activity in a person's nervous system are accompanied by subjective experience; they have a conscious "dimension." We don't know, and there is no reason to assume, that the pattern of activity *precedes,* rather than accompanies, the corresponding subjective experience. The naïve view that experience somehow gives us direct, although partial, access to what is going on in our brains stands up remarkably well to the available evidence.

Consider, for example, generalized states of consciousness. We may be awake or asleep. Awake, our attention may be focused or unfocused. Asleep, we may be having vivid and fantastic dreams during which our eyes undergo rapid movements; or we may be having more prosaic dreams, or not dreaming at all, with our eyes still. During these four psychologically distinct states, the electroencephalograph (a device that measures tiny changes in electric potential on the surface of the skull) records four distinct and distinctive patterns of brain waves (Figure 13.1). These patterns represent the integrated electrical activity of the whole brain (or at least of its surface).

What about specific conscious states? Is the brain's physical state different when its owner is looking at a picture than when he or she is listening to music? Would an observer who could monitor a brain's physical state be able to distinguish between occasions when its owner was solving a mathematical problem, recalling an earlier conversation, or trying to untie a knot in a fishing line? Would such an observer be able to tell whether the brain's owner was listening to music or to speech? The answer to all these questions is yes. There already exist nonintrusive devices that make it possible to examine the working brain and record levels of activity in different parts of it.

These devices exploit the fact that brain cells have to be constantly supplied with energy. The harder cells in a particular region of the brain are working, the greater their rate of energy consumption. Hence if the rate at which energy is supplied to different regions could somehow be recorded, we would be able to tell in which regions the cells were working hardest when the brain's owner was listening, looking, remembering, or thinking. Now, energy is supplied to brain cells (as to other cells) by blood, so one way of detecting variations in the energy supply would be to detect variations in the flow of blood in different regions. Scientists have been able to do this by injecting tiny quantities of a harmless radioactive substance into the blood, causing regions of high activity, and there-fore high concentration of the radioactive substance, literally to light up. An even more powerful technique, positron-emission tomography, or PET, exploits the fact that brain cells feed on glucose, the only fuel molecule that passes the brain–blood barrier. Glucose can be "tagged" with a radioactive element (fluorine-18) that decays by emitting positrons (antielectrons). These instantly encounter elec-trons, whereupon both particles disappear, producing a pair of high-frequency photons (gamma rays). With either technique the emitted radiation passes through the skull and is registered by a suitable recorder, whose output is analyzed by a digital computer that constructs a picture of brain activity.

Figure 13.2 shows PET scans of the human brain. The striking conclusion illustrated by such pictures is that specific mental tasks are associated with specific patterns of brain activity.

Patterns of brain activity may be far more complex than their subjective coun-terparts, however. This is shown most clearly by the remarkable and often-discussed observations of Roger Sperry on split-brain patients. In these patients the bridge of nerves connecting the right and left hemispheres of the brain has been cut for therapeutic reasons. Even in such patients, visual information is nor-mally sent to both hemispheres. The left half of the visual field *of each eye* proj ects onto an area in the right cerebral hemisphere; the right half of the visual field of each eye projects onto a similarly situated area in the left cerebral hemisphere. Sperry devised a simple experimental procedure that allowed the subject's left eye to receive light from only the left half of its visual field, and the right eye to receive light from only the right half of its visual field. In this setup, visual infor-mation from the left eye went to the right half-brain, and information from the right eye to the left half-brain. Sperry found that split-brain subjects could *name* objects presented to the right eye (and the left half-brain) but not objects presented to the left eye (and the right half-brain). These patients could *match* an object presented to the left eye—a spoon, say—with its counterpart in a heterogeneous collection, but they couldn't name it. Indeed, they insisted that they had seen nothing: no information had reached the verbal left hemisphere. Can someone be said to be *aware* of an experience he or she denies having?

Reality thus seems to consist of two parts: a part we can describe but can't directly experience, and a part we experience directly but can't fully describe. The part

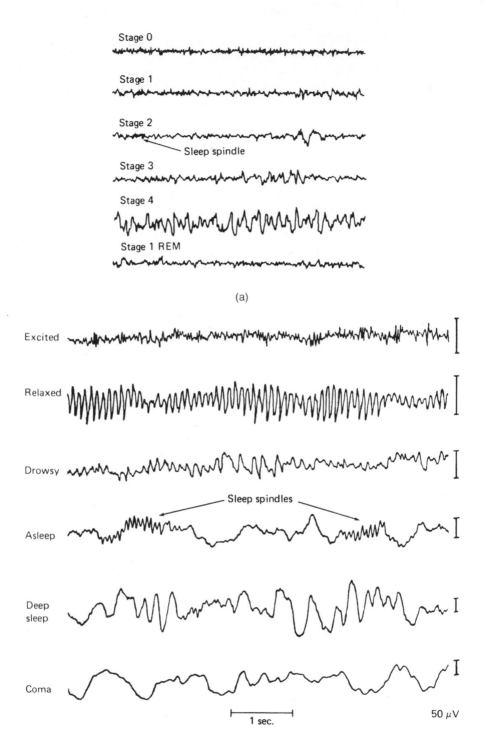

(a)

Stage 0

Stage 1

Stage 2

Sleep spindle

Stage 3

Stage 4

Stage 1 REM

Excited

Relaxed

Drowsy

Sleep spindles

Asleep

Deep
sleep

Coma

1 sec.

50 μV

(b)

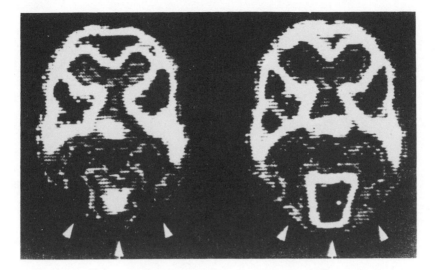

FIGURE **13.2** PET scans of the human brain when (left) eyes are closed and (right) eyes are open. The most conspicuous changes in the level of recorded activity occur in the visual cortex (lower part of the scan). (Photograph by M. E. Phelps et al., in *Neuronal Man: The Biology of Mind,* by Jean-Pierre Changeux, trans. Lawrence Garey. Copyright © 1985 by Pantheon Books. Reprinted with the permission of Pantheon Books, a division of Random House, Inc.)

we can describe—the external world—transcends individual human experience. Its deepest regularities are expressed in the universal language of mathematics. The part we experience directly isn't wholly indescribable, but it does contain an indescribable residue—a residue that makes one person's inner life distinct from another's.

Consciousness has obvious biological value. It enables an animal to respond creatively to novel and unforeseen challenges. It is the hallmark of behavior that is neither habitual nor reflexive nor instinctive. But if, as I have assumed, all events in the external world are governed by physical laws, *the subjective aspect* of consciousness has no functional role. This is true whether or not consciousness is fully reducible to physical phenomena. If subjectivity is an illusion, like the illusion of demonic possession, it is obviously nonfunctional; nonexistent demons

FIGURE **13.1** (a) Electroencephalograms (EEGs) recorded during periods of wakefulness and sleep. Stage 0: awake, eyes open. Stages 1–4: progressively deeper stages of sleep. Stage 1 REM: vivid dreaming, accompanied by rapid eye movements, during light (Stage 1) sleep. (b) EEGs recorded during different stages of arousal. ([a] From *A Manual of Standardized Terminology, Techniques, and Scoring System for Sleep Stages of Human Subjects,* edited by A. Rechtschaffen and A. Kales. Copyright © 1968. Courtesy of the National Institute of Neurological Disorders and Strokes. [b] From *Epilepsy and the Functional Anatomy of the Human Brain,* by W. Penfield and H. Jasper. Copyright © 1954 by Little, Brown and Company. Reprinted by permission)

are impotent. But even if consciousness has an irreducibly subjective component, as I believe it does, that component can't interact causally with events in the external world, because it isn't part of that world. Conceivably, a neurobiologist of the future may understand the workings of a bat's brain-mind as well as an electronics engineer understands the workings of a color television set; yet the biologist would know—and need to know—as little about "what it is like for a *bat* to be a bat" as the engineer would know about what it is like for a color television to be a color television.

The Evolution of Consciousness

Our present understanding of neurobiology stops short of explaining subjectivity. It offers no clues (that I can make out) either to how subjectivity might be annexed to the external world or, if it is irreducible, to why it exists. But neither does our present understanding of physics explain why nature has chosen to express its deepest laws in the language of mathematics. In general, science isn't very good at answering questions of the form: Why does so-and-so exist? It has had more success with questions of the form: How did so-and-so evolve? Even if we don't understand why subjectivity exists, we can still try to understand how the property of having an inner life came into existence in the course of biological evolution.

Surely it wasn't present in the earliest living organisms. Bacteria have complex molecular assemblies for processing information, but there is nothing in their physiology or behavior that would justify attributing an inner life to them. Nor is there any reason to suppose that sponges or plants have an inner life. But it is hard to imagine that we are the only possessors of inner lives. Apes and monkeys remind us too much of ourselves; we empathize too readily with our pets and even with some wild animals. The most plausible hypothesis seems to be that the property of having an inner life came into being at some stage in the evolution of animals. Can we go further and identify the biological basis of inner life?

Because subjective experience is mediated by the nervous system, one might suppose that all organisms that have nervous systems have an inner life. Perhaps the complexity of an animal's inner life is directly related to the complexity of its nervous system. Animals with very simple nervous systems—slugs and leeches, for instance—would have meager and virtually undifferentiated inner lives; those with more complex and highly differentiated nervous systems would have complex and highly differentiated inner lives.

This hypothesis is too simple, however. Consciousness can't be just nervous activity "viewed from the inside." A large portion of the activity of the human nervous system never impinges on consciousness. In particular, the autonomic nervous system, which regulates our internal economy, functions silently and in the dark. It is part of our bodies but not of our selves. This suggests a modification of the original hypothesis: consciousness is associated with the so-called somatic nervous system, which regulates an animal's interaction with its environment by collecting and processing sensory information and activating skeletal muscles.

But is it reasonable to assume that all animals are aware, to some degree, of sensory stimulation and muscular responses? In animals with simple nervous systems, a given stimulus may produce the same muscular response every time. Or the response may depend in a simple and predictable way on the state of the organism. For example, the response to a repeated stimulus may get progressively weaker (or stronger), or it may depend on the animal's state of arousal. In all these cases, the response is like a muscular reflex. Even an animal as high up on the evolutionary scale as the frog responds in a predictable, stereotyped way to specific features in its visual field. Is a frog aware of seeing something when its "bug detector" detects a bug? Or does this retinal event cause the frog's tongue to strike at its target without the intervention of inner experience, in the way that a flash of light causes the pupils of our eyes to contract? The second answer seems to me the more likely.

There is another obvious difficulty with the hypothesis that inner experience is associated primarily with nervous-system activity that mediates interactions between an animal and its environment. For us, at any rate, inner experience has as much to do with internal processes like thinking, remembering, and expecting as it does with sensing and acting. Thus consciousness is neither a necessary nor an exclusive concomitant of sensorimotor nervous activity.

Is there a kind of nervous activity of which consciousness *is* a necessary and exclusive concomitant? In other words, is there an objective criterion for consciousness? Given a sufficiently complete description of another person's brain events, such a criterion would enable us to say, *"Those* patterns of activity are accompanied by subjective experience." Of course, we still wouldn't be able to say what that experience was like, except by analogy with our own experience, but we would be able to tell that it was going on.

Perception as Construction

Perhaps the answer to a simpler question will provide a clue: What distinguishes the frog's way of seeing a bug from our way of seeing a bug? The frog's visual system is what workers in artificial intelligence call a feature detector: it extracts predetermined kinds of information from the changing pattern of light incident on the frog's retina. According to the theory sketched in Chapter 12, human visual perception works quite differently. It is a cyclic process in which the brain constructs, tests, and modifies perceptual hypotheses. Thus *in order to have a percept, we must construct it.* This immediately suggests that the conscious aspect of perception depends on, or is called forth by, the construction of perceptual hypotheses.

This view of perception has a natural place for mental imagery. It has long been known that retinal images give rise to representations of themselves in several distinct regions of the cerebral cortex (the thin outer layer of the main part of the brain). A point of light in the visual field of a monkey causes a small group of nerve cells to fire in each of these cortical regions. All the various cerebral representations are connected, directly or indirectly. Now, if perception involves the construction and testing of "perceptual hypotheses," it is tempting to suppose that the ''hypothesis'' for a visual scene is embodied in one (or more) of the scene's cerebral representations. According to the formula *perception = construction,* seeing the scene is the subjective experience that accompanies the construction of this cerebral representation.

Looking at a scene automatically initiates the perceptual cycle. But internal stimuli, such as the desire to recall the face of a friend or the desire to solve a problem, can also initiate the construction of cerebral representations, and thus, according to our formula, cause us to see images. The fact that such images are normally less vivid and detailed than those produced by light falling on the retina will eventually have to be explained, but is hardly surprising. More relevant to the present argument is the observation that some physical states—for example, sleep, extreme physical exhaustion, and states induced by hallucinogenic drugs—promote the production of vivid internally generated images and that internally produced images vary enormously in vividness and detail from person to person. Visual artists, mnemonists, and some religious mystics all in their separate ways have exceptional image-generating abilities.

The cyclic theory of perception applies to speech and music as well as to visual forms and colors. Everything I have said about visual images applies equally well to speech images and musical images. The quality and vividness of such images, too, vary enormously from person to person. Just as visual images guide the pen or brush of a painter, so musical images guide the fingers and bow of the violinist. All musicians hear ''music in the head''—sometimes to distraction. A conductor studying a printed score ''hears'' the music as he or she reads it; the composer ''heard'' it before committing it to paper. Once again, construction = perception.

This formula also accounts for the observation that a skill or a complex sequence of movements, once mastered, tends to become unconscious. A skilled cyclist doesn't have to concentrate on keeping his balance. A concert pianist performing in public focuses all her attention on the music; she is unaware of her fingers. Patterns previously constructed and laid down in memory don't have to be reconstructed.

Finally, the formula construction = perception offers a clue to the nature of dreaming. According to this formula, the pattern of cerebral excitation produced by the stimulation of a sensory surface is *not* accompanied by conscious awareness. Instead, awareness accompanies the *construction* of a matching or an anticipatory pattern of excitation. As we will discuss later, a precisely analogous account can be given for movement. Muscular contractions are initiated by ''primary'' patterns of cerebral excitation, but awareness accompanies the construction of matching or anticipatory patterns. In principle the constructive part of a perceptual cycle or its motor analogue could occur without sensory stimulation or movement; it could be initiated by events in the brain itself. Presumably this is what happens when we dream.

Psychiatrist J. Allan Hobson has recently developed a detailed psychophysiological theory of dreaming that is consistent with this general idea.[5] Hobson regards dreaming as a state of consciousness that shares with the waking state a property he calls "narratively coherent awareness." He explains this property, and the distinction between the dreaming and the waking states, as follows:

> The brain/mind organizes its percepts, thoughts, and emotions, of whatever provenance, in the form of a scenario. It extracts from and/or imposes upon sensorimotor data certain structures or frames that give them order. These include orientation (me, now, here), intention (going to the store, writing a letter), and tone (feels good, feels scary). In waking, these functions are relatively stable, in part because of interactions with a stable world and in part because of the activity of the brain/mind's own stabilizing mechanisms (aminergic modulation). In dreaming, both external-world and internal stabilization are lost.[6]

What the waking and dreaming states have in common according to this theory, which draws support from a variety of recent neurophysiological observations, are certain kinds of *constructive activity*. In the waking state, the products of this activity are regulated and controlled by external and internal inputs; in the dreaming state, the constructive activity persists, but regulation and control are relaxed.

The preceding discussion suggests that computational models of thought, with their emphasis on the manipulation of fixed symbols, are at best far too narrow. According to the view of mental processes advocated here, thought in its widest sense is a constructive process utilizing internal representations. These representations may be visual images, speech images, or musical images. People whose conscious thought is conducted largely in words find it difficult to imagine that "real" thinking can be carried out in any other way, just as some computer scientists have difficulty imagining that it can be other than disguised computation. The testimony of mathematicians and other visual thinkers, as well as the present theory, suggests that these views are too parochial.

Consciousness as an Emergent Property

Although the view of perception and cognition that I have been defending is not yet securely established, it does seem to be consistent with modern findings about how the brain works. Before turning to these findings, however, I want to say a little more in a philosophical vein about the problem of consciousness and the place of consciousness in the broader picture of the growth of order that I have been sketching.

Some readers will have noticed that I have made consciousness an *emergent* property of living organisms. The notion of emergence generally figures in mystical accounts of cosmic evolution. Most modern scientists and philosophers regard it as unnecessary and obscurantist. So what is it doing in what purports to be a scientific view of the growth of order?

Emergent theories posit that in the course of cosmic evolution there arise properties, structures, or processes that not only didn't exist earlier, but embody some new *and scientifically indescribable* element. So-called emergent evolution is hierarchical. Each new level, which embodies a new emergent principle, includes the preceding levels. For example, the philosopher Samuel Alexander posited five such levels: spacetime, matter, life, mind, and deity.

The countertheory to emergent evolution is radical reductionism, which asserts that all the properties of a complex structure or process are implicit in its components. Radical reductionism, as I have defined it, is clearly untenable. The properties of molecules *depend* on the properties of their constituent atoms but aren't implicit in them. Nor are the distinctive properties of living cells implicit in the properties of their constituent molecules. At each level of the biological hierarchy—molecules, molecular assemblies, organelles, cells, tissues, organs, organ systems, organisms, and communities—something qualitatively new appears. The new ingredient isn't scientifically indescribable, however. It is a new form of order. As we saw in Chapter 10, life itself is characterized by an "emergent" but perfectly intelligible property: reproductive instability. Throughout this book, I have argued that order, although seemingly less concrete than matter and energy, is just as fundamental a constituent of the external world.

The ability to construct internal representations of sensory stimuli, which (I have argued) underlies perception and cognition, is also an emergent property in the scientific sense. It greatly expands an animal's ability to generate useful information about its environment and to act on that information. Viewed objectively, internal representations are perfectly concrete entities, even though we can't yet characterize them precisely. But internal representations also have a subjective aspect: in certain circumstances we are *aware* of them. If my speculation that awareness accompanies certain kinds of constructive activity in the brain should turn out to be correct, we would in principle be able to predict *when* a person was having conscious experience. We might even be able to describe the experience quite precisely. But I think we still wouldn't know what it was like for *that person* to have the experience.

Nor does there seem to be any prospect of understanding *why* awareness should accompany a particular kind of activity in the brain. Functional explanations ("Feeling thirsty is adaptive because it causes me to try to meet my body's need for water") won't work because the functional role they attribute to awareness is already filled by the accompanying neural events. The pattern of neural excitation that accompanies (but neither causes nor is caused by) my feeling thirsty is what initiates and sustains my water-seeking behavior. If, as I have postulated, physical events are governed by universal physical laws, the subjective aspect of my thirstiness has neither consequences nor causes in the external world. What is adaptive isn't my subjective awareness of being thirsty but the pattern of neural excitation that accompanies that awareness. From a scientific standpoint, my private awareness is just a marker or tag that plays no functional role.

But surely, you may argue, consciousness *does* play an important part in human behavior. A person who lacked consciousness would behave like a robot or a zombie. Moreover, we know that impaired or heightened states of consciousness—the inability to experience pleasure or extreme sensitivity to pain—have profound effects on behavior. Such objections don't allow for the necessary connection between the objective and subjective aspects of mental states. A heightened sensitivity to pain must be accompanied (according to the theory I am advocating) by altered patterns of neural activity, and it is these patterns that affect behavior. Someone who lacked consciousness would indeed behave like a robot, if he could function at all, because his nervous system would be functionally impaired in just the ways that would deprive him of the capacity to perform actions normally accompanied by consciousness. Consciousness may be ''just'' a marker, but it is a *reliable* marker.

This view of consciousness suggests an answer to an old philosophical question: Would a nonbiological device, functionally identical with a human brain and furnished with appropriate sensory inputs, be conscious? If, as I have argued, consciousness is an emergent property of living organisms, there is no reason to suppose that it depends only on an organism's functional properties. I think it is more natural to suppose that it also depends on the physical properties of neurons and assemblies of neurons. If that is so, we might in principle be able to construct an *intelligent* robot—in limited ways this has already been done—but it would be impossible for a nonbiological device, however closely its behavior might simulate an intelligent animal's, to possess consciousness.

According to the view I advocate, consciousness is radically contingent. It is a brute fact that could not have been predicted. From the objective standpoint of natural science, it has no rhyme or reason. Hence it is emergent in a metaphysical, nonscientific sense. Is this view of consciousness mystical? The *American Heritage Dictionary* defines *mysticism* as a ''belief in the existence of realities beyond perceptual or intellectual apprehension but central to being and directly accessible by intuition.'' Consciousness is certainly a reality ''central to being and directly accessible by intuition.'' But it is not *beyond* perception; it is the stuff of perception. Nor is it entirely beyond intellectual apprehension, although we can't yet explain why it exists. But energy and electric charge are also realities whose existence we can't yet explain.

14

Brain and Mind

Chapters 12 and 13 presented arguments to support a constructivist account of thought and perception. In this chapter, I will try to connect this account with what is known about the brain and how it works.

A nervous system enables its possessor to collect and process information about the environment and to initiate coordinated muscular responses to that information. Our own nervous system enables us, in addition, to construct internal representations of the outside world, to remember past experiences, to formulate hypotheses about the future, and to act in the light of these hypotheses. All these functions are mediated by networks of specialized cells, *neurons,* that transmit electrical impulses to and receive electrical impulses from sensory surfaces, muscles, and other neurons.

Neurons

In the course of evolution, neurons have differentiated into many distinct types, each adapted to a specific subfunction, but they all share a body plan that enables them to fulfill the function of receiving and transmitting electrical signals. The common plan is shown schematically in Figure 14.1. The main body of a neuron, centered on the cell nucleus, has the same basic structure and performs the same kinds of functions as other animal cells. Within it, fuel molecules are broken down and the energy released in the process is stored in molecules of ATP. This stored energy is used to transcribe and translate appropriate segments of the DNA in the cell nucleus and to run "chemical pumps"—molecules embedded in the cell membrane that maintain the cell's chemical composition.

Dendrite

Cell Body

Axon Hillock

Myelin Sheath

Gap

Axon

Axon Terminal

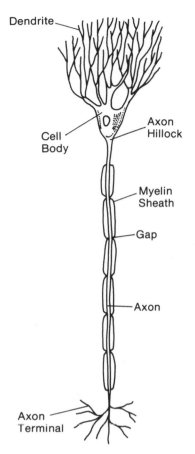

Figure **14.1** Basic body plan of a vertebrate neuron. The cell body contains the nucleus, mitochondria, and other cellular organelles. Voltage spikes originate in the cell body—specifically, in the axon hillock. Dendrites receive incoming signals from the terminals of other axons (not shown in this diagram) with which they synapse. Voltage spikes are transmitted along the axon, which is wrapped in a myelin sheath punctuated by narrow gaps. The axon terminals transmit outgoing signals to the dendrites of other neurons (not shown) with which they synapse. The diagram is not drawn to scale: the cell body is a few hundredths of 1 millimeter across; the distance between gaps in the myelin sheath is typically about 1 millimeter; the thickness of the sheath is a few thousandths of 1 millimeter; and axons reach 1 meter or more in length.

Growing out from the cell body in all directions like a fright wig are multiply branching processes called *dendrites* (literally, "parts of trees"). Nerve fibers from other neurons terminate on these branches as well as on the main cell body. In many, although not all, neurons, one of the excrescences slims down into a long, thin fiber—the *axon*—that connects the neuron to other regions of the nervous system. In vertebrates, long axons are embedded in insulating sheaths punctuated at regular intervals by gaps in the insulation, as illustrated in Figure 14.1. An axon may branch a few times. Each branch ends in a cluster of thinner, multiply branching fibers whose tips make contact with the dendrites and cell bodies of other neurons.

The basic design of neurons evolved before vertebrates and invertebrates diverged from their common ancestor. Among vertebrates, evolution fashioned more complex nervous systems by increasing the number of neurons and by encasing them in insulating sheaths. Invertebrates failed to invent insulation but did invent more complex, multipurpose neurons that helped them to cope with the severe space shortage that results from having an external skeleton. Apart from these differences, the nervous systems of a sea slug, a mouse, and a human being are composed of similar components that communicate in similar ways. They differ most obviously in the number of components. The nervous system of a sea slug contains about 100,000 neurons; that of a mouse, 5 or 6 million; our own, more than 100 billion.

These quantitative differences are mainly by-products of qualitative differences—differences in the way that the neurons are organized in circuits. Corresponding to the differences in behavioral complexity among sea slugs, mice, and humans, we expect to find differences in the organizational complexity of their nervous systems. We expect the nervous system of the mouse to have more levels of organization than that of the sea slug, and our own nervous system to have still more. In addition, we expect the mouse's nervous system to have more *developmental plasticity*—a greater capacity to modify genetically determined patterns of neuronal organization—than the sea slug's, and our own nervous system to have still more. As we will see, the available evidence supports these expectations. But first, we need to understand how neurons generate and transmit electrical impulses.

How Neurons Work

Because electrical signaling is a modern and highly sophisticated product of cultural evolution, it seems surprising at first sight that its biological counterpart arose very early in the history of life. On closer examination it seems less surprising. Electrical signaling requires (1) an electrically conducting medium, such as a copper wire, (2) a difference in voltage, or *electric potential,* as between the positive and negative terminals of a battery, and (3) a means of creating and transmitting changes in electric potential.

The first requirement is met automatically. An electrically conducting medium—gas, liquid, or solid—is one that contains mobile electrically charged particles, or ions. The body fluids of animals are good electrical conductors because they contain dissolved salts, which dissociate into positive and negative ions. Crystals of sodium chloride, for example, dissociate into positive sodium ions and negative chloride ions. Body fluids are salty because they evolved from seawater. Even today the body fluids of marine invertebrates are nearly indistinguishable from seawater in chemical composition. The body fluids of land animals resemble (less closely) dilute seawater.

A difference in electric potential—the second requirement for electrical sig-

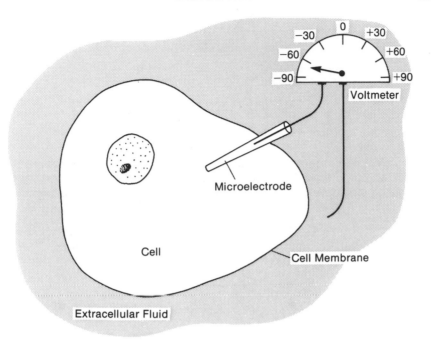

FIGURE 14.2 There is a difference of electric potential, or voltage, across the membrane bounding an animal cell. The interior of the cell is electrically negative relative to the fluid surrounding the cell. The microelectrode, one of the neurobiologist's principal tools, is a glass tube drawn out to a fine point and filled with an electrically conducting salt solution. The voltmeter, which measures differences in voltage between the electrodes, is calibrated in thousandths of 1 volt (millivolts).

naling is a normal feature of animal cells. As illustrated in Figure 14.2, the inside of a cell is like the negative terminal of a battery; the outside, like the positive terminal. The difference in electric potential, or voltage, of the cellular battery is typically several hundredths of a volt.

How does this difference in electric potential arise, and how it is maintained?

Cell membranes are made up largely of identical copies of a single kind of molecule, called a *phospholipid* (Figure 14.3). The head of a phospholipid is a strong acid, soluble in water, while the hydrocarbon tail is insoluble in water. The cell membrane is essentially a double layer of phospholipid molecules arranged as in Figure 14.4. In this orderly configuration—a two-dimensional liquid or liquid crystal—the "hydrophobic" tails of the phospholipids are shielded as much as possible from water molecules, while the "hydrophilic" heads maximize their contact with water molecules.

Phospholipids assemble spontaneously in bilayers, thereby creating order, in apparent violation of the second law of thermodynamics. In fact, the spontaneous assembly of a phospholipid bilayer, like the spontaneous folding of a polypeptide to form a functional protein, generates entropy. The increased order of the "fore-

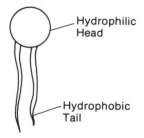

Figure 14.3 Schematic representation of a phospholipid molecule. The "polar head," represented by a circle, tends to associate with water molecules; it is "hydrophilic." The long hydrocarbon tails, represented by wavy lines, are not soluble in water; they are "hydrophobic."

ground'' molecule or molecular assembly is more than compensated by the increased disorder of the "background" water molecules.

The invention of the cell membrane opened the way for the evolution of life as we know it. It enabled free-living self-replicating molecules or colonies of molecules to control their immediate environment—an enormous advantage in the struggle for survival. But a cell membrane composed *entirely* of phospholipids would be useless. Because it would keep out fuel molecules and molecular building blocks, such a membrane would be a shroud rather than a shelter. To reap the benefits of sequestration, the first cells needed membranes that would allow fuel molecules and building blocks to get in but would not allow them to get out—membranes with selective channels. In modern cells these channels are formed by large, complex proteins and protein-containing molecules embedded in the membrane, like nuts in the frosting of a cake (Figure 14.5).

Figure 14.4 Phospholipid molecules in water spontaneously assemble into a double layer in which their hydrocarbon tails are shielded from contact with water molecules and their water-soluble heads maximize their contact with water molecules.

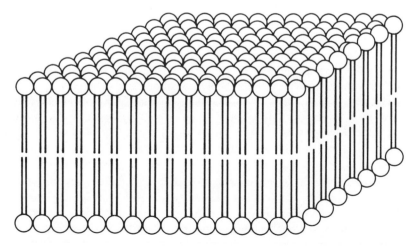

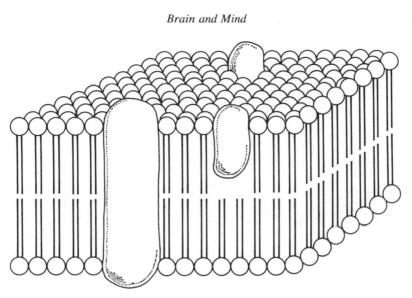

Figure 14.5 Embedded in the cell membrane are a variety of proteins and protein-containing molecules that allow the cell to communicate with and exchange matter with its environment. Some membrane proteins serve as channels that allow specific molecules to enter and leave the cell. Some of these channels are "gated"; that is, they are open only under certain conditions—for example, when the voltage across the membrane is greater than a certain value.

Among these molecules is the *sodium–potassium pump*, which transfers potassium ions from the outside of the cell to the inside, and sodium ions from the inside to the outside (Figure 14.6). Sodium–potassium pumps keep the concentration of potassium ions inside animal cells much higher than in the surrounding fluid, and the concentration of sodium ions much lower. (The caption to Figure 14.6 explains how the pump works.) In mammalian cells the ratios are typically between thirty and forty to one for potassium ions and between ten and twenty to one for sodium ions. Maintaining these ratios against the entropic tendency toward chemical uniformity is expensive. A substantial fraction of the fuel that animals burn is used to make the ATP that drives sodium–potassium pumps embedded in their cell membranes. (ATP molecules bind to sites on the inner surface of the pump.)

Cells derive a variety of benefits from the lopsided distributions of potassium and sodium ions. For example, the difference in sodium-ion concentration enables cells in the intestine and kidney to absorb glucose and amino acids from the surrounding fluid even when the concentrations of these molecules are lower than they are inside the kidney and intestine cells. The glucose and amino-acid molecules hitch rides on carrier proteins to which sodium ions have attached themselves. The sodium ions diffuse entropically from the sodium-rich fluid outside the cell into the sodium-poor cell interior, carrying the glucose and amino acids with them.

Experiments show that cell membranes are good at keeping sodium ions out,

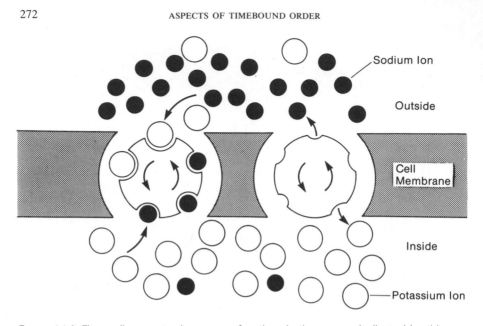

FIGURE 14.6 The sodium–potassium pump functions in the manner indicated by this mechanical model. The pump transfers sodium ions across the cell membrane from inside to outside, and potassium ions from outside to inside. For every three sodium ions that leave, two potassium ions enter. The pump is actually a molecular complex that binds sodium ions and ATP on its inner surface and potassium ions on its outer surface. Changes in the shape of the complex, powered by energy supplied by ATP, create channels that allow the sodium ions to leave the cell and potassium ions to enter it.

but that potassium ions pass back and forth quite freely. What, then, maintains the high concentration of potassium ions inside the cell against the entropic tendency toward uniformity?

It is true that sodium–potassium pumps embedded in the cell membrane keep bringing potassium ions into the cell, but experiments in which the sodium–potassium pumps are disabled by being deprived of ATP show that the rate at which the pumps recruit potassium ions is far slower than the calculated rate at which these ions would diffuse back into the surrounding fluid if nothing else was opposing the diffusion process. In fact, something else *is* keeping potassium ions trapped in the cell: an electric field across the cell membrane. This field pulls potassium ions into the cell at a rate that just balances the rate at which they diffuse from the potassium-rich interior to the potassium-poor exterior (Figure 14.7).

The electric field arises from the fact that the cell membrane is much less permeable to negative ions than to positive ions. The bulk of the negative ions in a cell are proteins to which one or more (usually two) electrons have attached themselves. These negative protein ions are much too big to get through the cell membrane. So when potassium ions leave the cell, in response to entropy's call, they are usually unable to take a negative ion with them. As a result, the departure of potassium ions creates a growing imbalance between positive and negative elec-

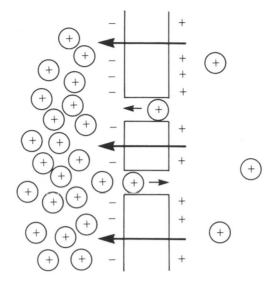

FIGURE 14.7 The cell membrane has permanently open channels through which potassium ions can pass in both directions. Nevertheless, the concentration of potassium ions is much higher inside than outside the cell. The tendency of potassium ions to diffuse out of the cell, down the concentration gradient, is balanced by an electric field (heavy arrows) that tends to push potassium ions back into the cell. This field is a consequence of the cell membrane's impermeability to negative ions. As the positive potassium ions diffuse out, they create a negative electric charge on the inside of the cell membrane. An equal density of positive charge arises on the other side of the cell membrane. This separation of negative and positive electric charge gives rise to the electric field.

tric charge inside the cell. As potassium ions diffuse outward, the interior of the cell becomes more negative and exerts a stronger pull on the escaping positive ions. Eventually a steady state is established in which the voltage drop across the membrane is just large enough to ensure that potassium ions enter and leave the cell at equal rates.

Thus the second requirement for electrical signaling, a voltage drop across the cell membrane, results from the fact that the membrane is much more permeable to positive ions than to negative ions. Its original function, presumably, was to maintain a high concentration of potassium ions inside the cell.

We come now to the property that distinguishes neurons from other cells: their ability to transmit electrical impulses. Neuronal signals and signaling have no analogues in the repertoire of the electrical engineer. Although nature's solution to the problem of long-distance communications is now understood very well, thanks largely to the work of Alan Hodgkin and Andrew Huxley at Cambridge University, no engineer would think of copying it.

The key events in the formation and propagation of an electrical impulse in a neuron are the opening and closing of *voltage-gated ion channels* in the cell membrane. Ion channels are formed by large molecules embedded in the cell membrane (Figure 14.5). They permit specific ions to enter and leave the cell. The membrane's permeability to a specific ion depends on how many open channels for that ion it contains. Cell membranes typically contain many permanently open potassium channels and relatively few permanently open channels for other ions. The cell membranes of neurons also contain channels for potassium, sodium, calcium, and chloride ions that are closed when the membrane voltage has its normal or "resting" value but begin to open as the membrane voltage becomes less negative. The permanently open ion channels control the cell membrane's resting voltage; the voltage-gated sodium channels subserve the production and propagation of voltage *changes*.

A nerve impulse is initiated by the sudden opening of sodium channels in the axon "hillock" (Figure 14.1). Leaving aside, for the moment, the causes of this event, let's consider its consequences. Because the concentration of sodium ions is much higher outside the axon than inside, sodium ions immediately rush in through the open channels. The part of the membrane that contains these open channels is actually more permeable to sodium ions than to potassium ions. Enough sodium ions therefore pass into the cell to neutralize and then reverse its negative electric charge, causing the voltage across the membrane to increase (in the region of open sodium channels) from about −70 millivolts to about +50 millivolts.

Although the opening of sodium channels causes a dramatic local change in the voltage across the cell membrane, it hardly alters the concentrations of sodium ions. Positive and negative electric charges are almost perfectly in balance, inside as well as outside the cell. A very slight excess of negative ions maintains the resting voltage of −70 millivolts, and the inflow of a relatively small number of positively charged sodium ions suffices to reverse the imbalance and create a positive voltage.

After about 1 millisecond, the sodium channels close spontaneously and electrical control reverts to the potassium ions, which diffuse out of the axon until the normal voltage across the cell membrane has been restored. (Actually, some additional potassium channels open up just before the sodium channels close.) Thus a sudden opening of sodium channels at one end of an axon produces a narrow voltage impulse or spike (Figure 14.8).

What happens to this voltage spike? An axon is like an insulated copper cable immersed in seawater. If you produce a voltage spike at one end of such a cable, it will propagate along the cable at nearly the speed of light, growing progressively weaker along the way. Axons, however, would make terrible cables. The interior of an axon is a relatively poor conductor of electricity, and the cell membrane is a relatively poor insulator. As a result, a voltage spike is considerably attenuated after traveling only 1 millimeter or so from its point of origin. Yet some axons transmit signals over distances of 1 meter or more. How is this possible?

The answer is that the voltage spike is regenerated along the way. As the spike travels into a neighboring region of the axon, the voltage there begins to increase (become less negative). This causes some of the voltage-gated sodium channels to open. Sodium ions rush in, causing the voltage to increase further. This, in turn, causes more channels to open. Thus there is positive feedback between the sodium channels and the voltage across the axon membrane. If the voltage increases above

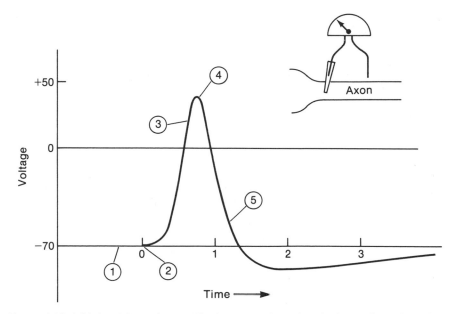

F<small>IGURE</small> **14.8** Initiation of a voltage spike in an axon. At time 1, the sodium channels are closed and the membrane voltage has its resting value (−70 millivolts in this example). At time 2, the sodium channels in a narrow region of the axon open, sodium ions rush in, and the voltage across the membrane climbs steeply. At time 3, somewhat less than 1 millisecond later, voltage-dependent potassium channels begin to open. The sodium channels close at time 4. The voltage-dependent potassium channels close at time 5. Thereafter, the voltage across the membrane returns slowly to its normal resting value. Because more than the normal number of potassium channels are open between times 4 and 5, the voltage dips below its resting value after time 5.

a certain threshold—about −60 millivolts—channels open at an exponentially increasing rate. Increasing the voltage is like increasing the depth of snow on a steep slope: beyond a certain value the result is an avalanche.

A voltage impulse travels along a neuron the way a spark travels along a fuse. The spark heats up and ignites the material just ahead of it. The voltage impulse opens sodium channels and causes an inflow of sodium ions in the region of the axon just ahead of it (Figure 14.9). Sodium channels that have just opened and closed need a brief recuperative period (a few milliseconds) before they can reopen. So a voltage impulse, again like a spark on a fuse, always travels outward from its starting point.

In vertebrates, long axons are wrapped in cells specialized to form thick insulating sheaths (Figure 14.1). These sheaths (called myelin) are interrupted at intervals of 1 millimeter or so by gaps, and the sodium channels are concentrated in the axon membrane that is exposed there. The voltage impulse is regenerated at the gaps, which act as relay stations. The myelin sheath allows an impulse to propagate between gaps with much less attenuation than it would experience in an

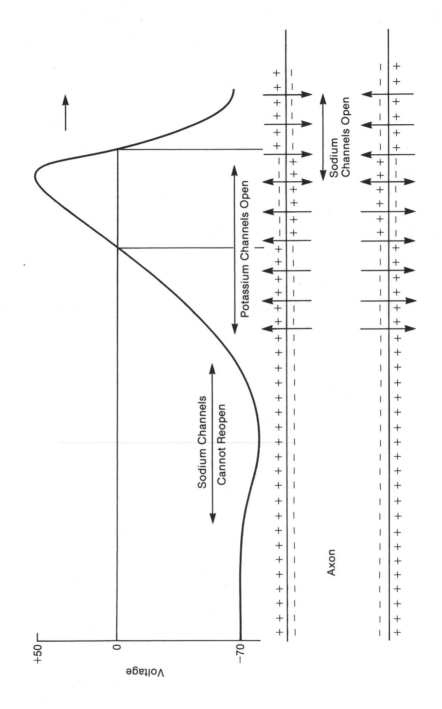

uninsulated axon of the same diameter. Thus myelinated axons conduct voltage impulses more efficiently and more rapidly than unmyelinated axons.

The invention of myelination played a key role in the evolution of vertebrates. Neuropsychologist Richard F. Thompson describes it in these terms:

> The evolution of myelinated nerve fibers in the vertebrates was a tremendous advance. Cells could be fast, yet small. Myelin is probably the primary factor that made possible the greatly increased complexity of the vertebrate brain. Invertebrates typically have many fewer neurons, which form specialized circuits that do particular things. If the circuits have to be fast, and some do so that the animal can eat and avoid being eaten, the neurons must be very large. With myelin the vertebrate brain could develop many more neurons that have less specialized functions because myelinated neurons are fast but still small in size. We have the lowly glial cell, which forms the myelin sheath around the nerve fiber, to thank for the ultimate development of the human brain.[1]

Synapses, Neurotransmitters, and Neuromodulators

Voltage changes are transmitted from sensory receptors to neurons, from neurons to other neurons, and from neurons to glands and muscle fibers. How does this transmission occur?

Axons and dendrites terminate in fibers tipped with tiny knobs. Viewed through an ordinary (light) microscope, these knobs seem to make direct contact with parts of other neurons. This observation, made around the turn of the century, suggested that neurons are in direct electrical contact like wires at a soldered joint. But other, less direct, kinds of experimental evidence suggested that specific chemical substances mediate the transmission of electrical signals from neuron to neuron.

The issue was settled in the 1950s when the electron microscope revealed the actual structure of neuronal junctions, or *synapses.* It turned out that some synapses are indeed direct electrical contacts, welding the neurons they join into a single electrical unit. Most junctions, however, are chemical synapses. At an electrical synapse the gap is less than 2 nanometers (two ten-millionths of a centimeter) wide. At a chemical synapse the width of the gap separating the presynaptic knob from the postsynaptic membrane is typically twenty times greater, but still

FIGURE 14.9 Propagation of a voltage spike along an axon. The inflow of positive sodium ions through momentarily open sodium channels in the cell membrane produces a local voltage spike that spreads, causing the voltage across adjacent regions of the cell membrane to become less negative. When the voltage in an adjacent region exceeds its threshold, sodium channels there open at an exponentially increasing rate, and a new voltage spike is produced. Channels that have just been opened cannot reopen immediately. Hence the voltage spike propagates away from its place of origin at the axon hillock. Because the voltage spike travels at constant speed along the axon, its profile at a given moment is the mirror image of the profile shown in Figure 14.8, which represents the time course of the voltage at a fixed point.

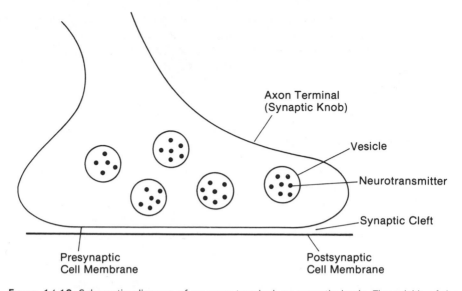

FIGURE 14.10 Schematic diagram of an axon terminal, or synaptic knob. The width of the gap separating the presynaptic and postsynaptic membranes (the synaptic cleft) is about one-tenth the wavelength of visible light. The molecules that act as chemical messengers across the gap are stored in small sacs, or vesicles.

less than one-tenth the wavelength of visible light (Figure 14.10). The knob contains vesicles filled with molecules called *neurotransmitters*, synthesized in the cell body. When a voltage impulse reaches one of these knobs, it opens voltage-gated channels in the cell membrane, allowing *calcium ions* to flow in. These ions initiate a complicated series of chemical reactions that permit the neurotransmitters in the vesicle to escape from the knob by causing the membranes that bound the vesicles to merge with the membrane that bounds the knob (Figure 14.11). Some of the neurotransmitters bind to matching receptors on the postsynaptic membrane, causing ion channels there to open. That is how signals are transmitted at chemical synapses.

The transmitted signals are of two distinct kinds: excitatory and inhibitory. The reception of an excitatory signal makes it easier for a neuron to fire; the reception of an inhibitory signal makes it harder. Whether a signal is excitatory or inhibitory depends on both the transmitter molecules and the receptors at the postsynaptic membrane. An excitatory signal opens sodium channels in the post-synaptic membrane, allowing positive sodium ions to enter the cell and causing the voltage across the membrane of the receiving neuron to become less negative. An inhibitory signal opens potassium or chloride channels, allowing positive ions to flow out or negative ions to flow in, thereby making the voltage across the membrane more negative. The magnitude and frequency of the chemical signals that a neuron initiates are determined by the combined effects of the excitatory and inhibitory signals that it has received in the immediate past.

Inhibitory signals play a key role in the functioning of the nervous system.

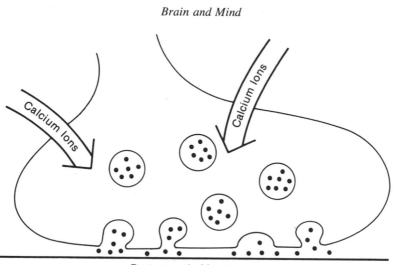

Postsynaptic Membrane

FIGURE 14.11 The arrival of a voltage spike at an axon terminal causes calcium-ion channels to open, allowing calcium ions to flow into the terminal. (Calcium-ion pumps maintain a low concentration of calcium ions inside the axon.) The calcium ions trigger a series of chemical reactions that cause the membranes bounding some of the vesicles to merge with the presynaptic membrane, allowing the neurotransmitters in the vesicles to enter the synaptic cleft. Some of the neurotransmitters then bind to specific receptors on the post-synaptic membrane.

When I lift a finger, one set of muscle fibers contracts and the opposing (antagonistic) set relaxes. Neurons that excite fibers belonging to the first set also excite neurons that inhibit the excitation of fibers belonging to the second set, and vice versa, as illustrated schematically in Figure 14.12. Inhibitory signals also make possible *negative-feedback loops,* a ubiquitous feature of neuronal circuitry. For example, the axon of a motoneuron (a neuron that excites a muscle fiber) has branches that transmit signals to neurons that send inhibitory signals back to the first neuron and to neighboring motoneurons (Figure 14.13). John Eccles, a pioneer investigator of synapses and neuronal circuits, describes the function of this circuitry in the following terms:

> When one group of motoneurones is firing strongly, it will exert a strong feedback inhibition . . . on the whole ensemble of motoneurones in their neighborhood no matter what they are doing. As a consequence there is suppression of all weak discharges. Only the strongly excited neurones survive this inhibitory barrage. In this way the actual motor performance is made much more selective by the negative feedback eliminating the stray weakly responding motoneurones that would frequently cause some disorder in the movement.[2]

In a later passage, Eccles characterizes the role of inhibition in more general terms:

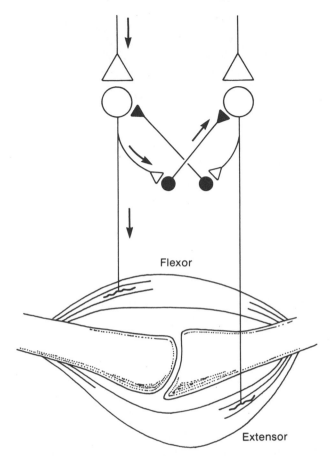

Flexor

Extensor

FIGURE 14.12 Reciprocal inhibition. Muscles at joints are arranged in antagonistic pairs. Contraction of the flexor reduces the angle between the bones that meet at the joint; contraction of the extensor increases the angle. In this highly simplified diagram, an excitatory signal to a flexor (arrow at top left) is transmitted by an axon that branches. One branch carries a voltage spike to a fiber of the flexor. The other branch synapses with a neuron (drawn in black) that makes an inhibitory input to the cell body of the neuron that transmits signals to the extensor. Thus excitation of the flexor automaticlaly makes it more difficult to excite the extensor, and vice versa. In this diagram, circles represent cell bodies, triangles represent axon terminals, and neurons that carry inhibitory signals are drawn in black.

> I always think that inhibition is a sculpturing process. The inhibition, as it were, chisels away at the diffuse and rather amorphous mass of excitatory action and gives a more specific form to the neuronal performance at every stage of synaptic relay. This suppressing action of inhibition can be recognized very clearly at higher levels of the brain—particularly in the cerebellum.[3]

When I lift my finger, I'm not aware of the complex interplay between neuronal excitation and neuronal inhibition that underlies this apparently simple act.

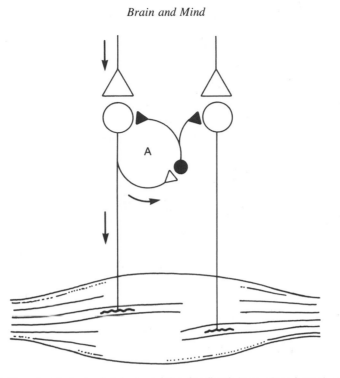

FIGURE 14.13 The circuit marked A is a negative-feedback loop. One branch of an axon carrying an excitatory signal to a muscle fiber transmits a signal to an inhibitory neuron (drawn in black), which inhibits the firing of the first neuron and of a neighboring neuron.

But there is one part of our lives where, if we stop to think about it, we can see at once that inhibition must play a critical role. I mean sleeping and dreaming. In sleep, our sensory thresholds are raised. We aren't aware of soft voices, dim lights, gentle pressure, or minor aches and pains. This commonplace observation suggests that sensory information isn't being relayed from the neurons that collect it to higher brain centers—or at least isn't being relayed as effectively as in the waking state. In 1978 neurophysiologist Otto Pompeiano showed experimentally that during sleep, sensory neurons are indeed inhibited by impulses that originate in the brain stem. (As discussed later, the brain stem is the seat of neuronal activity that regulates states of arousal like waking, sleeping, dreaming, and attention.) Again, in a dream I may feel myself running, swimming, or flying, but beyond the odd twitch, my muscles don't carry out the commands that my mind (and hence my brain) seems to be sending. In fact, skeletal muscles are in no condition to respond to commands: when dreaming begins they lose their normal resting tension and firmness, becoming flaccid and useless. Pompeiano's experiments showed why. Impulses originating in the brain stem during dreaming inhibit the motoneurons, preventing them from responding to commands issued by higher brain centers.

According to a widely accepted and now strongly corroborated hypothesis put forward in 1977 by J. Allan Hobson and Robert McCarley, waking and sleeping are controlled by two distinct populations of neurons in the brain stem. Recordings made with microelectrodes show that as the activity of one population waxes, the activity of the other wanes, and vice versa. When the first population is turned on, some of its members inhibit motoneurons, others prevent the transmission of sensory information to the brain, and still others excite muscles that produce rapid eye movements during vivid dreams. When the neurons of this population are turned on, they feed back excitatory impulses to themselves and to one another— an inherently unstable arrangement that, unopposed, would produce exponentially increasing activity. The same neurons, however, also excite neurons belonging to the second population, the population that is turned on during waking. *Those* neurons, when excited, *inhibit* the activity of neurons belonging to the first population and thus suppress, or at least temper, that population's potentially unstable behavior. Neurons belonging to the second population also tend to inhibit their own activity—an inherently stable arrangement. Thus the model neatly accounts both for the stability of the waking state (the fact that variations in our state of alertness, while awake, are relatively small, gradual, and controlled) and for the instability of sleep, with its frequent and erratic swings between deep dreamless sleep, vivid dreaming, and (especially in the elderly) intervals of wakefulness.

So far, we have discussed what happens at the simplest chemical synapses: the arrival of a voltage impulse at the knoblike terminal of a neuron initiates a sequence of events leading to the release of molecules into the synaptic cleft. When these molecules bind to receptors in the postsynaptic membrane they cause sodium, potassium, or chloride channels to open, thereby exciting or inhibiting voltage impulses in the postsynaptic neuron. The entire transaction, from the arrival of the first voltage impulse to the excitation or inhibition of a second impulse, takes about one-thousandth of a second. But at some synapses the binding of chemical messengers to the postsynaptic membrane has more complicated and longer-lasting effects. Instead of exciting or inhibiting a voltage impulse in the postsynaptic neuron, the binding of these molecules (called *neuromodulators*) initiates chemical reactions that change the way the neuron will respond to subsequent chemical signals. For example, neuromodulators may alter the permeability of specific ion channels. The neuronal populations that control stages of sleep and wakefulness presumably do so by modifying the excitability or "inhibitability" of large groups of neurons (primary sensory neurons, motoneurons, oculomotor neurons, and so on), although the cellular and molecular mechanisms involved are not yet fully understood.

A few instances of neuromodulation have, however, been studied in some detail. For example, neurobiologist John Dowling and his collaborators have studied the effects of dopamine in the retina of fish. They found that when dopamine molecules bind to "horizontal cells" in the fish retina (so called because they make connections between neurons that transfer information "vertically," that is, from the rods and cones toward the brain), they activate enzyme systems within these cells. The resulting changes in the observable properties of the horizontal cells include a decrease in their responsiveness to light incident on the retina and a reduction in the quantity of neurotransmitter they release when stimulated by a voltage impulse. Thus dopamine tends to reduce horizontal communications within the fish retina; it reduces the interaction between neurons carrying signals that originate at different points in the retina. Although only a small population of neurons in the fish retina releases dopamine, these neurons make contact with a large population of horizontal cells. The dopamine-releasing neurons therefore *regulate* the activity of horizontal cells.

Dowling and his colleagues found that the changes produced by dopamine in the fish retina last fifteen minutes or more. Other experiments have given evidence of even longer-lasting neuromodulatory changes. Such findings suggest that neuromodulation may be the cellular process that underlies learning and memory. Experiments by Eric Kandel and his collaborators support this idea. These experiments show that simple kinds of learning in the marine snail *Aplysia* are associated with changes in the efficacy of chemical transmission at specific synapses. Habituation of the withdrawal reflex—a gradual reduction in the snail's response to a repeated stimulus—results from a weakening of chemical transmission at particular synapses; sensitization—a sudden heightening of the snail's reflexive responses following an especially strong stimulus—results from a strengthening of chemical transmission at another set of synapses. Both habituation and sensitization can persist for long periods.

The Hierarchy of Neuronal Circuits

The circuits shown schematically in Figures 14.12 and 14.13 are called *local circuits.* A local circuit contains one or more *input fibers,* which carry signals into the circuit, and an *output neuron,* which carries signals out of the circuit. Most, although not all, local circuits also contain *interneurons,* whose connections lie entirely within the circuit. Interneurons may transmit either excitatory or inhibitory signals, and the three elements may be connected in many different ways besides those shown in Figures 14.12 and 14.13. The variety of possible local circuits is thus very great, and many of the possibilities are actually realized.

Local circuits form the bottom rung of a hierarchy of neuronal circuits. Neurobiologist Gordon Shepherd distinguishes four higher levels of neuronal organization. The second level comprises circuits that connect local circuits within the same region of the nervous system. "Their function characteristically appears to be to spread activity from a site, or to provide for antagonistic interactions between neighboring integrative units within a region."[4]

The third level of the neural hierarchy is made of circuits that connect different regions with one another.

These circuits are, in turn, embedded in more comprehensive systems, such as motor pathways that transmit commands to muscles, and sensory pathways that transmit information about the environment to the central nervous system.

Finally, at the fifth and highest level of organization are systems that mediate behaviors of the organism as a whole.

The hierarchy of neuronal circuits is fundamentally unlike military, ecclesiastical, and political hierarchies, in which commands flow in one direction only along lines that fan out from top to bottom, as in Figure 14.14a. The functional

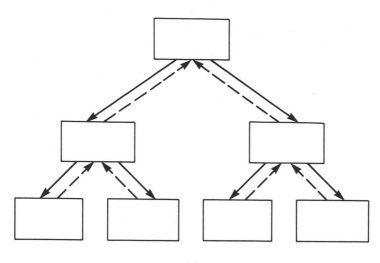

(a)

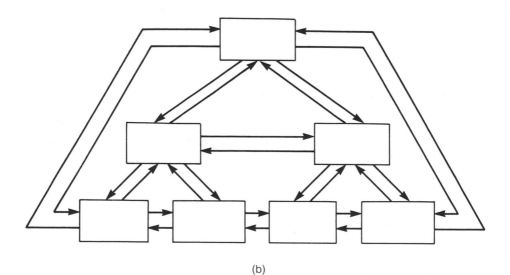

(b)

FIGURE 14.14 (a) Fragment of a military hierarchy. Solid lines indicate the flow of instructions; dashed lines, the flow of information. (b) Fragment of a neuronal hierarchy. There is direct communication between neurons on the same hierarchical level and between neurons on noncontiguous levels of the hierarchy. Neuronal signals may be either excitatory or inhibitory (Figures 14.12 and 14.13).

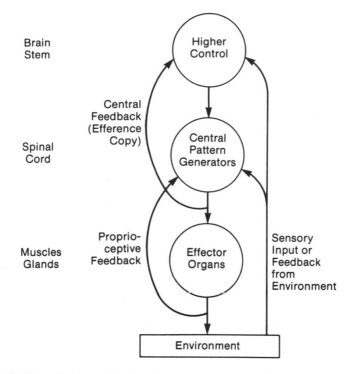

FIGURE 14.15 Hierarchical organization of neuronal circuits governing the control of movement. Note the nesting of feedback loops. (From *Neurobiology*, by Gordon M. Shepherd. Copyright © 1988, by Oxford University Press. Reprinted by permission of Oxford University Press)

organization of neuronal hierarchies is far more complex, as illustrated in Figure 14.14b. Information flows in both directions, and the lines of flow fan out in both directions. Feedback loops are present at every level and are nested one within another. Consider, for example, the motor pathway illustrated schematically in Figure 14.15. At every level, input signals are processed before being transmitted to lower levels and are modified by information fed back from lower levels. Moreover, although commands are passed along from higher to lower levels, they may also be transmitted directly between levels separated by several rungs.

Luria's Picture of the Brain's Functional Organization

Hierarchical structure manifests itself in the brain's *functional organization* as well as in the organization of neural circuitry. A. R. Luria, in *The Working Brain*, distinguishes three main functional units of the brain, each with its own system of anatomical structures. The three systems are strongly interconnected. Neural pathways carry information in both directions between every pair of systems, and every form of conscious activity involves all three working in concert.

The first functional unit regulates states of arousal: waking, sleeping, dreaming, and attention. The second unit subserves the reception, processing, and storage of sensory information. The third unit organizes conscious activity:

> Man not only reacts passively to incoming information, but creates *intentions,* forms *plans* and *programmes* of his actions, inspects their performance, and *regulates* his behaviour so that it conforms to these plans and programmes; finally, he *verifies* his conscious activity, comparing the effects of his actions with the original intentions and correcting any mistakes he made.[5]

All these activities are subsumed by the third of Luria's three functional units.

In the passage just quoted, Luria is talking about action, in the conventional sense, but his account of perception is exactly analogous:

> *Perception is an active process* which includes a search for the most important elements of information, their comparison with each other, the creation of a hypothesis concerning the meaning of the information as a whole, and the verification of this hypothesis by comparing it with the original features of the object perceived. The more complex the object perceived, and the less familiar it is, the more detailed this perceptual activity will be. Both the direction and character of these perceptual searches vary with the nature of the perceptual task, as is clearly shown by eye movements recorded during the examination of a complex object.[6]

Both passages epitomize the view discussed in Chapters 12 and 13 that *conscious activity is a constructive, cyclic process mediated by schemata.* Schemata serve to direct conscious activity, and the outcomes of conscious activity modify the schemata that direct them. In *The Working Brain,* Luria marshals an impressive array of neurological evidence to support this view.

Each of Luria's three functional units and each of the three sets of anatomical structures that subserve them is hierarchically organized.

The functional unit that mediates arousal and attention has its seat in a system of neurons named—perhaps by an admirer of Samuel Johnson—the *reticular formation.* (Dr. Johnson defined a net as "anything reticulated or decussated at equal distances, with interstices between the intersections.") The reticular formation is not in fact a nerve net, as it was once thought to be, but a very dense assembly of neurons. The cell bodies of these neurons form dense clumps or nuclei in the brain stem, and their axons fan out to all parts of the brain. Impulses transmitted by neurons of the reticular formation wake up the brain and also put it to sleep; they maintain and regulate "cortical tone." Other neurons carry messages from the cortex and the cerebellum back to the nuclei of the reticular formation, completing a feedback loop that enables the reticular formation to monitor activity in the higher brain centers and tailor its outputs to their needs.

The first functional unit is highly differentiated. For example, Luria distinguishes three broad classes of signals that activate the awakening or arousal apparatus: internal metabolic signals, such as thirst, hunger, and need for oxygen; external signals, like an alarm clock; and higher mental activity, such as antici-

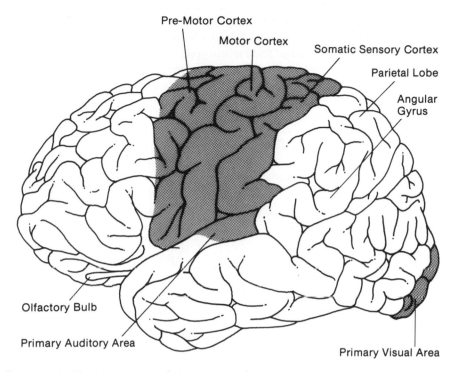

FIGURE **14.16** The human cortex (left hemisphere). Visual information gathered by rods and cones in the retina is projected onto (transmitted by neurons to) the primary visual area, which occupies an area at the back of the cortex. Also indicated in the diagram are the primary projection area for auditory information, the olfactory bulb, the motor and pre-motor cortexes, and the somatic sensory cortex.

pation. Each class of signals recruits the arousal apparatus through a distinctive set of pathways.

Luria divides the second unit, which receives, processes, and stores sensory information, into three anatomically distinguishable (if not actually distinct) zones that form a functional hierarchy. The first zone comprises the *projection areas* of the cortex. Visual information is projected from the retinas by means of a relay station to a specific area of the cortex (Figure 14.16), where it is mapped "topographically," as though the retinal image has been photographed on a rubber sheet that was then stretched in a nonuniform way. The central part of the retinal image, where receptors are most densely distributed and where vision is most detailed, gets the lion's share of the cortical map. This is partly because cortical neurons, unlike rods and cones in the retina, are distributed with uniform density. Since there are more receptors per unit area in the central part of the retina than in the periphery, they require proportionally more area in the cortical map. But in addition, and even more important, the mapping itself is nonuniform in a way that favors the central part of the retina and discriminates against the periphery, somewhat in the way a joke map of the world-as-seen-by-a-New-Yorker reduces Asia to the size of Queens and makes Staten Island bigger than Australia.

Analogously, auditory information is mapped topographically by pitch in its own projection area in a different part of the brain. Information that enables an animal to keep its balance, gathered by receptor organs in the inner ear, has its own projection area. Still another part of the cortex contains a topographic map of general sensory information (touch, pressure, pain) provided by receptors in the skin, skeletal muscles and joints, and internal organs. Here, too, regions of the sensory surface where receptors are most dense command the lion's share of the cortical map. Thus in the human cortex, the areas that correspond to the lips, mouth, and tongue, and to the hand—particularly the thumb—are greatly enlarged.

The second zone of the informational unit consists of *association areas*. Anatomically, these areas lie above and around the primary projection areas. They play an essential role in the processes—still not well understood—by which meaningful wholes are synthesized from the highly specific and localized outputs of the primary projection areas. Patients with lesions in the secondary visual zone suffer from what Freud called "visual agnosia." Unlike patients with lesions in the primary visual projection area, their visual field is intact; they are not partially or wholly blind. But they don't understand what they see; they see but they don't perceive.

> This is a typical case of a patient with such a lesion. This patient carefully examines a picture of a pair of spectacles. . . . "There is a circle . . . and another circle . . . and a stick . . . a cross-bar . . . why, it must be a bicycle?"[7]

As this example illustrates, such patients' ability to *reason* about what they see is unimpaired—indeed heightened, because they have to rely on it so heavily. The title story of Oliver Sacks's book *The Man Who Mistook His Wife for a Hat* describes a musician suffering from severe visual agnosia who learned to carry out complicated tasks dependent on visual recognition, such as dressing himself, by setting them to music.

The third zone of the informational unit plays an essential role in the construction of still higher-level schemata, including those that involve two or more sensory modalities. Anatomically, it is nestled conveniently between the secondary projection zones, like Switzerland between France, Italy, and Germany. Our information about its function, like most of the information on which Luria's picture rests, comes mainly from observations of patients with brain lesions.

> Patients with . . . lesions [of the third zone] experience difficulty in grasping the information which they receive as a whole; they cannot fit together the elements of incoming impressions into a single structure; they cannot convert the consecutive presentation of the elements of a new situation into the new quality of simultaneous perceptibility; they can no longer find their bearings in space and their attempt to do so is converted into a series of disconnected and fragmentary attempts at orientation.[8]

All the various zones belonging to Luria's second functional unit—the unit we have just been discussing, concerned with sensory information and its processing—lie in the back (posterior) part of the cortex. The third functional unit, which directs and regulates activity, occupies the front (anterior) part of the cortex. Luria divides it, too, into three anatomically and functionally distinct zones.

The first zone, called the *motor cortex,* is made up of neurons that send commands to the muscles through the spinal column. Each neuron activates a single muscle fiber, just as each neuron is a primary projection area of the second functional unit responds to a specific stimulus. Like the sensory projection zones, the motor cortex is topographically organized (Figure 14.17). The motor cortex, which sends commands to the muscles, and the adjacent somatosensory cortex, which receives sensory information from muscle, joint, and skin receptors, are strongly interconnected and form a single functional system. Their differentiation into more or less distinct systems is a relatively recent evolutionary development.

The second zone, the *premotor cortex,* plays an essential role in integrating individual muscle twitches into elementary organized movements—stepping, grasping, bending, and the like. Thus it is analogous to the secondary (association) zones of the informational unit, which subserve the construction of meaningful sensory patterns from abstract features. Just as patients with lesions of the secondary visual zone have difficulty recognizing patterns, so patients with lesions of the premotor cortex have difficulty performing skilled movements smoothly. And just as patients with lesions of the secondary visual zone can reason about patterns they can't recognize, so patients with lesions in the premotor cortex can string together sequences of voluntary movements but are unable to join these movements into smooth, seamless "melodies."

In the course of vertebrate evolution, the premotor cortex has expanded much more rapidly than the motor cortex. In marmosets the motor cortex is four times as large as the premotor cortex; in humans the ratio is nine to one in the opposite direction. Thus the capacity for performing complex movements has increased far more in the course of primate evolution than the basic repertoire of muscular contractions.

The third and highest zone of the functional system that organizes and directs activity occupies the *prefrontal* region of the cortex. This is the part of the brain that expanded most rapidly during hominid evolution. As a result, human brains differ most obviously from the brains of chimps and gorillas in the sheer size of their frontal lobes. It is therefore not surprising that the frontal lobes are implicated in all distinctively human behaviors—in prediction and planning, in learning complex skills, in voluntary acts, in creative activity of all kinds. Lesions of the frontal lobes do indeed impair just these abilities and behaviors.

The Neural Basis of Creative Activity

But what *is* creative activity? According to the view advocated in Chapters 12 and 13, creative activity is akin to perception: both involve the construction or modification of schemata. But what are schemata at the neuronal level of description?

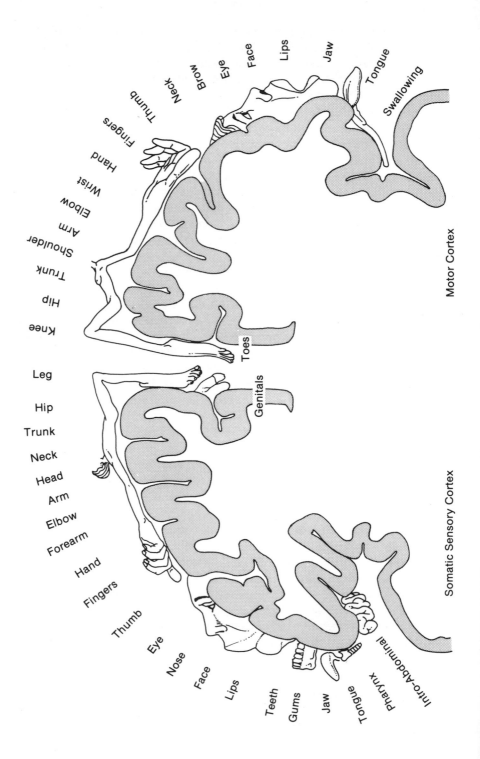

Motor Cortex

Somatic Sensory Cortex

Thumb
Neck
Brow
Eye
Face
Lips
Jaw
Tongue
Swallowing

Fingers
Hand
Wrist
Elbow
Arm
Shoulder
Trunk
Hip
Knee

Toes
Genitals

Leg
Hip
Trunk
Neck
Head
Arm
Elbow
Forearm
Hand
Fingers
Thumb
Eye
Nose
Face
Lips
Teeth
Gums
Jaw
Tongue
Pharynx
Intra-Abdominal

And how do they get modified? I will try to approach these questions, which don't yet have definitive answers, by discussing what may appear to be (but will prove not to be) an unrelated question: What accounts for the explosion of brain size during mammalian evolution, and in particular during primate evolution?

We humans don't, of course, have the biggest brains. The brains of elephants and whales are much larger. Nor do we have the largest ratio of brain weight to body weight: marmosets have higher ratios than ours. Is there an objective index of brain size that takes body size into account and yields a sensible answer (namely, that we are brainier than much smaller and much larger animals)? Roland Bauchot and Heinz Stephan[9] plotted the logarithm of brain weight against the logarithm of body size for species belonging to several behaviorally similar groups: lower insectivores, higher insectivores, lower monkeys, higher monkeys, great apes, and humans (Figure 14.18). In this plot, points representing members of the same group fall close to the same straight line, which accordingly represents the relation between brain weight and body weight for animals with similar behavioral repertoires. The straight lines for different groups turn out to have much the same slope. So we can "correct" brain weight for body weight; that is, we can refer all the brain weights to a standard body weight. The resulting "normalized" brain weight is lowest for the lower insectivores, higher for monkeys, and substantially higher for humans than for monkeys and apes. It's about the same for higher monkeys and apes.

This index doesn't, however, take into account *qualitative* changes in the architecture of the brain. As we have seen, *higher brain functions depend on neurons in the cerebral cortex* (or, to be anatomically more precise, the *neocortex*). This thin, bark-like outer layer, which in the human brain is highly convoluted and covers the two cerebral hemispheres like a mobcap, has expanded at a far higher rate than the rest of the brain. Between insectivores and higher monkeys, the "normalized" weight of the neocortex increased twentyfold; between higher monkeys and apes, it tripled; and between apes and humans, it tripled again— truly an explosive increase. What accounts for it? What selective advantage did this new part of the brain confer?

I suggest that the expansion of the neocortex made possible an evolutionary transition from *visual feature-detection* to *visual perception*. By "feature-detection" I mean the kind of processing performed by the frog's eye—an analysis of the visual stimulus that, after a fixed number of steps, leads to a decision about whether the stimulus contains certain predetermined features. By "percep-

FIGURE 14.17 The motor and somatic sensory cortexes (Figure 14.16) are topographically organized. Commands to muscles in a specific part of the body originate in a specific area of the motor cortex; sensory information from a specific part of the body comes to a specific part of the somatic sensory cortex. The cortical area devoted to a particular part of the body is proportional to the number of muscle fibers or sensory receptors in that part. The diagram shows the left somatic sensory area (which receives inputs from the right side of the body) and the right motor area (which issues commands to the left side of the body).

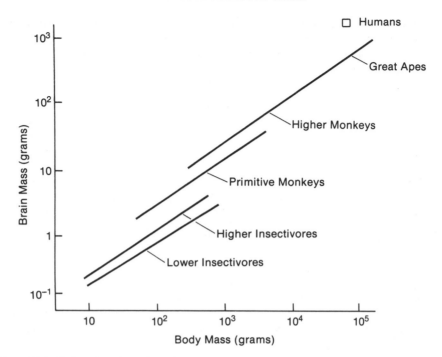

FIGURE **14.18** In a log–log plot of brain mass against body mass, the points for animals with similar behavioral repertoires fall close to a straight line with a slope of 2/3. Thus in such a group, the brain mass of an animal is proportional to the two-thirds power of its body mass. The coefficient in this relation increases with the complexity of the behavioral repertoire that defines the group. The coefficient is larger for higher insectivores than for more primitive insectivores (such as moles), larger for higher monkeys than for primitive monkeys (such as lemurs), and largest of all for humans. Interestingly, the coefficient isn't significantly higher for chimps, gorillas, and orangutans than for higher monkeys. (Based on data from Roland Bauchot and Heinz Stephan, *Mammalia* 33 (1969): 228–75; Heinz Stephan, "Evolution of Primate Brains: A Comparative Anatomical Investigation," in *The Functional and Evolutionary Biology of Primates,* ed. R. Tuttle. Copyright © 1972 by Aldine)

tion" I mean the kind of cyclic processing, involving the construction, testing, and modification of perceptual hypotheses, that we discussed in Chapter 12. Perception clearly has an enormous potential advantage over feature-detection. A feature-detecter's visual world is meager, stereotyped, and unalterable; a perceiver's visual world is *potentially* far richer, and its variety is not genetically constrained. The extent to which the perceiver's advantage is realized depends, however, on the brain's capacity to construct and modify perceptual hypotheses. What kind of neural structures confer this capacity?

Perceptual hypotheses are presumably embodied in more or less complex neural circuits; if they exist at all, there is nothing else they could be. Hence the capacity to construct and modify perceptual hypotheses is just the capacity to construct and modify neural circuits. What could be the biological basis for such a capacity?

Most body cells divide several or many times in the course of development. Blood cells, for example, are continually sloughed off and replaced by new ones. Bone and muscle regenerate themselves after injuries. Neurons are different. Once formed during embryogenesis, they never divide and can never be replaced. Thus the pattern of neurons laid down during early development—a pattern whose functional features are determined by the genes—is fixed and unalterable.

What *can* be altered are *patterns of neuronal connections.* Two related kinds of changes can occur. Although a dead neuron can never be replaced, a live neuron can grow new fibers, much as a tree grows new branches, and thus make new connections with other neurons. In addition, chemical synapses can be strengthened or weakened. In particular, neurobiologists have found that *some synapses are greatly strengthened by use.* Thus the nervous system does have the capacity to construct and modify neural circuits. So it has the capacity to construct, test, and store new perceptual and cognitive hypotheses.

How does it accomplish these tasks? With this question we have arrived at the frontiers of theoretical and experimental research in the biology of mind. In 1978 the immunobiologist and neurobiologist Gerald Edelman proposed a biologically realistic model for the construction, storage, and recall of perceptual hypotheses.[10] Although this model is still speculative, it does, I believe, contain important elements of the picture that will eventually emerge. Edelman's model, like an earlier qualitative suggestion by immunobiologist Niels Jerne, was inspired by an analogy between the nervous system and the immune system. Jerne and Edelman postulated that immune recognition (the recognition of foreign particles by antibody cells) and perception are active rather than passive processes—processes that generate order and information. The prototype of such processes is biological evolution, in which, as we discussed in earlier chapters, blind genetic variation creates a repertoire of genotypes on which natural selection then acts. An analogous strategy underlies the immune response (Figure 14.19). A large repertoire of antibody-producing cells is built up from a much smaller genetic repertoire by random processes that occur during early development. Each cell manufactures a specific antibody (a large globular protein) and carries some of the antibodies it manufactures on its surface. An antigen (foreign particle) may bind in varying degrees to several antibodies. An antibody-producing cell is stimulated to divide by the binding of antigens to the antibodies it carries on its surface. Antibody-producing cells whose antibodies recognize an invading antigen reproduce very rapidly and give rise to *clones,* remnants of which persist after the invaders have been eliminated. These remnants constitute the organism's immunological memory.

Edelman postulates that a *primary repertoire* of neural modules (representing, in the present terminology, perceptual hypotheses or fragments of such hypotheses) is laid down in the course of embryogenesis and early development. Each module contains between 50 and 10,000 neurons. Neurobiologist Vernon Mountcastle discovered that the motor cortex is made up of just such structural and functional units, which he called *cortical columns.* Later, David Hubel and Torsten Wiesel extended this finding to the visual cortex and discovered how the processing of visual information takes place within a column. Edelman postulates that his modules have a wide variety of internal connections but that their external

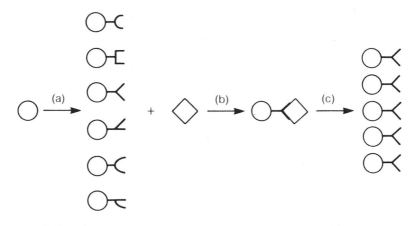

FIGURE **14.19** The clonal-selection theory of the immune response. (a) Generation of diversity: in the course of early development, cells carrying an immense variety of antibodies on their surfaces are generated. Each cell carries one specific antibody. (b) Recognition: a virus, bacterium, fungus, or foreign protein recognizes and binds a specific antibody (or a few specific antibodies). (c) Cloning: binding stimulates cells carrying the selected antibodies to reproduce (and to undergo certain internal changes).

connections are largely fixed. These modules are the analogues of antibody-producing cells. They are the units of perceptual function and the targets of selection.

The role of the antigen is taken by the perceptual stimulus—or more precisely, its cortical projection. Just as every antigen is recognized in varying degrees by several antibodies, so every stimulus evokes a response in several modules. The modules that "match" the stimulus best respond most strongly. They represent the perceptual hypothesis that is selected. Edelman posits that the response of a module causes alterations in the strength of synaptic connections in the neural pathway that mediates the response. A strong response facilitates future responses to similar stimuli. Thus perceptual stimuli leave enduring traces in the form of facilitated neural pathways. (This part of Edelman's model was actually put forward in 1949, long before the modifiability of synapses had been established experimentally, by neuropsychologist Donald Hebb in *The Organization of Behavior*.)

We can now give a partial answer to the question that set off our discussion of Edelman's model of perception: What kind of neural structures confer the capacity to construct, test, and modify perceptual hypotheses? Edelman suggests that the human brain has a large primary repertoire of cortical modules that act as *candidates for selection*—perhaps, he estimates, 1 million modules, each containing up to 10,000 neurons. There also must be an enormous number of modifiable neural pathways connecting these modules with primary projection areas. Moreover, the modules represent a single hierarchical level. The hierarchic construction of many-leveled neural configurations, representing complex percepts, will require additional large numbers of modifiable pathways. It seems clear that any model that, like Edelman's, depicts perception as a constructive process will require vast numbers of *uncommitted* neurons and synapses—candidates for selection. Because the brain's structure, including the basic pattern of connections between its neurons, is genetically predetermined, the candidates for selection must already exist at birth or shortly thereafter (as they do in the immune system). By modifying the strengths of existing neural connections, the process of perceptual construction selects new neural configurations from a vastly larger aggregate of candidates.

If, as I have suggested, perception replaced feature-detection early in mammalian evolution (or late in reptilian evolution), its advent would have created strong selection pressures favoring larger brain size—specifically, more neocortical area and, equally important, more connections within the new cortical areas and between them and other parts of the brain, especially the primary projection zones. These anatomical changes would have provided the substratum for an expanding perceptual capacity, probably allied with an expanded capacity for performing complex movements. It is encouraging for this hypothesis that, as I mentioned earlier, the evolutionary increases in the size of the mammalian neocortex have indeed taken place mainly in the secondary, association zones of the cortex and in the premotor cortex. These are the zones that provide the biological foundation for *sensorimotor intelligence.*

What about the most recent stage in brain evolution, the explosive growth of the frontal lobes during the emergence of *Homo sapiens?* As we have seen, this part of the neocortex plays a central role in mental processes that occur *at a higher hierarchical level* than that of sensorimotor intelligence. These processes—language and thought—depend on our ability to construct and manipulate *internal representations.* The biological promise of our hominid ancestors' dawning ability to construct inner models of the external world would have generated selection pressures comparable with those created by the advent of perception 70 million years or so earlier.

＊ ＊ ＊

The order-generating activity of the human mind and brain rivals, in some ways, that of the evolutionary process that gave birth to it. The products of cultural evolution are far less complex and diverse than those of biological evolution, but they have been brought forth during a far shorter period of time. There is another, perhaps more basic, way in which human creative activity seems to differ from the creative activity of biological evolution. Evolution is blind; it brings forth its fruits in darkness. Humans can foresee the future, however imperfectly. Even more, by inventing and pursuing goals, we *make* the future. Or so it seems. Is this view of the future as a work perpetually in progress—a work of which each of us is a coauthor—consistent with science? Or if not with science, with the speculative scientific view of the world I have sketched in the preceding pages? That is the question that launched this inquiry and to which we now return.

15

Chance, Necessity, and Freedom

To be fully human is to be able to make deliberate choices. Other animals sometimes have, or seem to have, conflicting desires, but we alone are able to reflect on the possible consequences of different actions and to choose among them in the light of broader goals and values. Because we have this capacity we can be held responsible for our actions; we can *deserve* praise and blame, reward and punishment. Values, ethical systems, and legal codes all presuppose freedom of the will. So too, as P. F. Strawson has pointed out, do "reactive attitudes" like guilt, resentment, and gratitude. If I am soaked by a summer shower I may be annoyed by my lack of foresight in not bringing an umbrella, but I don't resent the shower. I *could* have brought the umbrella; the shower just happened.

Freedom has both positive and negative aspects. The negative aspects—varieties of freedom *from*—are the most obvious. Under this heading come freedom from external and internal constraints. The internal constraints include ungovernable passions, addictions, and uncritical ideological commitments.

The positive aspects of freedom are more subtle. Let's consider some examples.

1. A decision is free to the extent that it results from deliberation. Absence of coercion isn't enough. Someone who bases an important decision on the toss of a coin seems to be acting less freely than someone who tries to assess its consequences and to evaluate them in light of larger goals, values, and ethical precepts.

2. Goals, values, and ethical precepts may themselves be accepted uncritically

or under duress, or we may feel free to modify them by reflection and deliberation. Many people don't desire this kind of freedom, and many societies condemn and seek to suppress it. Freedom and stability are not easy to reconcile, and people who set a high value on stability tend to set a correspondingly low value on freedom. But whether or not we approve of it, the capacity to reassess and reconstruct our own value systems represents an important aspect of freedom.

3. Henri Bergson believed that freedom in its purest form manifests itself in creative acts, such as acts of artistic creation. Jonathan Glover has argued in a similar vein that human freedom is inextricably bound up with the "project of self-creation." The outcomes of creative acts are unpredictable, but not in the same way that random outcomes are unpredictable. A lover of Mozart will immediately recognize the authorship of a Mozart divertimento that he happens not to have heard before. The piece will "sound like Mozart." At the same time, it will seem new and fresh; it will be full of surprises. If it wasn't, it wouldn't be Mozart. In the same way, the outcomes of self-creation are new and unforeseeable, yet coherent with what has gone before.

Although philosophical accounts of human freedom differ, they differ surprisingly little. On the whole, they complement rather than conflict with one another. What makes freedom a philosophical problem is the difficulty of reconciling a widely shared intuitive conviction that human beings are or can be free (in the ways discussed above or in similar ways) with an objective view of the world as a causally connected system of events. We feel ourselves to be free and responsible agents, but science tells us (or seems to tell us) that we are collections of molecules moving and interacting according to strict causal laws.

For Plato and Aristotle, there was no real difficulty. They believed that the soul initiates motion—that acts of will are the first links of the causal chains in which they figure. With few exceptions, modern neurobiologists have rejected the view of the relation between mind and body that this doctrine implies. They regard mental processes as belonging to the natural world, subject to the same physical laws that govern inanimate matter. The differences between animate and inanimate systems and between conscious and nonconscious nervous processes are not caused by the presence or absence of nonmaterial substances (the breath of life, mind, spirit, soul) but by the presence or absence of certain kinds of order. This conclusion is more than a profession of scientific faith. It becomes unavoidable once we accept the hypothesis of biological evolution, without which, as Theodosius Dobzhansky remarked, nothing in biology makes sense. The evolutionary hypothesis implies that human consciousness evolved from simpler kinds of consciousness, which in turn evolved from nonconscious forms of nervous activity. There is no point in this evolutionary sequence where mind or spirit or soul can plausibly be assumed to have inserted itself "from without." It seems even more implausible to suppose that it was there all along, although, as we saw earlier, some modern philosophers and scientists have held this view.

Karl Popper and other philosophers have tried to resolve the apparent conflict between free will and determinism by attacking the most sacred of natural science's sacred cows, the assumption that all natural processes obey physical laws.

In asserting that there *may* be phenomena that don't obey physical laws, these philosophers are obviously on safe ground. But the assumption of *in*determinism doesn't really help. A freely taken decision or a creative act doesn't just come into being. It is the necessary—and hence law-abiding—outcome of a complex process. Free actions also have predictable—and hence lawful—consequences; otherwise, planning and foresight would be futile. Thus every free act belongs to a causal chain: it is the necessary outcome of a deliberative or creative process, and it has predictable consequences.

Some physicists and philosophers have suggested that quantal indeterminacy may provide leeway for free acts in an otherwise deterministic Universe. Freedom, however, doesn't reside in randomness; it resides in choice. Plato and Aristotle were right in linking Chance and Necessity as "forces" opposed to design and purpose in the Universe.

Thus freedom seems equally inconsistent with determinism and indeterminism. Thomas Nagel has suggested that it isn't even possible to give a coherent account of our inner sense of freedom:

> When we try to explain what we believe which seems to be undermined by a conception of actions as events in the world—determined or not—we end up with something that is either incomprehensible or clearly inadequate.[1]

"The real problem," Nagel says, "stems from a clash between the view of action from inside and *any* view of it from outside." Yet the intuitive view of what it means to be free doesn't rest on introspection alone. We recognize other people's spontaneity and creativity even—or especially—when it is of such a high order that we can't imagine ourselves capable of it. We can apprehend the exquisitely ordered unpredictability of Mozart's music without beginning to be able to imagine what it would be like to compose such music. And even subjective impressions of freedom, unlike subjective impressions of pain or of self, aren't hard to describe. Consider the process of making a decision. Shall I do A or B? My head says A; my heart says B. I agonize. I try to imagine the consequences first of A, then of B. Suddenly, a new thought occurs to me: C. Yes, I'll do C. The essential aspect of such commonplace experiences is that their outcomes aren't determined in advance but are created by the process of deliberation itself, a process unfolding in time. All creative processes have this character.

Such processes, however, go on not only in people's subjective awareness but also in their brains. Conscious experience gives us a fragmentary and unrepresentative view of its underlying cerebral processes, but there is no reason to suppose that the view is *deceptive.* On the contrary, modern techniques of imaging brain activity, mentioned in Chapter 13, suggest that there is a high degree of structural correspondence between consciousness and brain activity. If, then, the outcome of a deliberative or creative process seems undetermined at the outset, if it seems to us that such processes create their outcomes, perhaps the reason is that the outcomes of the underlying cerebral processes *are,* in some objective sense, undetermined, *are,* in some objective sense, created by the processes themselves.

I will argue that the neural processes that give rise to subjective experiences of freedom are indeed creative processes, in the sense that they bring into the world kinds of order that didn't exist earlier and weren't prefigured in earlier physical states. These novel and unforeseen products of neural activity include not only works of art, but also the evolving patterns of synaptic connections that underlie the intentions, plans, and projects that guide our commonplace activities. Although consciousness gives us only superficial and incomplete glimpses of this ceaseless constructive activity, we are aware of it almost continuously during our waking hours. This awareness may be the source of—or even constitute—the subjective impression that we participate in molding the future.

Much of the argument that supports this view has already been given in earlier chapters. Let me now try to pull it together around the following three questions:

1. Do all law-abiding processes have predetermined outcomes?
2. What does it mean to say that a physical process creates its outcomes?
3. How is this kind of creativity related to creativity in contexts relevant to the problem of human freedom?

Law-abiding Processes Need Not Have Predetermined Outcomes

As we discussed in Chapter 2, events are determined by *laws* and *initial conditions*. The laws define what is possible; initial conditions, what is actual. The Earth's orbit is one of infinitely many allowed by Newton's laws of motion and gravitation: some of the allowed orbits are even rounder than ours, most are more elongated; some would take us closer to the Sun, others would carry us beyond the orbit of Pluto. The actual orbit is the outcome of processes determined by conditions in the primordial Solar System. Had these conditions been different, the planets and their orbits would also have been different. Thus physical determinism has two distinct aspects. Phenomena are determined because they are governed by laws. They are also *predetermined* to the extent that their initial conditions are specified.

Like nearly all scientists, I assume that physical laws brook no exceptions. But what does it mean to say that initial conditions are "specified"? Are they, too, "given," as are the laws? Let me rephrase the question: What kind of information, in addition to laws, would be contained in a complete description of the physical world? The customary answer has been: the values of physical quantities whose values are not determined by laws. For example, a complete description of a (hypothetical) universe governed by the laws of classical physics would contain information about the positions and velocities of all the particles at a single moment of time. Given that information, a sufficiently powerful computer would be able to compute the positions and velocities of all the particles at any other moment of time. This is the universe envisaged by Laplace and, with appropriate modifications to allow for the replacement of classical by quantal laws, by most modern scientists as well. The Laplacian or neo-Laplacian universe is deterministic in the strongest possible sense: the outcomes of all physical processes are predetermined.

Quantum physics seems at first sight to soften this strongly deterministic picture of the world, because the outcomes of some processes—radioactive decay, for example—aren't determined by either laws or initial conditions. But at a deeper level, the quantal picture is just as deterministic as the classical one. The evolution of quantum states, like the evolution of classical states, is determined by rigidly deterministic laws. If the initial quantum state of an isolated system is specified, the laws determine all earlier and later quantum states, exactly as in classical physics. The novel feature of quantal descriptions is that when a physical system is in a definite quantum state, measurements of some physical quantities don't have definite outcomes. But this kind of indeterminacy doesn't make quantal processes more "creative" than classical processes.

In Chapter 3 I argued that a complete description of the Universe need not—indeed *cannot*—include the values of all quantities whose values aren't determined by laws. How can a description that leaves out certain kinds of information be complete? The answer is that a complete description must omit not only information that is *supplied by* the laws, but also information that is *incompatible with* the laws. I proposed an additional law, or symmetry principle, which states that *a complete description of the Universe contains no information that serves to define a preferred position or direction in space.*

We saw that this postulate (the Strong Cosmological Principle) is tenable only if the underlying structure of the world is discrete. Discreteness is a recurrent theme in the account of physical reality sketched in the preceding pages. As we saw in Chapters 4 and 5, the permanence and stability of molecular, atomic, and subatomic systems are manifestations of discreteness. Discreteness is also presupposed by Boltzmann's definition of randomness, on which we based our discussion of timebound order.

A description of the Universe that is consistent with the Strong Cosmological Principle contains *only* statistical information. Of course, nonstatistical properties exist and can be *measured*. Geographers, geologists, and astronomers have gathered vast quantities of nonstatistical data about the Earth, the Solar System, the Galaxy, and distant galaxies. We use this information to test our theories and to suggest new theories. Without it we would surely have no theories. And yet, paradoxically, the kind of information that appears in atlases and astronomical catalogues couldn't appear in a complete description of the astronomical Universe. Just as we leave behind an aspect of subjective experience when we take up the objective point of view of natural science, so we leave behind the particular, nonstatistical data of astronomy and the earth sciences when we take up what is literally, in Thomas Nagel's phrase, "the view from nowhere," a view that excludes all reference to particular places.

We have now answered our first question: Do all law-abiding processes have predetermined outcomes? Outcomes are determined by laws plus initial conditions. They are undetermined to the extent that the initial conditions are unspecified. A complete description of a universe satisfying the Strong Cosmological Principle may contain, at the outset, very little information about initial conditions.

In What Sense Does a Physical Process Create Its Outcome?

To answer this question, we must examine the processes of cosmic evolution.

A theory of cosmic evolution requires initial conditions. The simplest initial conditions is that *the Universe began to expand from a purely random state—a state wholly devoid of order.* From this postulate, we can easily deduce the Strong Cosmological Principle. The inference hinges on the fact that none of our present physical laws discriminates between different points in space or between different directions at a point. (A physicist would say, "The laws are invariant under spatial translations and rotations.") This implies that no physical process can *introduce* discriminatory information. So if information that would discriminate between positions or directions is absent at a single moment, it must be absent forever. In short, if the Strong Cosmological Principle is valid at any single moment, it must be valid for all time.

If the Universe began to expand from a state of utter randomness, how did order come into being? Before reviewing our answer to this question, we have to recall how we dealt with the concept of order itself.

The two key ideas needed to formulate an adequate scientific definition of order were put forward by Ludwig Boltzmann. The first idea is the distinction between microstates and macrostates. Macrostates are groups of microstates, defined by their statistical properties. For example, the microstates of a gas may be assigned to macrostates defined by density, temperature, and chemical composition. Proteins may be assigned to macrostates defined by biological fitness. Boltzmann's second key idea was to identify the randomness or entropy of a macrostate with the logarithm of the number of its microstates. Supplementing this definition of randomness, we defined the order or information of a macrostate as the difference between its potential randomness or entropy (the largest value of the randomness or entropy consistent with given constraints) and the actual value. Thus maximally random macrostates have zero order and maximally ordered macrostates have zero randomness. According to these definitions, a physical system far removed from thermodynamic equilibrium (the macrostate of maximum randomness) is highly ordered. So is a protein whose biological fitness can't be improved by changes in its sequence of amino acids: it belongs to a very small subset of the class of polypeptides of the same length.

These definitions of randomness and order are important not just, or even primarily, because they lend precision to the corresponding intuitive notions in a wide range of scientific contexts. They are important primarily because they are adapted to theoretical accounts of the growth and decay of order. Boltzmann himself proved (under restrictive assumptions) that molecular interactions in a gas not already in its most highly random macrostate increase its randomness. In Chapter 8 we saw how the cosmic expansion generates chemical order (chemical abundances far removed from those that would prevail in thermodynamic equilibrium); in Chapter 9 we discussed the origin and growth of structural order in the astronomical Universe; and in Chapters 10 and 11 we saw how random genetic variation and differential reproduction generate the biological order encoded in genetic material.

Astronomical and biological order-generating processes are hierarchically linked in the manner discussed in Chapter 2. Each process requires initial conditions generated by earlier processes. For example, the first self-replicating molecules needed an environment that provided high-grade energy, molecular building blocks, and catalysts. High-grade energy was supplied, directly or indirectly, by sunlight, produced by the burning of hydrogen deep inside the Sun. To understand why hydrogen is so abundant, we have to go back to the early Universe, when the primordial chemical composition of the cosmic medium was laid down by an interplay between nuclear reactions and the cosmic expansion. Apart from hydrogen, the atoms that make up biomolecules (carbon, oxygen, and nitrogen are the most common) were synthesized in exploding stars far more massive than the Sun. So, too, were inorganic catalysts like zinc and magnesium. Finally, the emergence of an environment favorable to life as we know it resulted from planet-building processes, for which we still lack an adequate theory.

Although some of the specific order-generating processes we have discussed are speculative or controversial, the general principles underlying the emergence of order from chaos seem more secure. In particular, we can now understand why, in spite of the second law of thermodynamics, the Universe is not running down. The Second Law states that all natural processes tend to increase randomness. In an ordinary isolated system, the growth of randomness leads inevitably to a decline of order, because the sum of randomness and order is a fixed quantity. The Universe, however, is not an ordinary isolated system. Because space is expanding, the sum of randomness and order is not a fixed quantity; it tends to increase with time. Hence a gap may open up between the actual randomness of the cosmic medium and its maximum possible randomness. This gap represents a form of order. Chemical order (as evidenced by the prevalence of hydrogen) emerges when equilibrium-maintaining chemical reactions can no longer keep pace with the cosmic expansion. Structural order (in the form of astronomical systems) emerges when the uniform state of an expanding medium becomes unstable—that is, less than maximally random.

By making randomness an objective property of the Universe, the Strong Cosmological Principle also objectifies the timebound varieties of order, which consist in the absence of randomness. The infinitely detailed world picture of Laplace's Intelligence is devoid of macroscopic order. It contains no objective counterpart to astronomical or biological order. Laplace's Intelligence is an idiot savant. It knows the position and velocity of every particle in the Universe; but because this vast fund of knowledge (or its quantal counterpart) is complete in itself, there is no room in it for information about stars, galaxies, plants, animals, or states of mind. In this book I have argued that the external world—the world that natural science describes—is fundamentally different from the universe of Laplace and Einstein, which is given once and for all in space and time (or in spacetime). It is a world of becoming as well as being, a world in which order emerged from primordial chaos and begot new forms of order. The processes that have created and continue to create order obey universal and unchanging physical laws. Yet because they generate information, their outcomes are not implicit in their initial conditions.

The Strong Cosmological Principle has also enabled us to interpret quantum theory as an objective description of regularities underlying natural phenomena. In conventional interpretations of quantum theory, measurement is a primitive notion. The orthodox interpretation postulates that measurements cause discontinuous and unpredictable changes in the physical state of the system being measured. The instrumental interpretation regards quantum physics not as a description of physical reality but as a device for generating statistical predictions that relate the outcomes of measurements. Niels Bohr and his followers have argued that the very notion of a phenomenon is incomplete without a specification of how the phenomenon is to be measured or observed. Now, measurement is necessarily an irreversible process. Irreversibility, in turn, is a statistical phenomenon, according to theories that build on Boltzmann's statistical interpretation of thermodynamics. Physicists have usually taken it for granted that the probabilities that figure in these theories derive from human ignorance of the microscopic states of complex systems. Thus mainstream theories of irreversibility have a subjective cast, which is even more pronounced in the conventional interpretations of quantum theory, because of the prominent role these interpretations assign to measurement. In Chapter 3 I argued that the Strong Cosmological Principle allows us to give an objective interpretation of the probabilities that figure in Boltzmann's theory of entropy and entropy growth, and in Chapter 7 I suggested that conventional quantal descriptions should be regarded as fragments of more comprehensive descriptions that refer to infinite assemblies of identical systems satisfying the Strong Cosmological Principle. This point of view enabled us to interpret measurement within the framework of quantum theory and to resolve the measurement-centered paradoxes that arise in the orthodox interpretation.

Creative Processes

All order-generating processes may be said to be creative, but some seem to deserve the label more than others. For example, the evolution of chemical order in the early Universe seems less creative than the evolution of biological order. To gain insight into this difference, let's compare the evolution of a star cluster with the evolution of a biological population. Suppose we are given a statistical description of the cluster's initial state and asked to calculate its subsequent evolution. To do the calculation, we have to assign an initial position and velocity to each star. This can be done in many different ways that are consistent with the given statistical description of the initial state, and different assignments will yield different evolutionary trajectories. But if the number of stars is large, these evolutionary trajectories diverge very little, because each star responds to the combined attraction of all the others, and the combined attraction is insensitive to statistical fluctuations in the cluster's initial state.

Now consider a biological population. Suppose we knew everything that could in principle be known about the population's initial state, including the genotypes of all the organisms belonging to the population. Suppose we also had the ability to simulate on a supercomputer every relevant aspect of the evolutionary process.

Could we then predict what genotypes would be present in the population at some later time?

No—at least not for a population undergoing significant evolutionary change. The reason is that evolutionary outcomes are very sensitive to some of the random genetic changes brought about by mutation and genetic recombination. Suppose we could enumerate all the possible outcomes of every mutational and recombinational event and assign a probability to each of them. We would then be able, in principle, to construct a complete statistical description of our evolving population. This description would encompass a vast number of qualitatively distinct, multiply branching pathways, each with only a tiny probability of being realized (Figure 15.1). It would therefore contain very little information about the history of any given population. A prediction about the outcome of a horse race that assigns small and nearly equal probabilities of winning to each of a large number of entrants isn't very informative.

Biological evolution, therefore, not only generates order and information, but does so in an essentially unpredictable way. This, I suggest, is an essential element of every truly creative process. A creative process not only generates order, but does so in an essentially unpredictable way.

We don't yet fully understand the biological basis of creative human activity, but I find the analogy with biological evolution compelling. In Chapter 14 I suggested that higher mental processes are mediated by a cyclic process in which the brain constructs, tests, and modifies internal representations. It is tempting to speculate that the process by which internal representations are constructed has a strong random component, in addition to systematic components that are built up in the course of individual development and that constrain and channel the random component. The systematic components would play a role analogous to that of beta genes in the evolutionary theory sketched in Chapter 11. They would be responsible for the elements of an artist's work that we recognize as his or her individual style.

Creative human activity is unpredictable in the same way and for the same reasons that biological evolution is unpredictable. Unpredictability, however, is only one aspect of human freedom. We are free because we are, to a considerable extent, the authors of our own lives, and because every human life is something new under the Sun. That is what Democritus and Socrates believed; and if the picture I have sketched in this book is correct in its main outlines, it is also one of the lessons of modern science. Our awareness of the openness of the future and of our own ability to help shape it reflects a deep property of objective reality.

Science and Human Values

Two forms of reductionism are common among natural scientists.

Physical scientists who espouse the neo-Laplacian view of the world assert that meaning and value have no objective existence. Reality, they say, is a meaningless dance of elementary particles, whose orderliness resides entirely in the laws that govern the particles and their interactions; or if there is a pattern to the

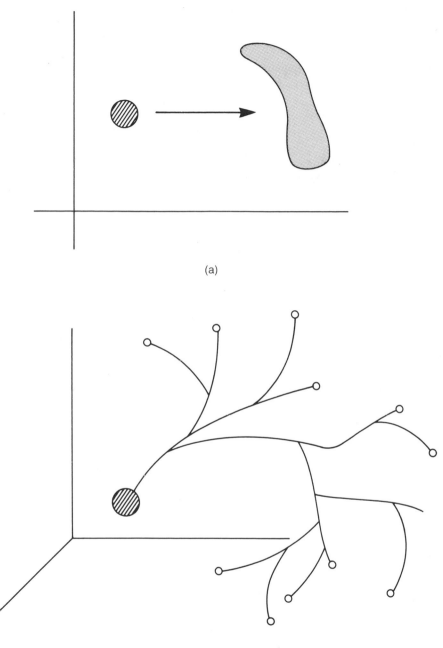

(a)

(b)

FIGURE 15.1 (a) A small statistical spread in the initial conditions of a star cluster gives rise to a moderate statistical spread in the cluster's measurable properties after, say, 5 billion years. (b) The evolution of a biological population may branch so strongly that statistical predictions, however refined, contain no useful information about the population's future state.

dance, it is one of inexorable decay—the universal tendency of order to dissolve into chaos.

Biologists are partial to another kind of reductionism. Social and political systems, ethical principles, religion, science, and art, say biological reductionists, all belong to the natural world. They are all products of cultural evolution, which is itself an extension of organic evolution. Hence cultural phenomena should admit biological explanations. Social Darwinism is a crude and now thoroughly discredited example of such explanation. Far more sophisticated and scientifically respectable are some current efforts to devise genetic explanations for cultural phenomena. The following passage from biologist Edward O. Wilson's *On Human Nature* conveys the spirit of this approach:

> In my opinion the key to the emergence of civilization is *hypertrophy,* the extreme growth of pre-existing structures. Like the teeth of the baby elephant that lengthen into tusks, and the cranial bones of the male elk that sprout into astonishing great antlers, the basic social responses of the hunter-gatherers have metamorphosed from relatively modest environmental adaptations into unexpectedly elaborate, even monstrous forms in more advanced societies. Yet the directions this change can take and its final products are constrained by the genetically influenced behavioral predispositions that constituted the earlier, simpler adaptations of preliterate human beings.[2]

In the neo-Laplacian worldview, the products of cultural evolution are less than real: the Laplacian Intelligence perceives only their constituent particles. In the sociobiological view of human culture, ethics, religion, art, architecture, music, and science are hypertrophied forms of "earlier, simpler adaptations of preliterate human beings." Despite the disparity between their starting points, the two approaches reach similar conclusions. The neo-Laplacian view denies the reality of cultural artifacts; the sociobiological view denies their essential novelty and richness.

The scientific worldview sketched in the preceding pages offers an alternative to reductionism in both its physical and its biological forms. It shows us that the Universe is more than a collection of elementary particles governed by immutable mathematical laws. Order and the processes that bring order into being lie at the heart of reality. Biological evolution, cultural evolution, and individual human lives not only are the most prolific sources of order in the known Universe, but also are creative. Because of them, the future is genuinely open.

It is often said that science doesn't supply values. It doesn't tell us that we should prefer altruism to greed, compassion to cruelty, knowledge to ignorance, truth to falsehood, life to death. It doesn't tell us how we should live our lives. All that is true. But I think it would be a great mistake to believe that science is irrelevant to human values. To begin with, as Jacob Bronowski argued in *Science and Human Values,* a commitment to a scientific view of the world presupposes and requires certain value judgments: that knowledge is better than ignorance, that truth is better than falsehood, that truth is knowable through reason and experience, that experience rather than authority or revelation is the final arbiter of truth.

A commitment to science entails a willingness to submit one's own beliefs and opinions to rigorous examination and an acknowledgment that what we can know of the truth is never final or complete.

But science can be relevant to human values and aspirations in another way. The ethical teachings of the great world religions derive much of their authority from more comprehensive worldviews. By explaining how we fit into the grander scheme of things, they provide a meaningful framework for human life. Ethics divorced from cosmology is a branch of philosophy, not a force in people's lives. Classical materialistic science offered an alternative to religious worldviews, but it was a dreary, unsatisfying, and unbelievable alternative. It is this Laplacian or neo-Laplacian picture that people usually have in mind when they reject the notion that science can help us to understand our place in the Universe. The worldview that modern science offers, as I have tried to describe it, is more interesting, more intellectually and emotionally satisfying, and, above all, truer. The old scientific worldview rested mainly on an important but unrepresentative fragment of natural science. The new scientific worldview embraces and reconciles quantum physics, macroscopic physics, molecular and evolutionary biology, and modern neuroscience. It invites renewed efforts to broaden and deepen our understanding of nature; at the same time, it inspires awe and wonder. And it assures us that there are no limits to what we and our descendants can hope to achieve and to become.

Notes

Chapter 1   The Unity of Science

1. Pierre Simon de Laplace, *Essai philosophique sur les probabilités* (Paris: Gauthier-Villars, 1821), p. 3 (translated by D. L.).

2. Albert Einstein, "The Religious Spirit of Science," in *Ideas and Opinions* (New York: Bonanza Books, 1954), p. 40.

3. The phrase "block universe" was used by William James, *A Pluralistic Universe* (Cambridge, Mass.: Harvard University Press, 1977), p. 140. James refers to "the block-universe eternal and without a history."

4. Arthur Stanley Eddington, *The Nature of the Physical World* (1929; Ann Arbor: University of Michigan Press, 1958), p. 294.

5. Willard Van Orman Quine, *Word and Object* (Cambridge, Mass.: MIT Press, 1960), p. 23.

6. Thomas Kuhn, *The Structure of Scientific Revolutions*, 2nd ed. (Chicago: University of Chicago Press, 1970), p. 206.

7. Albert Einstein, "The Method of Theoretical Physics," in *Ideas and Opinions*, p. 274.

8. Steven Weinberg, *The First Three Minutes*, 2nd ed. (New York: Basic Books, 1988), p. 119.

9. The following argument, invented by Georg Cantor, shows that the points in a line segment are not a discrete aggregate. Label the endpoints of the segment 0 and 1. Then each point in the segment corresponds to a real number between 0 and 1, and every real number between 0 and 1 corresponds to a definite point. Is the aggregate of real numbers between 0 and 1 discrete? Assume that it is. Then there must be a list of these numbers. If the numbers are represented by decimals, the list looks like this:

$$0.a_{11}a_{12}a_{13} \ . \ . \ .$$
$$0.a_{21}a_{22}a_{23} \ . \ . \ .$$
$$0.a_{31}a_{32}a_{33} \ . \ . \ .$$

where each entry is an integer from 0 to 9. This infinite list, by our assumption, contains all decimals less than 1. But we can easily construct decimals that aren't on the list. Any number $0.b_1b_2b_3 \ . \ . \ .$ (in which b_1 is different from a_{11}, b_2 is different from a_{22}, b_3 is different from a_{33}, and so on) can't be on the list because it differs in at least one place from every entry. We conclude that the aggregate of real numbers isn't discrete; the real numbers aren't listable or countable.

Chapter 2   Order and Randomness

1. See Eugene P. Wigner, "Events, Laws of Nature, and Invariance Principles," in *Symmetries and Reflections* (Cambridge, Mass.: MIT Press, 1967), p. 38.

2. Joseph Needham, "Evolution and Thermodynamics," in *Moulds of Understanding* (New York: St. Martin's Press, 1976), p. 173.

3. Erwin Schroedinger, *Science and the Human Temperament* (1935), reprinted as *Science, Theory and Man* (New York: Dover, 1957), p. 48.

4. James Clerk Maxwell, *Theory of Heat*, 10th ed. (London: Longmans, Green, 1891), p. 338.

5. Peter B. Medawar, "A Biological Retrospect," in *Pluto's Republic* (New York: Oxford University Press, 1982), p. 291.

6. You may recall that a free neutron decays into a proton, a positron, and an *antineutrino*. The neutrino is electrically neutral. Why does it have an antiparticle? The answer is that electric charge is one of several kinds of charge envisaged in the rule that states that a particle and its antiparticle carry opposite charges. The neutrino and the antineutrino (like the electron and the positron) carry opposite "leptonic charge." Leptonic charge, like electric charge, is conserved, according to currently well-established theories; that is, the combined leptonic charge of the particles in any isolated part of the Universe never changes. Nucleons carry an analogous charge, "baryonic charge," which is also conserved. Photons, however, carry no charge of any kind, so they have no antiparticle. When a particle and its antiparticle meet and disappear in a flash (or more accurately, two flashes) of light, matter disappears, but the total leptonic and baryonic charges don't change. Electric charge and energy also don't change.

Chapter 3   The Strong Cosmological Principle and Time's Arrow

1. Edwin P. Hubble, *The Realm of the Nebulae* (New Haven, Conn.: Yale University Press, 1936), in *Theories of the Universe,* ed. Milton K. Munitz (New York: Free Press, 1957), p. 289.

2. Richard P. Feynman, *Lectures on Physics* (Reading, Mass.: Addison-Wesley, 1963), vol. 1, p. 8.

3. For a modern philosophical discussion of this question, see Willard Van Orman Quine, "On What There Is," in *From a Logical Point of View* (Cambridge, Mass.: Harvard University Press, 1961), p. 1.

4. For a discussion of hidden order, see E. L. Hahn and R. G. Brewer, *Scientific American,* December 1984, p. 6.

Chapter 4   The Importance of Being Discrete

1. James Clerk Maxwell, "Molecules," in *The Scientific Papers of James Clerk Maxwell*, ed. W. D. Niven (New York: Dover, 1965), vol. 2, pp. 375–76.

Chapter 5   Seven Steps to Quantum Physics

1. Max Planck, *Naturwissenschaft* 31 (1943): 153.

2. Albert Einstein, *Annalen der Physik* 17 (1905): 132. The opening of this paper on the photon hypothesis parallels the opening of Einstein's paper on special relativity, published in the same year. In the paper on special relativity, Einstein remarks that Newton's theory treats all unaccelerated reference frames alike, whereas Maxwell's theory assumes that there is a preferred frame (in which light propagates with the same speed in all directions). Einstein then enunciates the principle of relativity: that no observation can define a preferred state of unaccelerated motion.

3. Quoted in Abraham Pais, "Einstein on Particles, Fields, and the Quantum Theory," in *Some Strangeness in the Proportion*, ed. H. Woolf (Reading, Mass.: Addison-Wesley, 1980), p. 197.

4. Pais, "Einstein on Particles, Fields, and the Quantum Theory," p. 197.

5. In his 1913 paper, Niels Bohr arrived by a less direct route at the rule requiring the angular momentum of an allowed circular orbit to be an integral multiple of $h/2\pi$. He began with the known empirical rule for the energy levels of hydrogen, $E = -C/n^2$, where n is an integer and C is an empirically determined constant. For very large values of n, the classical orbit is very large, so classical physics should apply to it. An electron in a circular orbit radiates light whose frequency is the same as the reciprocal of the orbital period. Bohr postulated that light of this frequency is emitted when the electron makes a transition from level n to level $n-1$. This requirement correctly predicts the constant C in the above formula for the energy levels. *Then* Bohr noticed that the allowed orbits are given by the angular-momentum rule. In a lecture on the spectrum of hydrogen delivered to the Physical Society in Copenhagen on December 20, 1913, Bohr didn't mention the angular-momentum rule. This suggests that he didn't at first attribute fundamental importance to it.

6. Niels Bohr, H. A. Kramers, and J. C. Slater, *Philosophical Magazine* 47 (1924): 785, in *Sources of Quantum Mechanics*, ed. B. L. van der Waerden (New York: Dover, 1968), p. 159. Commenting on Compton's theory of the scattering of light by electrons, these authors write, "By this process the electron acquires a velocity in a certain direction, which is determined, just as the frequency of the re-emitted light, by the laws of conservation of energy and momentum, an energy $h\nu$ and a momentum $h\nu/c$ being ascribed to each light-quantum. *In contrast to this picture*, the scattering of the radiation by the electron is, on our view, considered as a *continuous phenomenon* to which each of the illuminated electrons contributes through the emission of coherent secondary wavelets" (p. 173, my emphasis).

7. Albert Einstein, *Physilkalische Zeitschrift* 18 (1917): 121, in *Sources of Quantum Mechanics*, ed. B. L. van der Waerden (New York: Dover, 1968), p. 63.

8. Albert Einstein to Erwin Schroedinger, December 1950, *Letters on Wave Mechanics*, ed. K. Przibram, trans. M. J. Klein (New York: Philosophical Library, 1967), p. 39.

Chapter 6   Alice in Quantumland

1. Max Born, "On the Quantum Mechanics of Collisions," in *Quantum Theory and Measurement*, ed. J. A. Wheeler and W. H. Zurek (Princeton, N.J.: Princeton University Press, 1982), p. 52.

2. Born, "On the Quantum Mechanics of Collisions," p. 53.

3. In the text of his paper, Born says that the amplitude itself must represent this probability, but in a note added in proof, he writes, "More careful consideration shows that the probability is proportional to the square of the [amplitude]" (p. 54).

4. Born, "On the Quantum Mechanics of Collisions," p. 54.

5. Erwin Schroedinger, *Naturwissenchaften* 23 (1935): 807.

6. Eugene P. Wigner, "Two Kinds of Reality," in *Symmetries and Reflections* (Bloomington: Indiana University Press, 1967), p. 171.

7. Werner Heisenberg, *Daedalus* 87 (1958): 99.

8. Erwin Schroedinger, *What Is Life?* and *Mind and Matter* (Cambridge: Cambridge University Press, 1967), p. 99.

9. Andrew Osiander, quoted in *Ptolemy, Copernicus, Kepler* (Chicago: Encyclopaedia Britannica, 1939), p. 505. The preface was not signed, but scholars agree that it was written by Osiander, not Copernicus.

10. Eugene P. Wigner, "Interpretation of Quantum Mechanics," in *Quantum Theory and Measurement,* ed. J. A. Wheeler and W. H. Zurek (Princeton, N.J.: Princeton University Press, 1982), pp. 260–314.

11. See, for instance, Albert Einstein, "Reply to Criticisms," in *Albert Einstein: Philosopher-Scientist,* ed. P. A. Schilpp (Evanston, Ill.: Library of Living Philosophers, 1949), p. 683.

12. Niels Bohr, quoted in *Quantum Theory and Measurement,* ed. J. A. Wheeler and W. H. Zurek (Princeton, N.J.: Princeton University Press, 1982), p. 3.

13. Albert Einstein to Erwin Schroedinger, 9 August 1939, *Letters on Wave Mechanics,* ed. K. Przibram, trans. M. J. Klein (New York: Philosophical Library, 1967), p. 35.

14. Einstein to Schroedinger, 31 May 1928, *Letters on Wave Mechanics,* p. 31.

Chapter 7   The Strong Cosmological Principle and Quantum Physics

1. The reader who is familiar with quantum physics will have noticed that I have simplified the argument by talking about relative signs when I should have been talking about relative *phases*. In quantum theory, a and b are, in general, complex numbers, whose ratio a/b is a complex number that can be written in the form $R\exp(i\phi)$, where R and ϕ are real numbers i is the imaginary square root of -1, and exp denotes the exponential function. The real number ϕ is called the relative phase of a and b. If a and b are real numbers, the relative phase is zero if they have the same sign, and π radians (180°) if they have opposite signs. The mathematical objects that represent the microstates h and v also have a relative phase. The argument in the text hinges on the statement that the relative phases of h and v in the state of the assembly represented by row (4) are randomly distributed between zero and 2π radians.

2. A more complete discussion is given in David Layzer, "Quantum Mechanics, Thermodynamics, and Cosmology," in *Physics as Natural Philosophy,* ed. H. Feshbach and A. Shimony (Cambridge, Mass.: MIT Press, 1982), pp. 240–62.

3. Hugh Everett III, *Reviews of Modern Physics* 29 (1957): 454–62.

4. John A. Wheeler, quoted in *Quantum Theory and Measurement,* ed. J. A. Wheeler and W. H. Zurek (Princeton, N.J.: Princeton University Press, 1982), p. 185.

Chapter 8   Cosmic Evolution: The Standard Model

1. For an extended fictional treatment of this theme, see Jorge Luis Borges, "The Library," in *Ficciones,* ed. A. Kerrigan (New York: Grove Press, 1962).

2. Ludwig Boltzmann, quoted in *Physical Thought from the Presocratics to the Quantum Physicists: An Anthology*, ed. S. Sambursky (New York: Pica Press, 1975), p. 451.

3. Edwin P. Hubble, *The Realm of the Nebulae* (New Haven, Conn.: Yale University Press, 1936), in *Theories of the Universe*, ed. Milton K. Munitz (New York: Free Press, 1957), p. 289.

4. Arthur Stanley Eddington, *The Nature of the Physical World* (1929; Ann Arbor: University of Michigan Press, 1958), pp. 84–85.

5. Richard P. Feynman, *The Character of Physical Law* (Cambridge, Mass.: MIT Press, 1967), p. 116.

6. Is spatial nonuniformity compatible with the Strong Cosmological Principle? Yes. The principle says that a statistical *description* of the spatial distribution of mass must not discriminate between points in space, not that the distribution itself must be uniform. For example, the chance of finding a galaxy ten times as massive as the Milky Way within 1 million light-years of the Sun must be the same as the chance of finding such a galaxy within 1 million light-years of any other point in space.

7. Isaac Newton to Dr. Bentley, 1692, *Theories of the Universe*, ed. Milton K. Munitz (New York: Free Press, 1957), p. 211.

8. For a highly readable popular account of the standard (fireball) model, see Steven Weinberg, *The First Three Minutes*, 2nd ed. (New York: Basic Books, 1988).

Chapter 9   Gravitational Clustering and Structural Order

1. Despite their diversity, current hypotheses for the hot early Universe encounter a common difficulty. They predict a value for the so-called cosmological constant that is 120 powers of 10 (10^{120}) times as great as the largest value consistent with astronomical evidence. There have been many ingenious efforts to resolve this difficulty, but none of them is widely accepted by theorists. The difficulty doesn't arise in a Universe that expands from an initially cold state.

2. David Layzer, *Constructing the Universe* (New York: Freeman, 1984).

Chapter 10   Molecules, Genes, and Evolution

1. Plato, *Phaedo*, in *The Dialogues of Plato*, trans. Benjamin Jowett (New York: Bantam, 1986).

2. Henri Bergson, *L'évolution créatrice* (Paris: Quadrige/PUF, 1969), pp. 254–55 (this passage and those later in the chapter translated by D. L.).

3. Charles Darwin, *The Origin of Species*, 1st ed. (1859; Harmondsworth: Penguin, 1987), p. 456.

4. Quoted in Ernst Mayr, *The Evolutionary Synthesis*, ed. Ernst Mayr and William B. Provine (Cambridge, Mass.: Harvard University Press, 1980), p. 309.

5. Charles Darwin, *The Origin of Species*, 6th ed. (London: Macmillan, 1962), p. 242.

6. G. Ledyard Stebbins, *Darwin to DNA, Molecules to Humanity* (New York: Freeman, 1982), p. 46.

7. Quoted in Mayr, *Evolutionary Synthesis*, p. 309.

8. Werner Heisenberg, *Physics and Beyond* (New York: Harper & Row, 1972), p. 113.

9. Richard C. Lewontin, private communication.

10. Peter B. Medawar, "A Biological Retrospect," in *Pluto's Republic* (New York: Oxford University Press, 1982), p. 291.

11. Medawar, "Biological Retrospect," p. 291.

12. G. Ledyard Stebbins, *The Basis of Progressive Evolution* (Chapel Hill: University of North Carolina Press, 1969), pp. 119–20.

Chapter 11   Evolution and the Growth of Order

1. Jacques Monod, *Chance and Necessity* (New York: Vintage Books, 1971), pp. 116–17.

2. Thomas Robert Malthus, "Essay on the Principle of Population," in *The World of Mathematics,* ed. James R. Newman (New York: Simon and Schuster, 1956), p. 1192. Benjamin Franklin's essay "Observations Concerning the Increase of Mankind" was written in 1751. It had a polemical purpose. In the previous year, Parliament had passed an act that restricted the building of iron and steel works in the American colonies. Franklin argued that this act was opposed to England's best interests, which lay in maximizing the rate of growth of the colonial population. Industrialization would promote population growth, Franklin argued, by improving the standard of life, thus allowing people to marry earlier and have more surviving children. The essay was first published in 1755, and was reprinted several times. It is said to have had a considerable influence on English economic thought. The passage that Malthus paraphrases reads in its entirety as follows: "There is in short, no Bound to the prolific Nature of Plants or Animals, but what is made by their crowding and interfering with each other's Means of Subsistence. Was the Face of the Earth vacant of other Plants, it might be gradually sowed and overspread with one Kind only; as, for Instance, with Fennel; and were it empty of other Inhabitants, it might in a few Ages be replenish'd from one Nation only; as, for Instance, with Englishmen." Franklin estimated that the colonial population would double every twenty-five years, and Malthus adopted this estimate in his essay. I am indebted to Prudence Steiner for bringing Franklin's essay to my attention.

3. Miscopyings don't necessarily have lethal consequences. We can assign each unit of our self-replicating molecule an "index of reproductive importance," defined as follows. A text isn't reproductively viable if it contains two miscopied units with an index of $\frac{1}{2}$, three miscopied units with an index of $\frac{1}{3}$, and so on. If we now take N to be the sum of the indexes of all the units in the text, and call this quantity the effective length of the molecule, the preceding rule still holds: the longest reproductively viable molecule has effective length $1/f$, where f is the error frequency for copying a single unit. The effective length of a molecule is an approximate measure of the biological information it encodes. Segments of genetic material that play no functional role don't contribute to the effective length of a molecule or to the biological information it encodes.

4. In modern cells, the units that are linked together to form a molecule of DNA (deoxyribonucleotides) are manufactured from corresponding units of RNA (ribonucleotides) by a very elaborate process. This strongly suggests that RNA evolved first. Today, RNA plays several distinct roles. Molecules of RNA act as primers for the replication of DNA; they carry information from DNA to the ribosomes (which are themselves complexes of RNA and proteins); and, in the role of adaptors, they mediate the translation of genetic information into proteins. In some viruses, genetic information is stored and transmitted by RNA. Thus RNA is a very attractive candidate for the primordial self-replicator. Molecular biologist Leslie Orgel has shown that RNA molecules some tens of nucleotides long do in fact replicate themselves accurately in environments that contain activated building blocks and zinc ions, which act as nonspecific catalysts.

5. In RNA, the base uracil (U) replaces thymine. Uracil, like thymine, bonds snugly

with adenine to form a "rung" of the same width as that formed by cytosine and guanine.

6. Ivan Ivanovich Schmalhausen, *Factors of Evolution: The Theory of Stabilizing Selection,* ed. T. Dobzhansky (Philadelphia: Blakiston, 1949), p. 233.

7. D. D. Davis, *Fieldiana Memoirs (Zoology)* 3 (1964).

8. Quoted in Steven Stanley, *Macroevolution* (San Francisco: Freeman, 1979), p. 55.

9. A. C. Wilson, S. S. Carlson, and T. J. White, *Annual Reviews of Biochemistry* 46 (1977): 573–639.

10. David Pilbeam, "Human Origins," David Skomp Distinguished Lecture in Anthropology (Delivered at Indiana University, Bloomington, Indiana, 1986).

11. Charles Darwin, *The Origin of Species,* 6th ed. (London: Macmillan, 1962), p. 308.

12. Stanley, *Macroevolution,* p. 122.

13. Ernst Mayr, *Populations, Species, and Evolution* (Cambridge, Mass.: Harvard University Press, 1970), p. 374.

14. Richard C. Lewontin, private communication.

15. George Gaylord Simpson, *Fossils* (New York: Freeman, 1983), p. 167.

16. David Layzer, *American Naturalist* 115 (1980): 809.

17. D. T. Suzuki, A. J. F. Griffiths, J. H. Miller, and R. C. Lewontin, *An Introduction to Genetic Analysis* (New York: Freeman, 1986), p. 375.

Chapter 12   Language, Thought, and Perception

1. W. John Smith, *The Behavior of Communicating* (Cambridge, Mass.: Harvard University Press, 1977), pp. 357–58.

2. Alexander Romanovich Luria, *Lectures on Language and Cognition* (New York: Wiley, 1981), pp. 32–33.

3. Robin Fox, "The Cultural Animal," in *Man and Beast: Comparative Social Behavior,* ed. J. F. Eisenberg and W. S. Dillon (Washington, D.C.: Smithsonian Institution Press, 1971), pp. 263–96, quoted in E. O. Wilson, *Sociobiology* (Cambridge, Mass.: Harvard University Press, 1975), p. 560.

4. Melvin Konner, *The Tangled Web* (New York: Harper & Row, 1982), p. 8.

5. Frances Dahlberg, ed., *Woman the Gatherer* (New Haven, Conn.: Yale University Press, 1981), quoted in Konner, *Tangled Web,* p. 448.

6. Konner, *Tangled Web,* pp. 106–26, and references therein. See also R. Trivers, *Social Evolution* (Menlo Park, Calif.: Benjamin/Cummings, 1985); M. Daly and M. Wilson, *Sex, Evolution, and Behavior,* 2nd ed. (Boston: PWS Publishers, 1983).

7. Ulric Neisser, *Cognition and Reality* (San Francisco: Freeman, 1976), p. 67.

8. Neisser, *Cognition and Reality,* p. 66.

9. Peter C. Wason and Philip N. Johnson-Laird, *Psychology of Reasoning* (Cambridge, Mass.: Harvard University Press, 1972).

10. Luria, *Lectures on Language and Cognition,* pp. 207–9.

11. Noam Chomsky, *Aspects of the Theory of Syntax* (Cambridge, Mass.: MIT Press, 1965), p. 27.

12. Michael Polyani develops this theme at length in *The Tacit Dimension* (Garden City, N.Y.: Doubleday, 1966).

13. Roger Brown, *A First Language* (Cambridge, Mass.: Harvard University Press, 1973), p. 64.

14. Brown, *First Language,* p. 200.

Chapter 13   What Is Consciousness?

1. Thomas Nagel, *Mortal Questions* (Cambridge: Cambridge University Press, 1979), p. 196.

2. Nagel, *Mortal Questions,* p. 201.

3. For a discussion of this assertion, see Nagel, *Mortal Questions,* chaps. 12–14, which contain references to the philosophical literature. Representative philosophical essays that discuss the mind–body problem from the materialist perspective are collected in D. M. Rosenthal, ed., *Materialism and the Mind–Body Problem* (Englewood Cliffs, N.J.: Prentice-Hall, 1971). See also Daniel Dennett, *Brainstorms* (Cambridge, Mass.: MIT Press, 1981).

4. For a philosophical discussion of this view of mental states, see Thomas Nagel, *The View from Nowhere* (New York: Oxford University Press, 1986), chap. 3.

5. J. Allan Hobson, "Psychoanalytic Dream Theory: A Critique Based Upon Modern Neurophysiology," in *Mind, Psychoanalysis and Science,* ed. Peter Clark and Crispin Wright (Oxford: Basil Blackwell, 1988); J. A. Hobson, S. A. Hoffman, R. Helfand, and D. Kostner, *Human Neurobiology* 6 (1987): 157–64; J. Allan Hobson, *The Dreaming Brain* (New York: Basic Books, 1988).

6. J. Allan Hobson, private communication.

Chapter 14   Brain and Mind

1. Richard F. Thompson, *The Brain* (New York: Freeman, 1985), p. 74.

2. John Eccles, *The Understanding of the Brain* (New York: McGraw-Hill, 1973), p. 87.

3. Eccles, *Understanding of the Brain,* p. 89.

4. Gordon Shepherd, *Neurobiology,* 2nd ed. (New York: Oxford University Press, 1988), pp. 83–85.

5. Alexander Romanovich Luria, *The Working Brain* (New York: Basic Books, 1973), p. 80.

6. Luria, *Working Brain,* p. 240.

7. Luria, *Working Brain,* p. 116.

8. Luria, *Working Brain,* pp. 148–49.

9. Roland Bauchot and Heinz Stephan, *Mammalia* 33 (1969): 228–75; Heinz Stephan, "Evolution of Primate Brains: A Comparative Anatomical Investigation," in *The Functional and Evolutionary Biology of Primates,* ed. R. Tuttle (Chicago: Aldine, 1972), pp. 155–74. Both sources cited in Jean-Pierre Changeux, *Neuronal Man: The Biology of Mind,* trans. Lawrence Garey (New York: Pantheon, 1985), p. 41.

10. Gerald M. Edelman, "Group Selection and Phasic Reentrant Signaling: A Theory of Higher Brain Function," in *The Mindful Brain,* ed. Gerald M. Edelman and Vernon B. Mountcastle (Cambridge, Mass.: MIT Press, 1982), p. 51; Gerald M. Edelman, *Neural Darwinism: The Theory of Neuronal Group Selection* (New York: Basic Books, 1987).

Chapter 15   Chance, Necessity, and Freedom

1. Thomas Nagel, *The View from Nowhere* (New York: Oxford University Press, 1986), p. 110.

2. Edward O. Wilson, *On Human Nature* (Cambridge, Mass.: Harvard University Press, 1978), p. 89.

Index

*Page numbers in **boldface** contain definitions*